Thermal Evolution of the Tertiary Shimanto Belt, Southwest Japan: An Example of Ridge-Trench Interaction

Edited by

Michael B. Underwood
Department of Geological Sciences
University of Missouri
Columbia, Missouri 65211

SPECIAL PAPER

273

1993

Published by The Geological Society of America, Inc.
3300 Penrose Place, P.O. Box 9140, Boulder, Colorado 80301

Printed in U.S.A.

GSA Books Science Editor Richard A. Hoppin

Library of Congress Cataloging-in-Publication Data
Thermal evolution of the Tertiary Shimanto Belt, southwest Japan : an
    example of ridge-trench interaction / edited by Michael B.
    Underwood.
        p.   cm. — (Special paper / Geological Society of America ;
    273)
        Includes bibliographical references and index.
        ISBN 0-8137-2273-X
        1. Geology, Stratigraphic—Tertiary.   2. Geology—Japan—Shikoku
    Region.   I. Underwood, Michael B.   II. Series: Special papers
    (Geological Society of America) ; 273.
        QE691.T49   1993
        551.7'8'0952—dc20   92-43051
                                                                  CIP

Front cover: Oblique helicopter aerial photo of Cape Muroto, Kochi Prefecture, Shikoku Island, Japan. Note the well-developed marine terrace (150,000 years old) and the urban center of Muroto City. This region served as the principal site for the authors' research on the thermal evolution of the Tertiary Shimanto Belt. **Back cover—Left:** Middle Miocene (13 Ma) granitic rocks at Cape Ashizuri, Hata Peninsula, Kochi Prefecture, Shikoku. Widespread granitic intrusions, siliceous tuffs, and local mafic intrusions within the Outer Zone of Japan represent igneous manifestations of an anomalous forearc heating event that was triggered by interaction between the Shikoku Basin spreading ridge and the Shimanto accretionary prism. **Middle:** Folded turbidites of the Misaki assemblage (late Oligocene to early Miocene), Cape Muroto, Kochi Prefecture, Shikoku. These rocks, which make up part of the Nabae subbelt of the Shimanto Belt, were affected by a thermal overprint associated with intrusion of the Maruyama gabbro dikes at 14 Ma. **Right:** Fault truncation of tight, steeply plunging folds within the Gyoto structural domain, Cape Gyoto, Kochi Prefecture, Shikoku. These deformed turbidites are Eocene to early Oligocene(?) in age and form part of the Murotohanto subbelt of the Shimanto Belt.

10   9   8   7   6   5   4   3   2   1

Contents

Contents

Preface

This multidisciplinary study of the Shimanto Belt began as two simultaneous field-based investigations of structural fabrics on the Muroto Peninsula of Shikoku. Ph.D. dissertations ultimately were completed by Jim Hibbard (at Cornell University) and Lee DiTullio (at Brown University). In addition to the instrumental efforts of their faculty supervisors (Dan Karig and Tim Byrne), Asahiko Taira of the University of Tokyo provided scientific background on the regional and local geology, and invaluable logistical support; Ako thereby served as a critical liaison between the scientific communities of the United States and Japan. Laboratory studies of Shimanto thermal maturity (using the techniques of vitrinite reflectance and X-ray diffraction) followed at the University of Missouri as a logical outgrowth of the field work. Most of our initial objectives were rooted in tests of specific interpretations pertaining to the complex structural relations on the Muroto Peninsula, but the focus quickly expanded to include additional analytical techniques and additional field areas. As the project neared its completion, we benefited greatly from field trips and a free-wheeling exchange of scientific ideas at the International Conference on Accretionary Prisms, which was held in Muroto City in May 1991. Those of us who attended the Muroto conference will never forget the hospitality displayed by the citizens of Muroto City.

The following people served as formal reviewers for papers contained herein: Mark Brandon, R. M. Bustin, Jerry Clayton, Mark Cloos, Darrel Cowan, Rebecca Dorsey, Trevor Dumitru, Andy Fisher, Don Fisher, Martin Frey, Dan Karig, Richard Kettler, Hannan Kisch, Louie Liou, J. Casey Moore, Dan Orange, Kevin Shelton, Jim Sample, Jane Tribble, Asahiko Taira, and Chi Yuen Wang. We also acknowledge the assistance of GSA Book Editor, Richard Hoppin. For their financial contribution toward page charges, acknowledgment is made to the Donors of Petroleum Research Fund, administered by the American Chemical Society (Grant No. 19187-AC2 to Underwood).

The initial submission of papers for the volume began in March 1990, and the final revisions were completed in March 1992.

Geological Society of America
Special Paper 273
1993

Geologic summary and conceptual framework for the study of thermal maturity within the Eocene-Miocene Shimanto Belt, Shikoku, Japan

Michael B. Underwood
Department of Geological Sciences, University of Missouri, Columbia, Missouri 65211
J. P. Hibbard
Department of Marine, Earth, and Atmospheric Sciences, Box 8208, North Carolina State University, Raleigh, North Carolina 27695-8208
Lee DiTullio*
Department of Geology, Kochi University, Kochi 780, Japan

ABSTRACT

The primary purpose of this special publication is to show how Eocene through middle Miocene strata of the Shimanto Belt were affected by the Cenozoic geothermal regime of southwest Japan. This introductory paper synthesizes the regional and local geologic framework, as it applies to detailed studies completed on the Muroto Peninsula of Shikoku. In particular, we focus on the temporal and geometric relations between peak heating events and discrete stages in the deformation history. Rocks of the Shimanto Belt display the effects of a complicated history of polyphase folding, faulting, and cleavage formation. Interpretations of the structural geology follow conventional models for subduction-accretion via offscraping, underplating, and out-of-sequence thrusting. The positions of plate boundaries during the Paleogene phase of plate convergence in southwest Japan are somewhat difficult to substantiate, but that episode of tectonism may have been punctuated by subduction of the Kula-Pacific Ridge near the end of the Eocene epoch. The Neogene, on the other hand, was definitely a time of ridge-trench interaction. Geologic events of the middle Miocene (~15 Ma) are particularly noteworthy because they included anomalous near-trench magmatism, the cessation of backarc spreading in the Shikoku Basin, incipient collision between the Izu-Bonin and Honshu Arcs, the opening of the Sea of Japan, rapid rotation of crustal blocks in both northwest and southeast Japan, and the formation of high-rank coals within forearc-basin strata. The high ranks of organic metamorphism within the accretionary forearc, together with the anomalous basic and acidic volcanic and plutonic rocks, provide unambiguous and widespread evidence in favor of high geothermal gradients at relatively shallow depths. Furthermore, data from offshore extensions of the Shimanto Belt (Nankai accretionary prism) show that Holocene geothermal conditions are still unusually warm when compared to shallow levels of most "typical" subduction zones. Thus, much of the Cenozoic history of southwest Japan contradicts the paradigm of low-temperature, high-pressure metamorphism within subduction zones, and the exposed geology of the Muroto Peninsula provides some of the world's best examples of the thermal and structural effects of ridge-trench interactions.

---

*Present address: 326 12th St., Apt. 2L, Brooklyn, New York 11215.

Underwood, M. B., Hibbard, J. P., and DiTullio, L., 1993, Geologic summary and conceptual framework for the study of thermal maturity within the Eocene-Miocene Shimanto Belt, Shikoku, Japan, *in* Underwood, M. B., ed., Thermal Evolution of the Tertiary Shimanto Belt, Southwest Japan: An Example of Ridge-Trench Interaction: Boulder, Colorado, Geological Society of America Special Paper 273.

INTRODUCTION

The primary theme of this special publication is the relation between structural and thermal history within the Shimanto accretionary complex of southwest Japan (Fig. 1). One of the conceptual cornerstones in the field of accretionary-margin tectonics is the depression of isotherms within an overriding lithospheric plate as cold oceanic lithosphere is subducted (Ernst, 1974; Anderson and others, 1978; van den Beukel and Wortel, 1988). Abnormally low geothermal gradients are, no doubt, a prerequisite for the formation of a broad spectrum of blueschist-facies metamorphic terranes that exemplify many subduction complexes of the Alpine and circum-Pacific orogenic belts (Ernst, 1975, 1977; Cloos, 1985, 1986; Jayko and others, 1986). For example, the burial conditions associated with some of the high-pressure terranes and exotic blocks within the Franciscan Complex of California have been estimated at 5 to 10 kbar and 250°C or less (Cloos, 1982). Similarly, Jayko and others (1986) reconstructed Franciscan gradients ranging from roughly 7°C/km (Yolla Bolly terrane) to approximately 11°C/km (Pickett Peak terrane).

Not all subduction complexes have been exposed to the high-P, low-T conditions of burial inferred for Franciscan blueschists. The Sambagawa Belt of Japan, for example, records the effects of temperatures comparable to those of the Franciscan blueschists, but at significantly lower pressures (Ernst and others, 1970). Geothermal gradients for the Sambagawa were around 15 to 20°C/km based on temperature/pressure gradients of 40 to 65°C/kbar (Nakajima and others, 1977; Liou and others, 1987; Banno and Sakai, 1989; Tatasu, 1989; Wallis and Banno, 1990).

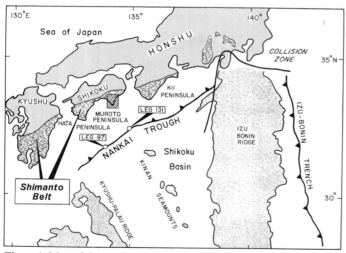

Figure 1. Map of the southwest Japanese Islands and important offshore tectonic and bathymetric features. Bathymetric highs are highlighted by stipple pattern. Note the locations of drill sites associated with DSDP Leg 87 and ODP Leg 131. Collision zone between the Izu-Bonin Arc (Iwo Jima Ridge) and the Honshu Arc is depicted by solid line without teeth. Major outcrop distributions of the Shimanto Belt are also highlighted. The principal study area for our investigations of Shimanto thermal structure is located on the Muroto Peninsula of Shikoku.

Even higher geothermal gradients (20 to 30°C/km) have been proposed for the underplated rocks of Kodiak Island, Alaska (Sample and Moore, 1987; Vrolijk and others, 1988), and for the Chugach accretionary complex of southeastern Alaska (Sisson and others, 1989). In addition to spatial variations in P-T gradients, the thermal regime along any single subduction margin can change considerably through time due to rearrangements of the governing plate boundaries. As an example, Underwood (1989) demonstrated that the geothermal gradient offshore northern California increased progressively through Cenozoic time as the Pacific-Farallon spreading ridge drew closer to the Franciscan subduction front (see also Zandt and Furlong, 1982; Heasler and Surdam, 1985).

Numerical models of thermal structure

It is widely accepted that the flux of heat from oceanic crust decreases as a simple function of crustal age (Lister, 1977; Parsons and Sclater, 1977; Yamano and Uyeda, 1988). Consequently, there is every reason to believe that the thermal structure of an accretionary prism will vary as a function of the age of subducting lithosphere at the deformation front. This hypothesis is supported, for example, by the computer models of Wang and Shi (1984), which graphically demonstrated that maximum temperatures within an accretionary prism should rise significantly (roughly 100°C) by allowing a mere 25 mW/m^2 of conductive heat transfer across the base of the wedge (Fig. 2). In reality, heat flow equals 95 to 100 mW/m^2 for oceanic crust that is 25 m.y. in age, and values for 120-m.y. crust are 43 to 46 mW/m^2 (Lister, 1977; Parsons and Sclater, 1977). Therefore, the effects of heat conduction through the prism should be quite pronounced wherever evenly moderately young oceanic crust enters a subduction zone.

Many numerical models have been developed to illustrate how thermal structure is controlled by changing boundary conditions (Fig. 2). Two of the most important variables, in addition to the age of the subducting slab, are the subduction rate and the angle of subduction (Wang and Shi, 1984; James and others, 1989; Dumitru, 1991). In addition, heat-flow data collected along individual marine transects show that geothermal gradients drop off markedly as a function of distance arcward of the subduction front; this wedging of isotherms, as illustrated in Figure 2E, occurs because of the combined effects of cooling the slab as it is subducted and thickening the deformed sediments in the overlying accretionary wedge by underplating and out-of-sequence thrusting (e.g., Davis and others, 1990; Langseth and others, 1990). Interpretations of data from heat-flow transects are complicated further by the effects of fluid advection (Reck, 1987; Moore and others, 1987, 1988; Davis and others, 1990; Langseth and others, 1990; Fisher and Hounslow, 1990). Warm water typically is liberated from deeper levels of the prism because of both mechanical compaction and diagenetic alteration of hydrous minerals such as smectite (Vrolijk, 1990). Shallow-level temperature gradients, therefore, can be very erratic in detail because of transient flow along fault-controlled and/or facies-controlled fluid conduits (Fisher and Hounslow, 1990).

Thermal effects of ridge-trench collision

Numerical simulations help illustrate the most extreme departures away from the "normal" scenario of cold oceanic subduction, particularly where an active spreading ridge comes into close proximity to a trench (DeLong and others, 1979; James and others, 1989). Near-trench volcanism is but one of several possi-

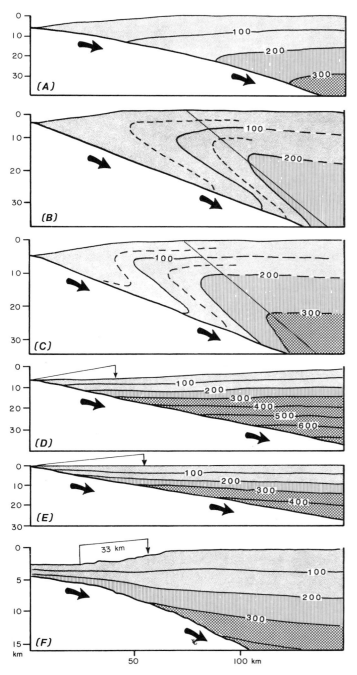

ble geologic manifestations of a ridge-trench collision (DeLong and Fox, 1977; Marshak and Karig, 1977; DeLong and others, 1978; Forsythe and Nelson, 1985; Cande and Leslie, 1986; Forsythe and others, 1986; Byrne and Hibbard, 1987; Nelson and Forsythe, 1989). Studies by Cande and others (1987) of the Chile Rise collision zone show that heat flow near the toe of the accretionary prism is actually higher than predicted in the model of DeLong and others (1979). Calculated values of heat flow range from a low of 62 mW/m^2 north of the ridge-trench intersection to a high of 343 mW/m^2 adjacent to the rift valley. As Cande and others (1987) pointed out, the plan-view intersection angle between the ridge segments and the subduction front exerts considerable control over regional geothermal patterns. One might infer that the region of elevated heat flow would be most confined along strike if an active spreading ridge is subducted at a right angle to the deformation front. The Chile Rise, on the other hand, is aligned nearly parallel to the subduction front, yet each successive spreading segment to the north of the collision zone steps away from the trench. Thus, the area of the subduction complex corresponding to extremely high heat flow is still fairly restricted.

Heat flow in subduction zones

The Oregon-Washington subduction margin is another well-studied example where young crust is being subducted (Kulm and Fowler, 1974). In this case, the Gorda-Juan de Fuca Ridge system is aligned at a higher angle to the subduction front (compared to offshore Chile), but the ridge-transform segments step successively towards the trench. Perturbations in heat flow caused by this geometry appear to be widespread (though not as

Figure 2. Comparison of models A to F showing thermal structure within accretionary prisms. Horizontal scales have been modified so that all are approximately equal; vertical scales, however, differ for each model. Model A is from van den Beukel and Wortel (1988), with key boundary conditions set as follows: age of subducting slab A = 70 Ma and subduction rate R = 80 km/my. Model B is from Wang and Shi (1984); R = 10 km/my and no heat flux is allowed across the base of the wedge. Model C is also from Wang and Shi (1984), with a heat flux of 25 mW/m^2 allowed across the base of the wedge; heat flow of this magnitude is unrealistically low for oceanic crust <120 Ma in age, but temperatures near the rear base of the wedge increase by nearly 100°C compared to Model B. Model D, from Dumitru (1991), shows an iteration that maximizes the prism temperatures by setting A = 4.2 Ma, R = 20 km/my, and basal shear stress = 10 MPa; note that the 300°C isotherm intersects the base of the prism less than 50 km from the deformation front, at a depth of approximately 9 km. Model E, from James and others (1989), also maximizes temperatures by setting A = 1.5 Ma and R = 75 km/my; the 300°C isotherm intersects the base of the prism roughly 55 km inboard of the deformation front at a depth of approximately 10 km. Model F, from Davis and others (1990), is based on projections of near-surface heat-flow data onto a multichannel seismic-reflection profile from the northern Cascadia margin; critical boundary conditions are A = 6 Ma and R = 45 km/my. Note that the 300°C isotherm intersects the base of the wedge only 33 km inboard of the deformation front at a depth of approximately 7.5 km.

extreme in magnitude as offshore Chile); numerous sea-floor measurements show peak values at the toe of the Cascadia slope in excess of 150 mW/m^2 (Korgen and others, 1971; Shi and others, 1988; Davis and others, 1990).

A final submarine example that is quite pertinent to our subaerial investigation of the Shimanto Belt is the Nankai Trough, located offshore Shikoku, Japan (Fig. 1). Temperature measurements from shallow heat-flow probes and DSDP (Deep Sea Drilling Project) boreholes, together with calculations based on the depths of gas-hydrate bottom-simulating reflectors (BSR), show that the Holocene thermal regime there is relatively warm (Yamano and others, 1982, 1984; Kinoshita and Yamano, 1986; Ashi and Taira, this volume). Typical geothermal gradients on the landward slope of Nankai Trough range from 46 to 50°C/km (in DSDP boreholes) to 41 to 66°C/km (from BSR analyses), with the expected landward decrease in heat flow. An even higher gradient (approximately 110°C/km) was documented recently at Site 808 of the Ocean Drilling Program (ODP), which penetrated the toe of the prism offshore Cape Muroto (Taira and others, 1991; Yamano and others, 1992). This temperature gradient decreases slightly with depth (largely because of down-hole increases in thermal conductivity), and the data from Site 808 agree fairly well with models based solely on the conductive flux of heat from young oceanic crust that formed by back-arc spreading in the Shikoku Basin (Yamano and others, 1984; Kinoshita and Yamano, 1986; Nagihara and others, 1989; Yamano and others, 1992). Comparative values of shallow-level geothermal gradients from other subduction zones are shown in Table 1. This compilation shows that the present-day conditions in Nankai Trough are not really that unusual.

Tectonothermal history of southwest Japan

Two fundamental questions arise from the overview above. First, have the effects of relatively high geothermal gradients been preserved in the Cenozoic rock record of southwest Japan, thereby signaling a departure away from the "blueschist-norm" of subduction zones? Our studies (as fully described in the following papers in this volume) clearly show that the answer to this question is "yes." Second, how far back in geologic time can we trace the abnormal geothermal regime of Nankai Trough? This second question is particularly critical because if an asthenospheric window opens up in the aftermath of a ridge collision (as a consequence of triple-junction migration), then the expected thermal effects should be restricted in both time and space (Dickinson and Snyder, 1979; Zandt and Furlong, 1982; Furlong, 1984). Similarly, the ridge-collision models reveal that the thermal effects should decrease rapidly through time as spreading ceases and the oceanic lithosphere cools (DeLong and others, 1979). In fact, based on recent computer simulations, James and others (1989) and Dumitru (1991) have suggested that the subducting oceanic crust must be quite young (less than 2 m.y. old) to produce an appreciable thermal anomaly. It should be noted, however, that these recent models were designed to simulate conditions within the middle of a forearc region (i.e., 80 to 100 km inboard of the

subduction front), and higher geothermal gradients might be expected closer to the prism toe (Fig. 2). Nevertheless, it remains paradoxical to explain how abnormally high heat flow can be maintained over broad expanses of a subduction margin for extended periods of time (i.e., 10 to 15 m.y.), unless the offending spreading ridge persists as a magmatic heat source throughout the same time span.

Most of the papers in this volume focus on Eocene-Miocene accreted rocks within the subaerial Outer Zone of Japan (Fig. 1). We stress at the beginning that the history of interaction between the Japanese accretionary margin and subducting plates of the Pacific Basin has been long and very complicated (Taira and others, 1980, 1982, 1988). For example, the Miocene tectonic events alone include the cessation of sea-floor spreading in the Shikoku Basin, the opening of the Sea of Japan, the rotation of

TABLE 1. SHALLOW-LEVEL GEOTHERMAL GRADIENTS IN MODERN SUBDUCTION ZONES

Locality	Gradient (°C/km)	Method*	References†
Eastern Aleutians	28	BT	1
	32	GH	1
Central Aleutians	21–23	GH	2
Northern Cascadia (Vancouver/Washington)	50	GH	3
	40–80	HFP	4
Southern Cascadia (Oregon)	40–80	HFP	4
	110	HFP	5
Northern California (Eel River Basin)	55	GH	6
Mexico	22–40	BT	7
	22–31	GH	8
Nicaragua	30–35	GH	8
Costa Rica	30–44	GH	8
Peru	42–52	BT	9
Chile Rise collision	60–350	HPF	10
Barbados	28–105	BT	11
	33–80	HFP	12
Japan Trench	24–36	BT	13
Nankai Trough	46–50	BT	14
	41–66	GH	8
	110	BT	15

*BT = DSDP/ODP borehole temperature; GH = gas-hydrate method; HFP = heat-flow probe.
†1 = Kvenvolden and von Huene (1985); 2 = McCarthy and others (1984); 3 = Davis and others (1990); 4 = Shi and others (1988); 5 = Korgen and others (1971); 6 = Field and Kvenvolden (1985); 7 = Shipley and Shepard (1982); 8 = Yamano and others (1982); 9 = Suess and others (1988); 10 = Cande and others (1987); 11 = Fisher and Hounslow (1990); 12 = Langseth and others (1990); 13 = Langseth and Burch (1980); 14 = Kinoshita and Yamano (1986); 15 = Taira and others (1991).

large crustal blocks in both southeast and northwest Japan, a possible change from subduction of the fused Pacific/Kula Plate to subduction of the juvenile Philippine Sea Plate, widespread anomalous near-trench magmatic activity, and the incipient collision between the Izu-Bonin Arc and the Honshu Arc (Seno and Maruyama, 1984; Niitsuma and Akiba, 1985; Otofuji and Matsuda, 1987; Niitsuma, 1988; Taira and others, 1989). Any interpretations of local tectonothermal processes must take these regional events into account.

The field investigations completed on the Muroto Peninsula (Fig. 3) were designed to document deformation pathways and to interpret the local tectonostratigraphic evolution (Hibbard and Karig, 1987; Hibbard, 1988; DiTullio, 1989; DiTullio and Byrne, 1990; Hibbard and others, 1992). Laboratory analyses of thermal maturity followed in an attempt to test some of the structural interpretations (DiTullio and others, this volume; Hibbard and others, this volume; Laughland and Underwood, this volume). At the same time, structural and stratigraphic correlations have improved between the Shimanto onshore geology and younger off-

shore extensions of the accretionary prism, as recorded by seismic-reflection data and DSDP/ODP boreholes (Taira, 1985; Kagami and others, 1986; Niitsuma, 1988; Okuda and Honza, 1988; Taira and others, 1988, 1989; Moore and others, 1990). Similarly, the plate-tectonic reconstructions of eastern Asia and the northwest Pacific Basin have undergone significant revisions and refinements (Engebretson and others, 1985; Kimura and Tamaki, 1986; Lonsdale, 1988; Jolivet and others, 1989; Maruyama and others, 1989; Otsuki, 1990; Hibbard and Karig, 1990a). All of these scientific developments have enhanced our ability to interpret the paleothermal structure as it was superimposed on the accretionary stratigraphy of southwest Japan. The remainder of this introductory paper establishes the observational background into which we place those interpretations.

GEOLOGIC BACKGROUND OF MUROTO PENINSULA

The regional geologic framework of the Muroto study area has been summarized by Taira and others (1982, 1988, 1989).

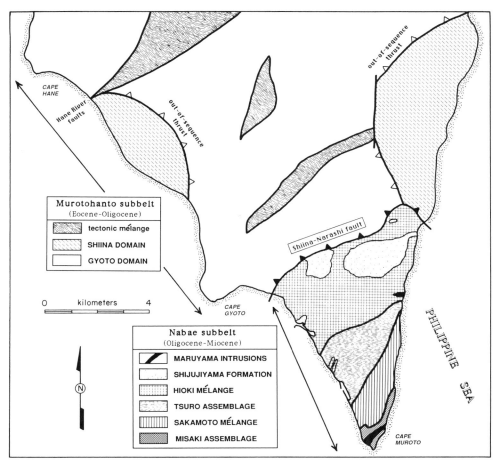

Figure 3. Tectonostratigraphic units within the Tertiary Shimanto Belt, Muroto Peninsula, Shikoku. Structural interpretations and stratigraphic nomenclature for the Murotohanto subbelt (Eocene-Oligocene) are from DiTullio (1989) and DiTullio and Byrne (1990). Analysis of the Nabae subbelt (Oligocene-Miocene) was completed by Hibbard (1988) and Hibbard and others (1992).

The Shimanto Belt (or Supergroup, in the conventional strati-graphic sense) extends from the Boso Peninsula to the Ryuku Is-lands (Fig. 1), and these rocks are separated from predominantly Mesozoic accreted terranes of south-central Japan (to the north) by the Butsuzo Tectonic Line. In essence, all of the structural and stratigraphic relations within the Shimanto Belt can be inter-preted within the context of progressive accretion-related deformation, over a time span of roughly 100 m.y. On Shikoku Island, the Shimanto Belt has been further divided into various subbelts, largely on the basis of fossil ages. The youngest of these tectonostratigraphic units, termed the Nabae subbelt (Taira and others, 1980), crops out on the southern tip of the Muroto Penin-sula (Fig. 3). These late Oligocene to early Miocene strata consist mostly of marine sedimentary deposits. The Nabae subbelt is separated from Eocene and early Oligocene strata of the Muroto-hanto subbelt to the north by a steeply dipping boundary known as the Shiina-Narashi fault (Fig. 3). Collectively, the Eocene through Miocene rocks comprise the Upper Shimanto Group, and Cretaceous sequences farther to the north are placed strati-graphically within the Lower Shimanto Group. Three discrete stages of Cenozoic tectonic deformation have been recognized on the Muroto Peninsula (Table 2). The rough equivalents of stages 1 and 2 occurred earlier in the Murotohanto subbelt than in the Nabae subbelt, but stage-3 deformation evidently affected the entire study area simultaneously.

Murotohanto subbelt

The Murotohanto subbelt was subdivided by DiTullio and Byrne (1990) into two structural domains containing relatively coherent strata, plus several fault-bounded units of polymictic mélange (Fig. 3). Age control for these rocks comes from mollusc shells, radiolaria, and calcareous nannoplankton. The mélange zones are characterized by a well-indurated matrix of black-gray argillite that displays a strong, anastomosing, spaced cleavage. Blocks within the mélange include oceanic rock types such as pillow basalt, basaltic tuff, chert, and hemipelagic mudstone (Taira and others, 1988). Fossils from the black-shale matrix are Eocene in age, as are fossils from the blocks of hemipelagic mudstone (Taira and others, 1982; Okamura and Taira, 1984). Coherent sequences of turbidites make up the remainder of the section; these strata are mostly early to late Eocene in age and consist of arkosic sandstone, shale interbeds, rare thin ash beds, and a minor facies of debris flows and slump folds. One fossil locality at Cape Gyoto contains late Eocene to early Oligocene radiolaria (Sakai, 1988), and fossils within equivalent rocks on the Hata Peninsula (Fig. 1) are as young as about 33 Ma (Taira and others, 1980; Agar and others, 1989). The full extent of early Oligocene strata on the Muroto Peninsula remains uncertain. The turbidite section has been subdivided into two structural domains (Gyoto and Shiina). Two separate deformation events (D_1 and

TABLE 2. DEFORMATION STAGES IN THE TERTIARY SHIMANTO BELT, MUROTO PENINSULA, SHIKOKU, JAPAN

Rock Unit	Stage 1 Features	Stage 2 Features	Stage 3 Features
Murotohanto subbelt (Eocene–early Oligocene)			
Oceanic mélange	Mélange foliation	Not documented	Not documented
Shiina domain	Clastic dikes Slump folds Mesoscale imbrication	E-trending faults/folds (F_1) E-striking cleavages (S_{1a}, S_{1b}) Out-of-sequence thrust NE-striking cleavage (S_{2a}?)	Spaced high-angle faults Muroto flexure Shiina-Narashi fault
Gyoto domain	Clastic dikes Slump folds Mesoscale imbrication Layer-parallel extension	Layer-parallel extension (?) Out-of-sequence thrust NE-trending folds (F_2) NE-striking cleavages (S_{2a}, S_{2b})	Spaced high-angle faults Muroto flexure Shiina-Narashi fault
Nabae subbelt (late Oligocene–middle Miocene)			
Hioki/Sakamoto	Mélange foliation Bedding-extension faults	Pressure-solution cleavage Mesoscale folds Thrust faults	Muroto flexure Shiina-Narashi fault Meso-scale high-angle faults Maruyama intrusions
Tsuro/Misaki	Isoclinal, sheath, and lobate folds	Pressure-solution cleavage Mesoscale folds Megascale folds Thrust faults	Muroto flexure (?) Mesoscale high-angle faults Maruyama intrusions
Shijujiyama	None	Megascale folds Homoclinal tilting	Shiina-Narashi fault Muroto flexure Mesoscale high-angle faults

D_2) have been assigned to stage 2, and the interpreted divergence in structural history between the domains is based on a discordance in the orientations of the respective stage-2 fabrics (DiTullio and Byrne, 1990).

Shiina domain. Slump folds, seaward-vergent imbrication, and clastic dikes are the principal manifestations of stage-1 deformation within the Shiina domain (Table 2). Turbidite beds strike dominantly to the east, in response to regional-scale seaward-vergent faulting and folding (D_1). This D_1 regime of north-south shortening (early stage 2) was also responsible for the formation of two sets of pressure-solution cleavage. The S_{1a} cleavage is axial planar to the east-trending F_1 folds, whereas the S_{1b} cleavage transects the F_1 folds. Unlike the Gyoto domain (described below), seaward-verging folds did not develop in the Shiina domain during the latter part of stage 2; a local northeast-striking cleavage has been mapped, however, and this cleavage appears to be the only time-equivalent to the D_2 fabrics of the Gyoto domain (i.e., correlative with S_{2a}). The absence of D_2 folds in the Shiina domain may be related to the fact that the strata were already tightly folded during the D_1 event; in addition, rocks of the Shiina domain probably were faulted into a lower level of the accretionary prism during late stage-2 deformation, and this deeper zone evidently was removed from the regime of subhorizontal shortening (DiTullio and Byrne, 1990).

Gyoto domain. Strata within the Gyoto domain (Fig. 3) maintain a characteristic strike to the northeast, rather than east-west. As in the Shiina domain, stratal-disruption features assigned to stage 1 include nontectonic (slump) folds, seaward-vergent mesoscale thrust faults, and clastic dikes. The clastic dikes indicate that the shortening direction during stage 1 was dominantly north-south. In addition, however, the Gyoto domain contains early phase fabrics produced by layer-parallel extension. This initial phase of deformation began prior to lithification, or at least before significant compaction of the sediments had been completed, and DiTullio and Byrne (1990) attributed the early stratal disruption to sediment offscraping during subduction-accretion. Layer-parallel extension may have continued into stage 2 of the deformation history, which is interpreted as a phase of subsequent intraprism shortening. Stage 2 did not produce D_1 folds or cleavage in the Gyoto domain, but instead culminated in younger (D_2) northeast-trending megascopic folds (F_2) and two sets of northeast-striking pressure-solution cleavage. The S_{2a} cleavage is axial planar, whereas S_{2b} transects the F_2 folds. These later D_2 features collectively define the prevailing structural grain of the Gyoto domain. The transition from stage-1 to stage-2 deformation, in summary, included a pronounced change in the direction of shortening, from north-south to northwest-southeast (with respect to the present-day geographic coordinates).

Nabae subbelt

The Nabae subbelt, as introduced by Taira and others (1980), is herein subdivided into three major tectonostratigraphic elements (Fig. 3). Most of these rocks apparently formed within

the span of late Oligocene to middle Miocene time; the internal biostratigraphy within this part of the Shimanto section is not very precise (Hibbard and others, 1992), and Eocene radiolarians have been reported from some blocks within the mélange (Sakai, 1988). An alternative to our nomenclature and structural subdivision also has been proposed by Sakai (1988), with the spatial limit of coherent Eocene rocks placed farther to the north. Most of the rocks that we assign to the Nabae subbelt are deep-marine sedimentary deposits (turbidites and hemipelagites) that exhibit such a highly complicated structure that a "normal" stratigraphy cannot be discerned. Consequently, Hibbard and others (1992) suggested using the name "Nabae complex" to encompass these intensely deformed units.

Nabae complex. Figure 3 shows that the Nabae complex contains two belts of mudstone-rich mélange (Hioki and Sakamoto mélanges) and two belts of comparatively coherent turbidites and hemipelagites (Tsuro and Misaki assemblages). Three stages of deformation can be recognized in the Nabae complex, in the same gross sense as the stages recognized in the Murotohanto subbelt (Table 2). The two mélanges are lithologically alike, and structural analysis supports the idea of a subsurface connection (Hibbard and others, 1992). Unlike the Eocene mélanges to the north (which are dominated by blocks of basalt and pelagic sedimentary rocks), phacoids within the Hioki and Sakamoto mélanges consist mostly of clastic sedimentary rocks that are indigenous to the coherent units of the Nabae complex; the remaining lithologies include carbonate, small fragments of vein quartz, and rare mafic volcanic rocks (pillow basalt). The inclusions are engulfed in an intensely deformed matrix of multicolored mudrock. The primary structural grain strikes to the northeast, and this crude foliation is the collective expression of an anastomosing scaly cleavage, pinch-and-swell fabrics, aligned phacoids, and bedding-extension faults. These early (stage-1) structures were overprinted by an intermediate phase of deformation (stage-2) that included folding, thrust faulting, and cleavage development (Table 2).

The Tsuro assemblage (Fig. 3) comprises variegated, silica-rich hemipelagic mudstone, rhythmic interbeds of sandstone turbidites and shale, massive sandstone, and minor pebbly mudstone. Strata strike east-northeast and dip moderately to steeply to the south-southeast; many of the beds are overturned. The Tsuro assemblage likely represents a coarsening-upward succession of abyssal-plain and trench-wedge deposits. The lithologies of the coherent Misaki assemblage (Fig. 3) are similar but somewhat less varied, consisting mostly of sandstone-shale interbeds, together with conglomerate, massive sandstone, mudstone, and minor limestone. Misaki strata also strike to the northeast and dip to the southeast; deposition of these sediments probably occurred within a trench-wedge setting, as well. Early phase (stage-1) structures within both of the "coherent" assemblages consist of folds with isoclinal, sheath, and lobate geometries. Stage-2 deformation produced mesoscopic and macroscopic folds (asymmetric and landward vergent), telescoping along thrust faults (oriented subparallel to fold axes), and axial-planar pressure-solution cleavage (Hibbard and Karig, 1987; Hibbard and others, 1992).

Shijujiyama Formation.
The Shijujiyama Formation (Fig. 3) is a distinctive stratal succession that displays a more intact character than surrounding units within the Nabae complex. Lithologies include massive sandstone, massive mudstone, conglomerate, and a near-basal volcanic member composed of pillow lava, massive basalt, mafic breccia, and volcanic conglomerate. The volcanic rocks have not been dated by radiometric techniques, but interbeds of the massive mudstone contain early Miocene microfossils and mollusc fragments (Ishikawa, 1982; Sakai, 1988). Chemical analyses show an "island arc" petrochemical signature, as defined particularly well by REE patterns (Hibbard and Karig, 1990b). Bedding attitudes within the main outcrop area (Fig. 3) delineate a single regional-scale syncline, and north-northwest dipping homoclinal structures have been mapped in the smaller outliers. The pervasive early-stage and intermediate-stage mesoscale folds and cleavages that typify the Nabae complex are conspicuously absent in the Shijujiyama Formation. The origin of these nearly undeformed strata is controversial. Sakai (1981, 1988) interpreted the section as a huge (kilometer-scale) olistolith within an olistostrome belt that stretches across the entire Outer Zone of southwest Japan. In contrast, Hibbard and others (1992) have argued that the Shijujiyama Formation represents a trench-slope-basin environment that was constructed above an accreted mélange basement, analogous to those described by Moore and Karig (1976).

Maruyama intrusive suite.
A suite of mafic rocks intrudes the Nabae complex along the east coast and the southern tip of the Muroto Peninsula. Included in this suite are two major bodies of layered gabbro and numerous stocks and dikes of fine-grained basalt. Igneous ages, as determined by both Rb/Sr and K/Ar radiometric methods, cluster at approximately 14 Ma (Hamamoto and Sakai, 1987; Hibbard and Karig, 1990a, b). These dates coincide closely with the ages of other "Outer Zone" plutons that intrude the Shimanto Belt elsewhere (Shibata, 1978; Hibbard and Karig, 1990a). The Maruyama basalts, moreover, can be classified chemically within the context of mid-ocean ridge basalts (MORB). REE patterns are transitional between normal (N-type) and enriched (E-type) MORB (Hibbard and Karig, 1990b).

Contacts and deformation history

Gyoto-Shiina contact.
The structural differences between the Gyoto and Shiina domains (Fig. 3) can be explained in terms of a divergence in the respective deformation pathways during an intermediate stage of the deformation history. Because both domains show the effects of similar stage-1 deformation histories, DiTullio and Byrne (1990) assumed that they were initially accreted to the margin of Japan together. The Gyoto domain, however, then escaped the D_1 event of regional folding and cleavage (early in stage 2) and experienced only the subsequent D_2 folding and cleavage (later in stage 2). The contact between the domains,

therefore, is interpreted as a major out-of-sequence thrust fault that was activated during the middle of stage 2 (Fig. 4B). Locally, the Gyoto domain structurally overlies units of oceanic mélange, which in turn overlie the Shiina domain. According to the interpretation of DiTullio and Byrne (1990), trench turbidites of the Gyoto domain were thrust beneath the Shiina domain during stage 1; once placed in this deeper structural position within the Eocene accretionary prism, Gyoto strata were isolated from the first episode of stage-2 folding but still subjected to layer-parallel extension. When the shortening direction changed from north-south to northwest-southeast (a consequence, presumably, of a fundamental realignment in plate motions), the out-of-sequence thrust developed, thereby carrying the Gyoto rocks up into higher levels of the accretionary prism (Fig. 4B); the D_2 deformation fabrics then were superimposed.

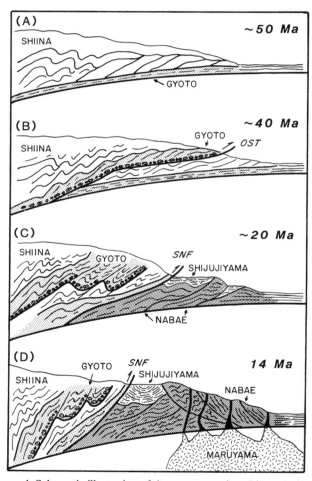

Figure 4. Schematic illustration of the tectonostratigraphic evolution of the Tertiary Shimanto Belt on the Muroto Peninsula of Shikoku. OST, out-of-sequence thrust (Shiina-Gyoto domain boundary); SNF, Shiina-Narashi fault (Murotohanto-Nabae subbelt boundary). See text for descriptions of critical stages in the depositional and deformation history. Based largely upon the structural interpretations of DiTullio and Byrne (1990) and Hibbard and others (1992).

Nabae complex. The nature of the original contact between the Tsuro assemblage and the Hioki/Sakamoto mélanges has been obscured by subsequent late-stage, high-angle (stage 3) faulting. Where best exposed, the contact appears to be a folded thrust fault; based on the presence of related recumbent folds, Hibbard and others (1992) have suggested northwest-directed emplacement of the Tsuro assemblage over the mélange. The Hioki and Sakamoto mélanges evidently merge and form a continuous substrate to the Tsuro thrust sheet, and windows of mélange are exposed along the southwest coastal section of the Tsuro assemblage (Fig. 3). Farther south, the contact between the coherent Misaki assemblage and the Sakamoto mélange is best regarded as a structural transition; the contact zone includes structural slices of coherent turbidites and mudrocks that have been sheared into the mélange, and the boundary is overprinted by a strong intermediate-stage cleavage. Because of these complications, the structural position of the Misaki assemblage, with respect to the Sakamoto mélange, remains uncertain.

Basal Shijujiyama contact. Shijujiyama strata definitely rest above the Hioki mélange, and the contact marks a dramatic change in both the deformation style and depositional facies. However, because of the combined effects of poor inland exposure and an apparent structural gradation between the two units, the origin of the contact remains unambiguous. The boundary has been interpreted as an olistolith surface (Sakai, 1981, 1988), as a fault (Taira and others, 1980), and as a tectonized unconformity separating slope-basin deposits from underlying Hioki mélange (Fig. 4; Hibbard and others, 1992). The unconformable relationship is consistent with descriptions of broadly coeval strata located elsewhere in the Outer Zone; analogous stratigraphic units include the Kumano Group on the Kii Peninsula to the east (Tateishi, 1978; Chijiwa, 1988; Hisatomi, 1988) and the Misaki Group near Cape Ashizuri (Hata Peninsula) of western Shikoku (Taira and others, 1982; Sakai, 1988).

Shiina-Narashi fault. The Shiina-Narashi fault (Hibbard and others, 1992) is clearly the most fundamental structural break within the Tertiary Shimanto Belt of Muroto (Figs. 3, 4C). The fault is high-angle near the surface, cuts across both structural domains of the Murotohanto section, and truncates Oligocene-Miocene strata of both the Hioki mélange and the Shijujiyama Formation. Although the fossil control is not definitive, there may be a mid-Oligocene hiatus across the subbelt boundary. The fault was recognized in the field because of pronounced changes in structural fabric. Rocks to the north display a homogeneous, continuous, almost slaty cleavage, whereas rocks in the Nabae subbelt are generally polished and slickensided near the fault, with slickenlines showing highly variable orientations. Farther away from the fault, the matrix of the Hioki mélange displays a crude, anastomosing, scaly fabric, and the Shijujiyama Formation is comparatively undeformed. Thus, the fault not only separates rocks of differing age, it also juxtaposes rocks with fundamental differences in structural style and deformation history. Based upon the earlier reconnaissance work of Mori and Taguchi (1988), there is also a significant change in the rank of organic

metamorphism across the fault. Although the north side of the fault clearly is up, the absolute slip direction remains uncertain. We believe that the fault is a rotated, north-dipping and possibly, listric thrust (Fig. 4), but in the absence of reliable kinematic indicators, the alternative of normal displacement along a south-dipping fault cannot be eliminated.

Muroto flexure. Stage-3 deformation affected all rock units within the Tertiary part of the Shimanto Belt, presumably at the same time. The most important structure to develop during stage 3 is a major north-trending cross fold, termed the Muroto flexure (Hibbard and Karig, 1990a). Within the Murotohanto subbelt, the Muroto flexure appears to maintain a subhorizontal axis (Fig. 5); this interpretation is favored because regional-scale stage-2 folds on the west coast of the peninsula display subhorizontal axes, whereas folds of the same generation on the east coast have had their axes rotated from subhorizontal to subvertical plunges (DiTullio and Byrne, 1990). In addition, the axial zone of the flexure appears to coincide with a high-angle reverse fault along which some differential uplift of the Eocene section has occurred (Fig. 5). Based upon the map pattern illustrated by DiTullio and Byrne (1990), the Gyoto-Shiina out-of-sequence thrust maintains a westward dip on the east coast and reverses to an eastward dip on the west coast; evidently, this warp of the thrust also occurred during stage 3. Smaller high-angle faults (e.g., Ozaki, Flat Rock, Kabuka, Shimizu) offset strata along the east coast within the Shiina structural domain (Fig. 5), and these faults probably splay off the axial zone of the Muroto flexure. Similar high-angle faults have been mapped within both domains on the west coast (Fig. 5).

The Muroto flexure appears to tighten toward the south. According to Hibbard and Karig (1990b), the plunge of the fold axis also steepens toward the south, and the flexure rotates the pervasive strikes of both bedding and mélange fabric of the Hioki mélange and Shijujiyama Formation along the east side of the peninsula. In addition, mafic igneous rocks of the Maruyama intrusive suite (Fig. 3) are concentrated near the axis of the Muroto flexure, and, based on structural and paleomagnetic arguments presented by Hibbard and Karig (1990b), the 14-Ma magmatic emplacement event (Fig. 4) occurred during or after the formation of the fold. The flexure is likewise responsible for the pronounced bend in the surface trace of the Shiina-Narashi fault (Figs. 3, 5). Finally, there is widespread evidence of cogenetic displacement and precipitation of quartz veins on a pervasive system of high-angle, mesoscale faults and tension fractures throughout the Nabae subbelt (Hibbard and others, 1992).

Neotectonic activity. The Outer Zone of southwest Japan is generally regarded as a nonactive fault province, but the peninsular region of Shikoku has experienced great earthquakes, uplift and tilting of coastal terraces, and low-angle thrusting events (Sakai, 1987). Mori and Taguchi (1988) showed that the magnitude of Holocene uplift increases toward the southern tip of Cape Muroto, and the average rate of uplift has been 2 cm/100 years over the past 200,000 years (Yoshikawa and others, 1964). Two groups of active faults have been mapped, and Sakai (1987) has

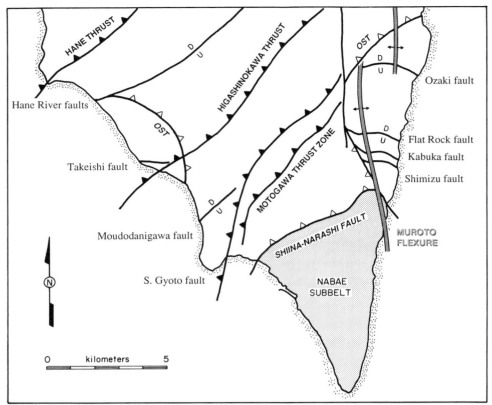

Figure 5. Map of late-stage structural features and active Quaternary thrust faults along the Muroto Peninsula of Shikoku. Surface traces of the Muroto flexure (hinge zone) and the Shiina-Narashi fault are from Hibbard and Karig (1990b). Quaternary thrust faults were mapped by Sakai (1987). High-angle faults are from Taira and others (1980) and DiTullio and others (this volume).

attributed this fault activity to compressional crustal deformation and shear fractures caused by subduction of the Philippine Sea Plate and progressive uplift of the accretionary prism.

Faults of the first group are south-vergent thrusts that strike northeast (Fig. 5), roughly parallel to the dominant structural grain of the Shimanto Belt. Fault offsets have been documented in the upper Pliocene beds of the Nobori Formation near Cape Gyoto (Sakai, 1987). It is worth noting that Sakai (1987, 1988) placed the southern limit of the early Tertiary (Murotohanto) subbelt along the Motogawa thrust zone (Fig. 5), which intersects the coast just north of the Shiina-Narashi fault boundary, as mapped by Hibbard and others (1992). The Quaternary thrust faults also can be traced offshore using seismic-reflection data; the thrusts connect with zones of drape folds that affect shelf to continental-slope deposits of Pliocene to Quaternary age. The second group of active faults consists of near-vertical conjugate sets with normal senses of displacement; these faults are oriented discordant to the Shimanto trend, and they offset beach-pebble beds and alluvial deposits on uplifted coastal terraces.

TYPES OF PALEOTHERMAL ANOMALIES

The contact relations and deformation history, as summarized above, provide the observational and interpretive framework for our investigation of paleothermal structure within the Shimanto Belt. Past studies have shown that the relative timing of peak heating within any orogenic belt, particularly with respect to specific stages of deformation, can be established by determining whether discordance exists between megascopic structural geometries and paleothermal gradients. For example, a fault contact must be considered "post-metamorphic" if the paleothermal structure is truncated and marked contrasts in organic rank or metamorphic grade are obvious across the boundary. As illustrated in Figure 6, several types of geometric and temporal relations are possible between the thermal structure and the rock structure.

The simplest configuration of thermal structure is produced by a uniform increase in temperature with stratigraphic depth (Fig. 6A). This type of gradient is rare within subduction com-

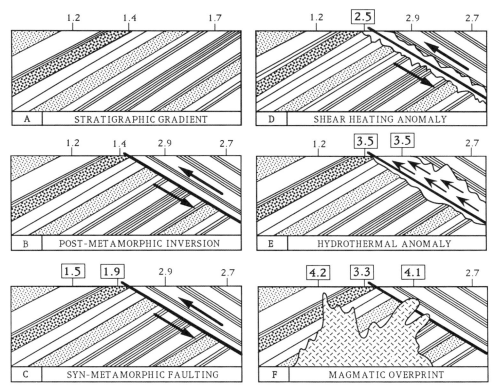

Figure 6. Types of thermal-maturity anomalies in terranes affected by thrust faulting. Numbers refer to hypothetical values of mean vitrinite reflectance. Figure 6A shows a stratigraphic gradient that has been tilted but remains intact; in highly deformed terranes (i.e., mélanges), gradients of this type may change with respect to the depth of structural burial rather than the primary depositional stratigraphy. Figure 6B illustrates intact gradients within both the hanging wall and footwall of a thrust; these gradients were imparted prior to faulting and never reset. Figure 6C displays a thrust system in which rocks of the footwall have been reset due to heat conduction across the base of the warmer hanging wall; increases in thermal maturity within the footwall depend largely upon depth below the thrust. In Figure 6D, shear heating during fault dislocation produces a narrow aureole of anomalous values in the immediate vicinity of the thrust; both hanging wall and footwall can be affected. Figure 6E depicts a system of hydrothermal circulation along a highly permeable fault zone; the spatial pattern of this anomaly will depend on the ability of hot fluids to penetrate into the hanging wall. Figure 6F exhibits reset thermal maturities in both hanging wall and footwall due to magmatic intrusions; the local rank of organic metamorphism under these circumstances will be dictated by distance from an intrusive contact, but widespread magmatic activity also should be accompanied by a regional-scale increase in heat flow.

plexes because of the effects of progressive deformation. Nevertheless, temperature gradients can be reconstructed with respect to structural position, even if the original depositional stratigraphy has been destroyed completely. If a major phase of deformation within the accretionary prism postdates thermal maturation, then isothermal contours (e.g., as defined by values of vitrinite reflectance) should be folded into antiformal and synformal geometries and/or truncated by faults; moreover, the thermal structure within one structural domain may be strongly discordant to the pattern defined by an adjacent tectonostratigraphic unit.

Thermal-maturity inversion occurs in a thrust belt when hanging-wall strata are uplifted and thrust over less mature strata of a footwall (Fig. 6B). A pronounced increase in thermal maturity above such a fault requires the hanging wall both to have reached maximum temperature and to have cooled prior to thrusting (i.e., post-metamorphic faulting); otherwise, the footwall will be subjected to the effects of conductive heat transfer across the fault surface, and thermal-maturity indicators will be reset. With synmetamorphic faulting (Fig. 6C), emplacement of a hot hanging wall over cooler footwall strata triggers a temporary saw-tooth inversion of the geothermal gradient, followed by thermal relaxation (Oxburgh and Turcotte, 1974; Brewer, 1981; Shi and Wang, 1987). The extent to which the footwall's thermal structure is altered by this heat transfer is governed by several factors, such as the temperature contrast between the fault blocks, thermal conductivity, rate of slip, total time of thrusting, distance moved, thickness of the thrust sheet (i.e., depth of structural burial), and the prior temperature history of the footwall. In addition, the regional-scale thermal structure of a thrust belt will be affected by the number of individual faults and the stacking order of the thrust sheets (Davy and Gillet, 1986).

In theory, shear heating anomalies can be produced during slip events on faults (Cardwell and others, 1978; Scholz, 1980; Johnston and White, 1983). Under these circumstances, thermal perturbations may develop along the margins of both fault blocks (Fig. 6D). Some shear heating aureoles are limited to the scale of centimeters (e.g., Bustin, 1983), and they are most likely to occur near geometric asperities such as lateral ramps. Several important parameters influence the magnitude of shear heating, including slab thickness, distance and rate of movement, coefficients of friction for both fault blocks, shear strength, shear strain rates, and time intervals over which the shear-derived heat is applied. Appreciable shear heating at shallow levels of an accretionary prism seems unlikely because of the relatively low values of sediment shear strength.

It is becoming increasing apparent that both diffuse fluid advection and focused fluid flow can alter the shallow-level thermal regime of an accretionary prism (Kulm and others, 1986; Moore and others, 1987, 1988; Reck, 1987; Moore, 1989; Fisher and Hounslow, 1990). As the prism sediments dewater through mechanical compaction and clay diagenesis, warm fluids migrate toward the sea floor and move preferentially along zones of high permeability, such as faults or sections of sand. The water-temperature anomalies actually recorded at the sea-floor vent sites are minute, on the order of tenths of 1°C (e.g., Le Pichon and others, 1987b), so preservation in the rock record of this type of shallow-level activity is unlikely. Nevertheless, at greater depths, the effects of hydrothermal activity on both organic and inorganic metamorphism can be pronounced (Fig. 6E). Borehole samples from continental hydrothermal systems (e.g., Barker and others, 1986) and outcrops within accreted terranes (e.g., McLaughlin and others, 1985; Underwood and others, 1988; Vrolijk and others, 1988) both exhibit the effects of overprints caused by hot migrating fluids.

Figure 6F illustrates two hypothetical accreted terranes that were thermally overprinted (after amalgamation) by heat emanating from a pluton. Under these circumstances, isothermal contours cut across the terrane boundary and form a concentric pattern whose geometry, in detail, is dictated by the size and the shape of the intrusion. Anomalies in both organic and inorganic metamorphism also can be caused by conduction away from smaller sills and dikes. As a general rule, values of vitrinite reflectance will increase above background over a distance equal to approximately two times the thickness of the intrusion (Dow, 1977; Bostick and Pawlewicz, 1984). Because many magmatic events are rapid compared to the time required for vitrinite reflectance to equilibrate with a given temperature field, the maximum

paleotemperature at the boundary of an intrusion generally will be underestimated if one calculates the "equilibrium" value using any of the available models for organic metamorphism (e.g., Waples, 1980; Barker and Pawlewicz, 1986; Sweeney and Burnham, 1990). Estimates of peak rock temperature near intrusions, therefore, should be calculated using the results of short-term laboratory experiments, such as those summarized by Bostick (1979). In addition to these local effects, however, widespread magmatism should be accompanied by regional-scale increases in heat flow. Thus, if the vertical geothermal gradient increases substantially and uniformly along strike, absolute values of vitrinite reflectance should increase even if the overall geometry of the thermal structure does not mimic the shape of an intrusion.

REGIONAL CONTEXT OF SHIMANTO THERMAL MATURITY

There are many important geologic events and controversial issues that have direct bearing on interpretations of paleothermal and structural data from the Tertiary Shimanto Belt. In order to appreciate fully the thermal history of the Muroto Peninsula, our results must be placed within the proper regional context, both as defined by data from broadly coeval rocks throughout Japan and as defined by regional plate-tectonic reconstructions. Additional insights may be gained by comparing the rock record to offshore data from the Nankai Trough. As summarized below, data from a wide variety of sources show that large areas of southwest Japan have been sites of abnormal geothermal activity during much of late Cenozoic time.

Tertiary plate reconstructions

Reconstructions of early Tertiary lithospheric plate motions in the western Pacific Basin are highly speculative because most of the direct evidence (e.g., from sea-floor magnetic anomalies) has been subducted (Fig. 7). Most models for the early Eocene epoch show lithosphere of the Pacific Plate being subducted beneath Japan (Engebretson and others, 1985; Seno, 1985; Lonsdale, 1988; Jolivet and others, 1989). One of the major uncertainties of the Eocene reconstructions, however, involves the position of the western extension of the spreading boundary between the Kula and Pacific Plates, prior to the fusion of the two plates at about 44 to 43 Ma (Fig. 7A). DiTullio and Byrne (1990) have attributed variations in structural fabrics and the out-of-sequence thrusting event within the Murotohanto subbelt to a change in convergence direction at about 42 Ma. Conceivably, this reorganization was caused by a change from Kula subduction to subduction of the fused Kula-Pacific lithosphere. If a segment of the Kula-Pacific Ridge crest had been in fairly close proximity to the Shimanto subduction front in late middle Eocene time (Fig. 7C), and juvenile lithosphere from the Kula-Pacific spreading had entered the Shimanto subduction zone just before or just after the demise of the Kula-Pacific Ridge, then this plate-tectonic

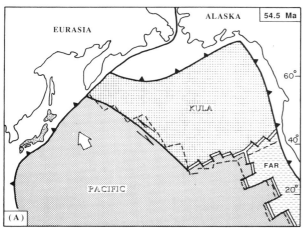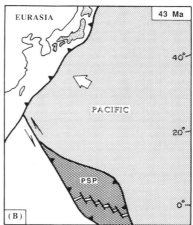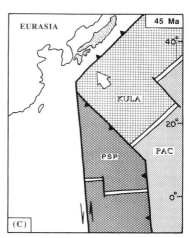

Figure 7. Comparison of plate-tectonic reconstructions for the North Pacific basin and southwest Japan during late Eocene time. Model A is from Lonsdale (1988), with the reconstruction set at 54.5 Ma; an earlier configuration of the Kula-Pacific spreading ridge (at 57 Ma) is shown by the dashed line. Fusion of the Kula-Pacific boundary occurred at 44 to 43 Ma. Note that the southwestward extent of the Kula Plate is poorly constrained in this model, but that Eocene subduction of the Pacific Plate beneath the Japanese margin is inferred. Model B, from Jolivet and others (1989), also depicts subduction of the Pacific Plate beneath Japan during late middle Eocene time (43 Ma). The final model (from DiTullio, 1989), shows segments of the Kula-Pacific Ridge in close proximity to the Shimanto subduction front just prior to the demise of spreading (45 Ma). Subduction of young crust associated with this ridge system may have been responsible for relatively high geothermal gradients and late stage 2 thermal overprints within the Murotohanto subbelt. PSP, Philippine Sea Plate; PAC, Pacific Plate; FAR, Farallon Plate.

regime certainly would have influenced the paleothermal structure of the Eocene accretionary prism. One independent piece of evidence in favor of young crust entering the subduction zone during late Eocene time is the progressive narrowing in the age gap between basaltic rocks (inferred oceanic basement) and overlying pelagic and turbidite sediments (inferred trench-wedge) within the Shimanto Belt; the time gap for Eocene sediments is less than 10 Ma (Taira, 1985).

The Pliocene to Quaternary phase of subduction beneath southwest Japan has involved the Philippine Sea Plate (Ranken and others, 1984). As outlined by Hibbard and Karig (1990a), reconstructions of the Philippine Sea Plate and related plate boundaries become increasingly speculative as one goes back into early Miocene and Oligocene time. Several controversial points merit comment, the first of which is the timing of cessation of back-arc spreading in the Shikoku Basin. The most recent interpretation of marine magnetic anomalies suggests that the final episode of volcanic activity (perhaps due to north-south spreading) took place between 14 Ma and 12 Ma (Chamot-Rooke and others, 1987). On the other hand, the oldest sedimentary rocks recovered with igneous basement at ODP Site 808, near the abandoned axis of the spreading ridge, are approximately 13.6 Ma in age (Taira and others, 1991). Other debated aspects of the regional tectonic reconstructions include the mechanism of opening of the Sea of Japan and its effect on the amount of block rotation in southwest Japan (e.g., Otofuji and others, 1985a; Kimura and Tamaki, 1986; Lallemand and Jolivet, 1986; Faure

and Lalevee, 1987; Celaya and McCabe, 1987; Otofuji and Matsuda, 1987). Another point of contention involves the amount and direction of migration of the Izu-Bonin Arc (i.e., the anchored slab versus retreating trench scenarios of Seno and Maruyama, 1984).

For the purposes of our paleothermal study, perhaps the most important element of these Miocene reconstructions would be the initial arrival time of hot, young, Shikoku Basin lithosphere at the Shimanto subduction front. Beyond the issue of timing, we would expect the thermal anomaly imparted by this subduction regime to be expressed most dramatically at the intersection point between the Shikoku Basin spreading ridge and the trench, but the position of the collision zone remains uncertain. For example, the plate-tectonic model of Jolivet and others (1989) shows subduction of the Pacific (rather than Philippine Sea) Plate beneath the proto Nankai Trough through early Miocene time, and the central spreading axis of the Shikoku Basin is placed well to the southwest of the island of Shikoku during the critical time period of 15 Ma to 12 Ma (Fig. 8A). In contrast, the block-rotation model of Otsuki (1990) fixes the Izu-Bonin Arc to the northeast, offshore Honshu, and migrates the spreading axis of Shikoku Basin to the southwest; according to this reconstruction, the position of the ridge crest at 15 Ma was east of the Kii Peninsula (Fig. 8C). The reconstruction of Seno (1985) places the ridge axis immediately offshore central Shikoku at 18 Ma. Finally, a modification of the Seno (1985) model proposed by Hibbard and Karig (1990a) shows crust of the Pacific Plate sub-

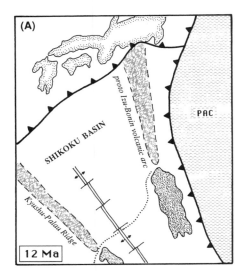
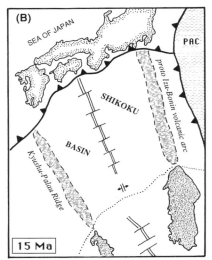
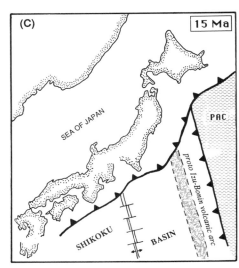

Figure 8. Comparison of plate-tectonic reconstructions for southwest Japan during middle Miocene time. PAC = Pacific Plate. Dotted line in the Shikoku Basin represents the approximate limit of crust that has been subducted since the time of the reconstruction. Model A, which places the collision zone between the Shikoku Basin spreading ridge and the Shimanto subduction front to the southwest of the Muroto Peninsula (12 Ma), is from Jolivet and others (1989). According to Model B, from Hibbard and Karig (1990a), the subduction of the juvenile Philippine Sea Plate was initiated at about 15 Ma. Widespread near-trench magmatism and high heat flow were triggered by this change (earlier subduction involved old crust of the Pacific Plate). Note that the ridge-trench collision zone is located near its present-day position, near Cape Muroto. In Model C, from Otsuki (1990), the collision zone at 15 Ma is located farther to the northeast near the Kii Peninsula. The position of the collision zone should correspond to the most intense zone of magmatic activity and overprints of thermal maturity.

ducted during Oligocene through early Miocene time, with a strike-slip boundary separating the Pacific Plate from the Shikoku Basin; subduction of the Philippine Sea Plate then began at about 15 Ma, with the crest of the back-arc spreading ridge fixed off-shore the Muroto Peninsula for the remainder of Cenozoic time (Fig. 8B). The structural configuration and timing of the Muroto flexure, combined with synchronous MORB magmatism, are cited as the prime evidence for locating the collision zone near the southeastern edge of Shikoku Island (Hibbard and Karig, 1990b).

Paleomagnetic data and block rotations

Any interpretation of structural trends within the Shimanto Belt must take into account the paleomagnetic evidence for rapid clockwise rotation of southwest Japan about a subvertical axis during the Miocene epoch. This rotation of 42 to 56° was concurrent with the fanlike opening of the Sea of Japan (Otofuji and Matsuda, 1983, 1984, 1987; Hayashida and Ito, 1984; Otofuji and others, 1985a) and also coincided with the counterclockwise rotation of northeast Japan (Otofuji and others, 1985b). Additional components of local rotation were superimposed on this regional trend in response to the collision of the Izu-Bonin Arc with the Honshu Arc (Hyodo and Niitsuma, 1986; Itoh, 1988; Kikawa and others, 1989; Takahashi and Nomura, 1989) and/or the differential rotation of separate crustal blocks (Itoh, 1986).

The timing of regional rotation is constrained tightly by both radiometric dates on volcanic rocks and fossil control extracted from interbeds of sediment. For example, Hayashida (1986) showed that the inception of coherent block rotation is recorded within the Setouchi Miocene Series at a stratigraphic boundary between the late early Miocene and the early middle Miocene (15.1 Ma). Because the overlying Setouchi volcanic rocks (which are younger than 14 Ma) retain magnetizations that are in the present-day north-south orientation, the rotational motions evidently terminated abruptly (within about 1 Ma of their inception). These data are in close agreement with the estimates of Otofuji and others (1985a), who used regional paleomagnetic compilations to argue for a rotational climax at 14.9 Ma with a transient duration of only 0.6 m.y. Similar results were obtained by Nakajima and Hirooka (1986), who pinpointed the interval of rotation for northwest Honshu between 16.0 ± 1.8 Ma and 15.4 ± 2.0 Ma.

The notion of rapid clockwise rotation of southwest Japan and segmentation of rotational provinces is further supported by analyses of the Neogene stress field (Takeuchi, 1986). Tsunakawa (1986) used radiometric data and changes in the strike directions of dike swarms to infer a major shift in the regional stress field of southwest Japan at about 15 Ma; the stress field changed at that time from a north-south orientation of maximum horizontal tension to a regime of east-west horizontal tension (relative to present-day coordinates). Additional shifts evidently occurred at 12 Ma (with a return to north-south horizontal tension), at around 8 to 9 Ma and at 1 Ma (Tsunakawa, 1986).

Although the temporal coincidence between block rotations and the opening of the Sea of Japan seems certain based on paleomagnetic and radiometric data, objections to the fanlike model of opening have been voiced, and other geometries and mechanisms have been proposed (Lallemand and Jolivet, 1986; Jolivet, 1987; Celaya and McCabe, 1987). One objection is related to the apparent lack of field evidence for contemporaneous deformation of exposed rock units at the time of the bending event. Recent analyses demonstrate, however, that Neogene fault motions were coupled with synrotational bends in some of the most fundamental regional-scale structural trends of southwest Japan (Faure and Lalevee, 1987). Furthermore, the formation of so-called megakink folds in the Shimanto Belt and other pre–middle Miocene rock units is now well documented. The Muroto flexure (Fig. 5) is but one example of these "megakinks" (Hibbard and Karig, 1990a, b), and similar structures have been mapped in the Outer Zone of Kyushu, the Hata Peninsula of eastern Shikoku, and the Kii Peninsula of Honshu (Yanai, 1986, 1989; Murata, 1987; Kosaka and others, 1988; Kano and others, 1990). These large folds (up to 100 km or more in width) display vertical or subvertical axes of rotation, are variable in shape, and are commonly accompanied by strike-slip faults; the structures have been attributed to horizontal compression at shallow levels of the crust during the opening of the Sea of Japan.

Miocene igneous activity

The occurrence of anomalous near-trench magmatic activity during middle Miocene time must be regarded as one of the most important and enigmatic events in the Neogene evolution of southwest Japan (Fig. 9). High regional heat flow certainly would be expected as a by-product of magmatic activity, thereby affecting the thermal structure of the entire forearc. Shibata and Nozawa (1967) and Hibbard and Karig (1990a) have presented summaries of the radiometric and zircon fission-track data extracted from the various igneous bodies that cut across and/or overlie rocks of the Shimanto Belt. This volcanic and plutonic activity was widespread, short-lived, and nearly simultaneous across the entire strike of the Neogene accretionary prism. Volcanic flows and intrusions affected strata on the islands of Honshu, Shikoku, and the southern part of Kyushu (Fig. 9). Furthermore, the occurrence of related submarine igneous masses is suggested by acoustic data and magnetic anomalies (Miyake and Hisatomi, 1985). Radiometric dates for the onshore igneous rocks range from about 15 Ma to 12 Ma, but most of the data cluster at approximately 14 Ma (Fig. 9). Thus, most of the activity was synchronous with the clockwise rotation of southwest Japan (Otofuji and others, 1985a).

The igneous rock types of the Outer Zone and their chemical compositions are diverse and include the following: MORB-type basalts of the Maruyama igneous suite of the Muroto

Peninsula (Hibbard and Karig, 1990b); basalts of island-arc affinity that are interbedded within the basal Shijujiyama Formation on the Muroto Peninsula (Hibbard and Karig, 1990b); the voluminous rhyolitic pyroclastic rocks and granite porphyries of the Kumano and Ohmine units on the Kii Peninsula (Murata, 1984; Chijiwa, 1988); the Shionomisaki abyssal volcanic/intrusive association (basalt, hornblende dolerite, cumulate gabbros, etc.) on the Kii Peninsula (Miyake, 1985, 1988; Miyake and Hisatomi, 1985); the Osuzuyama acid volcanic rocks and numerous coeval granitic plutons of southeast Kyushu (Oba, 1977; Nakada, 1983); and the Ashizuri granitic bodies of the Hata Peninsula on the southwest side of Shikoku (Oba, 1977). Furthermore, Torii and Ishikawa (1986) described the occurrence of local intrusions of high-magnesian andesite from the Kii Peninsula (Fig. 9); these rocks also were intruded at about 15 Ma, just prior to the rotation of southwest Japan. The chemical disparities between all of these coeval igneous suites, the crude zoning (with S-type magma bodies generally to the south), the occurrences of xenoliths, plus the presence of intermediate zones of migmatite, collectively support the contention of magma contamination by felsic materials (e.g., pelitic rocks) as mafic melts from a deep crustal source moved upward through the Shimanto accretionary prism (Oba, 1977; Nakada and Takahashi, 1979; Nakada, 1983; Terakado and others, 1988). The Nd and Sr isotopic variations documented by Terakado and others (1988) suggest incorporation of lower crustal materials for the I-type magmas, whereas the S-type magmas seem to require mixing with considerable amounts of sedimentary materials. The plutonic rocks of the Outer Zone also yield initial $^{87}Sr/^{86}Sr$ values that are relatively high (>0.706), which is consistent either with the concept of contamination of a deep-seated magma by sialic crustal material or with the generation of the main part of the melt within the continental crust (Shibata and Ishihara, 1979).

It is worth noting that the anomalous Miocene magmatic activity was not restricted to the Outer Zone of southwest Japan. For example, traces of rhyolitic tuff yielding K/Ar dates of 15 to 16 Ma crop out locally to the north of the Median Tectonic Line, within the Setouchi Zone (Matsuda and others, 1986). In addition, a more-widespread phase of volcanism occurred in that region during late middle Miocene time (Fig. 9). Those volcanic rocks, collectively known as the Setouchi volcanic belt, are characterized by unusual high-magnesian andesites, which occur together with basalts and dacitic/rhyolitic pyroclastic flows (Tatsumi and Ishizaka, 1982). K-Ar and fission-track age determinations place the timing of intermediate to basic volcanism at about 12 Ma, whereas the Setouchi silicic volcanism began slightly earlier at about 14 Ma. Geochemical characteristics suggest that the andesitic and basaltic magmas were generated independently at different depths from partial melting of a mantle source, with little crustal contamination or fractionation (Tatsumi, 1982; Ishizaka and Carlson, 1983).

The Setouchi volcanic rocks in some areas overlie sedimentary rocks of Setouchi Miocene Series, and those sedimentary strata have been dated precisely using microfossils. As outlined

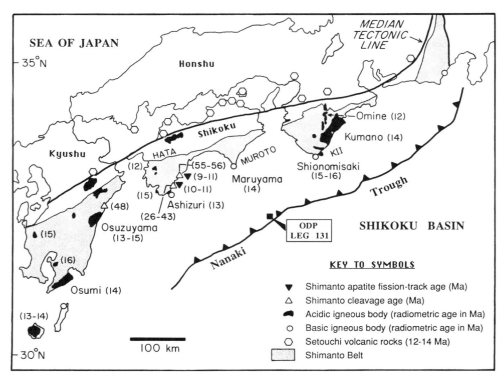

Figure 9. Map showing the distributions of anomalous basic and acidic igneous bodies of middle Miocene age within the Outer Zone of Japan, together with complimentary radiometric age dates (Ma). Occurrences of the somewhat-younger Setouchi volcanic series are shown north of the Median Tectonic Line. Also depicted are control points and corresponding absolute K-Ar ages (Ma) of cleavage formation within the Shimanto Belt, plus uplift/cooling ages (Ma) based on apatite fission tracks. Data for this figure were obtained from the following sources: Shibata and Nozawa (1967), Oba (1977), Miyake (1985), Hamamoto and Sakai (1987), Terakado and others (1988), Agar and others (1989), Hibbard and Karig (1990a), and Mackenzie and others (1990).

previously, detailed paleomagnetic data show that the rotation of southwest Japan began during the deposition of the Setouchi Miocene Series, at roughly 15 Ma (Hayashida, 1986). Thus, whereas the anomalous igneous activity within the Outer Zone occurred prior to and/or during rotation, the Setouchi phase of volcanism occurred after the rotation of southwest Japan had reached its culmination.

One long-standing explanation for the occurrence of igneous activity in the Outer Zone involves the development and subsequent migration of a ridge-trench-trench triple junction (Marshak and Karig, 1977). The situation was more complicated than that, however. The initiation of magmatism almost certainly coincided with the opening of the Sea of Japan to the north (Otofuji and others, 1985a; Otofuji and Matsuda, 1987). Furthermore, according to Hibbard and Karig (1990a), this same time period saw a fundamental change in the nature of plate boundaries to the south, as subduction of the Pacific Plate was replaced by subduction of the Philippine Sea Plate along the proto–Nankai Trough. Thus, near-trench magmatism likely was triggered by the arrival of very young lithosphere formed by back-arc spreading in Shi-

koku Basin. Some of the focal points of magmatism may have occurred within the collision zone, as defined by the intersection of the ridge crest and the trench. However, there is no evidence to support the idea of a progressive shift of the magmatic front either to the northeast or to the southwest, as would be expected with the strike-parallel migration of a ridge-trench-trench triple junction.

Tertiary coal fields of Japan

The spatial and stratigraphic distribution of Tertiary coal fields in Japan has been summarized by Aihara (1989). These coals are important because they provide accurate measures of peak burial temperatures within several different tectonic settings. The two most extensive zones of coal accumulation are associated with the Hidaka forearc basins of Hokkaido Island and the back-arc basins on the northwest side of Kyushu Island (Fig. 10). Maturation gradients within the Hidaka forearc terrane are in no way unusual. The maximum values of mean vitrinite reflectance (%R_m) range from 0.5%R_m to 0.9%R_m, and corresponding maximum burial depths are 3 km or less; the estimated paleogeothermal gradients vary between 16°C/km and 24°C/km (Aihara, 1989). In contrast, the back-arc coal basins of Kyushu (Fig. 10) were intruded by arc-related magmas during Paleogene time and, therefore, they contain higher rank coals, with $\%R_m$ values as high as $2.5\%R_m$ at depths of 3 to 4 km. Measurements of fluid-inclusion homogenization temperatures, using veins in siliciclastic interbeds, confirm the fact that paleotemperatures rose to levels as high as 250°C (Aizawa, 1990). Background paleogeothermal gradients calculated for northwest Kyushu range from 50 to 70°C/km and local anomalies are even higher because of the effects of contact metamorphism (Aihara, 1989).

Within the Outer Zone of Japan, the late early Miocene to early middle Miocene Kumano Group unconformably overlies Shimanto strata on the Kii Peninsula (Fig. 10). This sedimentary succession is comparable in many respects to the Shijujiyama Formation on the Muroto Peninsula. Coal-bearing intervals are common within the Kumano Group, and deposition evidently occurred within a forearc-basin setting (Chijiwa, 1988). The Kumano Group was both intruded by and covered unconformably by acidic magmas yielding radiometric and zircon fission-track dates of about 14 to 15 Ma (e.g., Shibata and Noz-awa, 1967; Tagami, 1982). Mean maximum vitrinite reflectance values ($\%R_{max}$) for the Miocene coals range from 0.8% to as high as 6.0%, and this corresponds to an equivalent range of 0.8 to $5.0\%R_m$ when expressed in terms of mean random reflectance (see Stach and others, 1982, for the relation between $\%R_m$ and $\%R_{max}$). Stratigraphic gradients in vitrinite reflectance are remarkably high, ranging from about 0.3 to $0.6\%R_{max}/100$ m (Chijiwa, 1988). Using the somewhat conservative time-temperature model of organic metamorphism proposed by Bostick (1974), Chijiwa (1988) calculated a paleogeothermal gradient of 80 to 100°C/km; if the $\%R_m$-temperature conversion of Barker (1988) is used instead, then the typical paleothermal gradient increases to approximately 140°C/km. Regardless of the exact paleotemperatures, this region obviously was a site of bona fide geothermal activity during the middle Miocene. Presumably, underlying rocks of the Shimanto Belt were likewise affected.

Metamorphism within the Shimanto Belt

Overall, the inorganic mineral phases within the Shimanto Belt define zones of zeolite facies, prehnite-pumpellyite facies, and greenschist-facies metamorphism (Toriumi and Teruya, 1988). These authors estimated the maximum burial conditions to be 3 to 5 kbar and 200 to 300°C, based upon the equilibrium limits of key index minerals. However, the occurrence of pumpellyite in the absence of lawsonite suggests that burial pressures were never significantly above 3 kbar (Liou and others, 1985). The precision of paleotemperature estimates can be improved from analyses of organic matter, as summarized below.

Kyushu. Aihara (1989) demonstrated that the level of thermal alteration of dispersed organic matter is unusually high within both Cretaceous and Tertiary portions of the Shimanto Belt exposed on the southeast side of Kyushu (Fig. 10; see Nishi, 1988, for appropriate geologic descriptions of these rocks). The youngest fossils extracted from this part of the Shimanto are early Miocene (Aquitanian) in age (Sakai, 1988). Cleavage formation within the early to middle Paleogene portion of the prism has been dated by Mackenzie and others (1990) at 48.4 ± 1 Ma using K/Ar methods (Fig. 9). Mean vitrinite reflectance values range from about 1.0 to $4.0\%R_m$. Although it is difficult to obtain accurate estimates of stratigraphic thickness because of intense stratal disruption (including the possible formation of olistostromes), average maturation gradients appear to be on the order of $0.1\%R_m/100$ m (Aihara, 1989). Using the $\%R_m$-temperature correlation of Barker (1988), this crudely defined $\%R_m$ gradient corresponds to a paleotemperature gradient of approximately 65°C/km. Some of the Shimanto sections show virtually no change in organic metamorphism over thicknesses of 3.5 km; this type of isothermal pattern is characteristic of overpressured stratigraphic intervals (Law and others, 1989) and zones of advective hydrothermal activity (Barker and others, 1986).

It is also important to note that late Miocene to Pliocene strata of the Miyazaki Group rest unconformably above the Shimanto Belt of southeast Kyushu (Sakai, 1988), and these post-

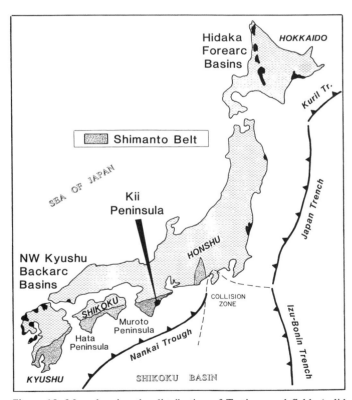

Figure 10. Map showing the distribution of Tertiary coal fields (solid black pattern) in the Japanese Islands (modified from Aihara, 1989). Areas of the Shimanto Belt are highlighted by the stippled pattern. Collision zone between the Honshu Arc and the Izu-Bonin Arc is shown by the dashed line. See text for descriptions of coal rank within each region.

orogenic sediments are submature, with maximum $\%R_m$ values of 0.5% or less (Aihara, 1989). Therefore, uplift and subaerial erosion of the high-rank Shimanto rocks must have occurred prior to the Tortonian (about 11 Ma). The Miyazaki Group displays its own vertical gradient in organic metamorphism, similar in slope to that of the Shimanto, such that values of $0.5\%R_m$ (roughly equivalent to 75°C) are found at present-day depths of only 500 m. The minimum $\%R_m$ values at the surface are 0.4% (50°C), which means that the Pliocene-Pleistocene geothermal gradient retained by rocks on southeast Kyushu was approximately 50°C/km.

Shikoku. Prior to our investigation, the largest compilation of paleothermal data from the Shimanto Belt was produced by Mori and Taguchi (1988), who completed several regional transects along the Muroto Peninsula of Shikoku. Their study documented a consistent pattern of increasing organic metamorphism southward, over a range of about 0.7 to $5.0\%R_m$. Following Barker (1988), this level of organic metamorphism corresponds to a temperature range of about 110 to 315°C. The southward increase in $\%R_m$ values is repeated at least four times in separate fault-bounded sections, and these sections include both Cretaceous and Tertiary strata. Regional-scale block rotation and differential uplift during movement on listric thrusts is the most likely explanation for this pattern. Sharp breaks are evident across some of the major faults and tectonostratigraphic boundaries, and anomalously high values were also noted in the immediate vicinity of igneous intrusions.

The Shimanto Belt, where exposed on the Hata Peninsula of Shikoku (Fig. 9), has been described by Agar (1988, 1990) and Cowan (1990). To our knowledge, vitrinite-reflectance measurements have not been completed on these rocks, but recent analyses by Agar and others (1989), using fission-track and K-Ar methods, showed that the Hata Peninsula has been affected unevenly by Tertiary heating, with thermal maturity increasing toward the granitic intrusions at Cape Ashizuri. Evidently, some of these Shimanto rocks were never exposed to temperatures above 110 to 120°C (i.e., all of the fission tracks in some detrital apatite grains were not annealed). Apparent cleavage ages for Maastrichtian samples of the Lower Shimanto Group vary between 43 and 56 Ma, whereas cleavages within the Paleogene section formed between 18 and 34 Ma (Agar and others, 1989). Subsequent cooling below the apatite partial-annealing window occurred well after the formation of cleavage, however, at around 9 to 11 Ma. Moreover, the fission-track ages indicate that a single phase of rapid uplift and cooling affected the entire Cretaceous through Oligocene section during late Miocene time.

Kii Peninsula. Little is known about the degree of organic metamorphism within Shimanto strata exposed on the Kii Peninsula of Honshu (see Kumon and others, 1988, for pertinent geologic descriptions). Two borehole gradients (from Suzuki and others, 1982, in Aihara, 1989) show unusually high values of vitrinite reflectance, ranging from about 2.5 to $4.0\%R_m$ at the surface to nearly $7.0\%R_m$ at depths of less than 1 km. The estimate for maximum paleotemperature, therefore, is about 350°C.

These Shimanto strata obviously were affected by the same generation of igneous activity and high heat flow that was responsible for the high-rank coals of the Miocene Kumano Group (Chijiwa, 1988).

Nankai Trough

Several types of thermal data have been obtained from offshore extensions of the Shimanto accretionary complex. Vitrinite-reflectance data from borehole samples of the Deep Sea Drilling Project (Leg 87, located offshore Hata Peninsula) are somewhat difficult to interpret because of mixing between indigenous (first-cycle) material, higher-rank grains recycled from uplifted coals, and organic particles of uncertain origin. The maximum value of (indigenous) mean reflectance is $0.40\%R_m$; this value (equivalent to a peak temperature of approximately 50°C) occurs at a subbottom depth of 731 m at Site 582, which penetrated the turbidites of Nankai Trough (Mukhopadhyay and others, 1986). At Site 583, located at the toe of the accretionary wedge, the maximum $\%R_m$ value is only 0.28% at a depth of 395 m (Mukhopadhyay and others, 1986). Borehole temperature gradients for Sites 582 and 583 are 46°C/km and 50°C/km, respectively (Kinoshita and Yamano, 1986), and recent data from ODP Site 808, located immediately offshore Cape Muroto, suggest a gradient of approximately 110°C/km (Taira and others, 1991; Yamano and others, 1992). Shallow-level geothermal gradients calculated from the depths of gas-hydrate reflectors are consistent with the borehole data, ranging from 41°C/km to a maximum of 66°C/km (Yamano and others, 1982; Kinoshita and Yamano, 1986; Ashi and Taira, this volume). Near-surface heat-flow measurements show a concentration of higher values (>130 mW/m^2) along the axis of Nankai Trough, and the zone of high heat flow (>100 mW/m^2) penetrates approximately 15 km landward of the deformation front (Ashi and Taira, this volume). Heat flow for most of the middle to upper trench slope ranges between 45 and 70 mW/m^2 (Kinoshita and Yamano, 1986; Ashi and Taira, this volume).

The thermal data cited above prove that the Holocene regime of Nankai Trough is relatively warm, at least within the upper kilometer of the sedimentary section near the deformation front. These elevated geothermal gradients can be explained, in part, by examining the evolution of the adjacent Shikoku Basin. Bathymetric features, magnetic anomalies, and DSDP drilling recoveries collectively support the contention that oceanic crust formed through sea-floor spreading behind the Izu-Bonin volcanic arc, and this extinct spreading ridge (highlighted bathymetrically by the Kinan seamounts) is aligned at a high angle to the Shikoku subduction front (Watts and Weissel, 1975; Klein and Kobayashi, 1980; Seno and Maruyama, 1984; Chamot-Rooke and others, 1987; Le Pichon and others, 1987a). The last phase of back-arc spreading (which may have been directed north-south) evidently ended about 12 Ma (Chamot-Rooke and others, 1987).

Heat flow from oceanic crust that is 12 m.y. in age should reach 136 to 145 mW/m^2, according to the empirical curves of Parsons and Sclater (1977) and Lister (1977). Consequently, appreciable heat still is being conducted from the igneous crust and through the overlying basin sediments. Heat-flow contours are aligned parallel to the trench rather than parallel to either the dominant group of magnetic anomalies of Shikoku Basin (Shih, 1980) or the trend of the Kinan seamount chain. Thus, the high heat flow cannot be attributed solely to the subduction of the extinct spreading ridge; the insulating effects of a thick turbidite wedge in Nankai Trough also must be influential (i.e., the "heat rebound" theory of Yamano and others, 1984). Rigorous testing of this hypothesis by numerical modeling suggests that hydrothermal circulation in the basement layer and transient fluid advection through the accreted sediments also play important roles near the toe of the prism (Nagihara and others, 1989; Yamano and others, 1992; Ashi and Taira, this volume).

SUMMARY

In recent years, a great deal of attention has been directed toward studies of shallow-level fluid circulation, fluid venting at the sea floor, and near-surface heat flow in Holocene accretionary prisms (e.g., Kulm and others, 1986; Boulegue and others, 1987a, b; Le Pichon and others, 1987b, 1991; Moore and others, 1987, 1988, 1991; Shi and others, 1988; Henry and others, 1989; Moore, 1989; Carson and others, 1991). Our paleothermal study of the Tertiary Shimanto Belt, therefore, is particularly timely because it focuses on a range of depths within the accretionary-prism environment that cannot be sampled by ocean drilling. Both regional-scale and local-scale studies completed elsewhere prove that precise paleotemperature distributions can be reconstructed within ancient accretionary terranes (e.g., Underwood and Howell, 1987; Underwood and others, 1988). The papers that follow, particularly DiTullio and others (this volume) and Hibbard and others (this volume), provide detailed documentation of the relations between Shimanto deformation history and the thermal history of southwest Japan.

As a first-order issue, our results certainly prove that accretionary subduction margins are not always sites of depressed geothermal gradients and low-temperature, high-pressure metamorphism (Laughland and Underwood, this volume). The paradigm used to account for blueschist-facies conditions, therefore, is not universally applicable. We demonstrate, moreover, that the paleothermal structure of the Shimanto Belt is not a direct consequence of the primary temperature distribution of the Eocene-Miocene accretionary prism. Instead, virtually all of the paleotemperature gradients are artifacts of subsequent thermal overprints that were triggered by the subduction of young oceanic crust and/or ridge-trench collision(s).

On a finer scale, this multidisciplinary investigation tests several interpretations and hypotheses that are specific to the geologic evolution of the Muroto Peninsula. Among the local issues, we address the following: (1) What was the relative timing of peak heating with respect to the three stages of deformation defined for the Murotohanto subbelt, particularly the formation of cleavage? (2) Is the out-of-sequence thrust within the Muroto-

hanto subbelt (Shiina-Gyoto contact) a premetamorphic, syn-metamorphic, or post-metamorphic fault? (3) Have the early stages of deformation within the Nabae subbelt been overprinted by late-stage heating? (4) Can the slope-basin model for the Shijujiyama Formation be proven correct (i.e., is the relation between cooler basin fill over warmer accreted basement preserved); conversely, has the original temperature gradient across the contact been destroyed by thermal overprinting or post-metamorphic faulting? (5) Were strata of the Nabae subbelt affected, either directly or indirectly, by the intrusion of mafic magmas during middle Miocene time, and if so, how widespread was the thermal effect? (6) Are discordant thermal relations evident across any of the major stage-3 faults (including the Shiina-Narashi fault), thereby signifying post-metamorphic displacement? (7) Is there any paleothermal expression of differential uplift or vertical offset across the limbs of the Muroto flexure within either the Murotohanto or the Nabae subbelt? (8) Can the locations of paleothermal anomalies be used to resolve some of the uncertainties and ambiguities regarding the position of critical plate-tectonic elements, such as the locus of collision at 15 Ma between the proto–Nankai Trough and the central spreading ridge of the Shikoku Basin? (9) When and how did uplift and cooling of the Shimanto Belt occur?

ACKNOWLEDGMENTS

Funding to Underwood was provided by the National Science Foundation (Grant EAR 87-06784). Acknowledgment is also made by Underwood to the Donors of The Petroleum Research Fund, administered by the American Chemical Society, for partial support of this research (Grant No. 19187-AC2). Field work on the Muroto Peninsula by Hibbard and DiTullio was funded by National Science Foundation Grants EAR 85-09461 and EAR 87-20743, two Harold T. Stearns Fellowships from the Geological Society of America, and a grant from Sigma Xi. Dan Karig, Tim Byrne, and Matt Laughland played instrumental roles in our joint investigation of the Shimanto Belt, as reflected by other papers in this volume. Finally, we thank Asahiko Taira and J. Casey Moore for sharing their vast knowledge of local and thematic issues, and for providing insightful reviews of the manuscript.

REFERENCES CITED

Agar, S. M., 1988, Shearing of partially consolidated sediments in a lower trench slope setting, Shimanto Belt, SW Japan: Journal of Structural Geology, v. 10, p. 21–32.
——, 1990, The interaction of fluid processes and progressive deformation during shallow level accretion: Examples from the Shimanto Belt of SW Japan: Journal of Geophysical Research, v. 95, p. 9133–9147.
Agar, S. M., Cliff, R. A., Duddy, I. R., and Rex, D. C., 1989, Accretion and uplift in the Shimanto Belt, SW Japan: Journal of the Geological Society, London, v. 146, p. 893–896.
Aihara, A., 1989, Paleogeothermal influence on organic metamorphism in the neotectonics of the Japanese Islands: Tectonophysics, v. 159, p. 291–305.
Aizawa, J., 1990, Paleotemperatures from fluid inclusions and coal ranks of

carbonaceous material of the Tertiary formations in northwest Kyushu, Japan: Journal of Mineralogy, Petrology, and Economic Geology, v. 85, p. 145–154.

Anderson, R. N., DeLong, S. E., and Schwarz, W. M., 1978, Thermal model for subduction with dehydration in the downgoing slab: Journal of Geology, v. 86, p. 731–739.

Banno, S., and Sakai, C., 1989, Geology and metamorphic evolution of the Sambagawa Belt, Japan, in Daly, J. S., Cliff, R. A., and Yardley, B.W.D., eds., Evolution of metamorphic belts: Geological Society of London Special Publication 43, p. 519–532.

Barker, C. E., 1988, Geothermics of petroleum systems: Implications of the stabilization of kerogen thermal maturation after a geologically brief heating duration at peak temperature, in Magoon, L. B., ed., Petroleum systems of the United States: U.S. Geological Survey Bulletin 1870, p. 26–29.

Barker, C. E., and Pawlewicz, M. J., 1986, The correlation of vitrinite reflectance with maximum temperature in humic organic matter, in Buntebarth, G., and Stegena, L., eds., Paleogeothermics: New York, Springer-Verlag, p. 79–93.

Barker, C. E., Crysdale, B. L., and Pawlewicz, M. J., 1986, The relationship between vitrinite reflectance, metamorphic grade, and temperature in the Cerro Prieto, Salton Sea, and East Mesa geothermal systems, Salton Trough, United States and Mexico, in Mumpton, F. A., ed., Studies in diagenesis: U.S. Geological Survey Bulletin 1578, p. 83–95.

Bostick, N. H., 1974, Phytoclasts as indicators of thermal metamorphism, Franciscan assemblage and Great Valley sequence (upper Mesozoic), California: Geological Society of America Special Paper 153, p. 1–17.

——, 1979, Microscopic measurement of the level of catagenesis of solid organic matter in sedimentary rocks to aid exploration for petroleum and to determine former burial temperatures—A review, in Scholle, P. A., and Schluger, P. R., eds., Aspects of diagenesis: Society of Economic Paleontologists and Mineralogists, Special Publication 26, p. 17–43.

Bostick, N. H., and Pawlewicz, M. J., 1984, Paleotemperatures based on vitrinite reflectance of shales and limestones in igneous dike aureoles in the Upper Cretaceous Pierre Shale, Walsenburg, Colorado, in Woodward, J., Meisner, F. F., and Clayton, J. L., eds., Hydrocarbon source rocks of the greater Rocky Mountain region: Denver, Rocky Mountain Association of Geologists, p. 387–392.

Boulegue, J., Benedetti, E. L., Dron, D., Mariotti, A., and Letolle, R., 1987a, Geochemical and biogeochemical observations on the biological communities associated with fluid venting in Nankai Trough and Japan Trench subduction zones: Earth and Planetary Science Letters, v. 83, p. 343–355.

Boulegue, J., Iiyama, T. J., Charlou, J. L., and Jedwab, J., 1987b, Nankai Trough, Japan Trench and Kuril Trench: Geochemistry of fluids samples by submersible "Nautile": Earth and Planetary Science Letters, v. 83, p. 363–375.

Brewer, J., 1981, Thermal effects of thrust faulting: Earth and Planetary Science Letters, v. 56, p. 309–328.

Bustin, R. M., 1983, Heating during thrust faulting in the Rocky Mountains: Friction or fiction?: Tectonophysics, v. 95, p. 309–328.

Byrne, T., and Hibbard, J., 1987, Landward vergence in accretionary prisms: The role of the backstop and thermal history: Geology, v. 15, p. 1163–1167.

Cande, S., and Leslie, R., 1986, Late Cenozoic tectonics of the southern Chile Trench: Journal of Geophysical Research, v. 91, p. 471–496.

Cande, S. C., Leslie, R. B., Parra, J. C., and Hobart, M., 1987, Interaction between the Chile Ridge and Chile Trench: Geophysical and geothermal evidence: Journal of Geophysical Research, v. 92, p. 495–520.

Cardwell, R. K., Chinn, D. S., Moore, G. G., and Turcotte, D. L., 1978, Frictional heating on a fault zone with finite thickness: Geophysical Journal of the Royal Astronomical Society, v. 52, p. 525–530.

Carson, B., Holmes, M. L., Umstattd, K., Strasser, J. C., and Johnson, H. P., 1991, Fluid expulsion from the Cascadia accretionary prism: Evidence from porosity distribution, direct measurements, and GLORIA imagery: Philosophical Transactions of the Royal Society of London, Series A, v. 335, p. 331–340.

Celaya, M., and McCabe, R., 1987, Kinematic model for the opening of the Sea of Japan and the bending of the Japanese Islands: Geology, v. 15, p. 53–57.

Chamot-Rooke, N., Renard, V., and LePichon, X., 1987, Magnetic anomalies in the Shikoku Basin: A new interpretation: Earth and Planetary Science Letters, v. 83, p. 214–228.

Chijiwa, K., 1988, Post-Shimanto sedimentation and organic metamorphism: An example of the Miocene Kumano Group, Kii Peninsula: Modern Geology, v. 12, p. 363–387.

Cloos, M., 1982, Comparative study of mélange matrix and metashales from the Franciscan subduction complex with the basal Great Valley sequence, California: Journal of Geology, v. 91, p. 291–306.

——, 1985, Thermal evolution of convergent plate margins: Thermal modeling and reevaluation of isotopic Ar-ages for blueschists in the Franciscan Complex of California: Tectonics, v. 4, p. 421–433.

——, 1986, Blueschists in the Franciscan Complex of California: Petrotectonic constraints on uplift mechanisms, in Evans, B. W., and Brown, E. H., eds., Blueschists and eclogites: Geological Society of America Memoir 164, p. 77–93.

Cowan, D., 1990, Kinematic analysis of shear zones in sandstone and mudstone of the Shimanto Belt, Shikoku, SW Japan: Journal of Structural Geology, v. 12, p. 431–441.

Davis, E. E., Hyndman, R. D., and Villinger, H., 1990, Rates of fluid expulsion across the northern Cascadia accretionary prism: Constraints from new heat flow and multichannel seismic reflection data: Journal of Geophysical Research, v. 95, p. 8869–8889.

Davy, Ph., and Gillet, Ph., 1986, The stacking of thrust slices in collision zones and its thermal consequences: Tectonics, v. 5, p. 913–930.

DeLong, S., and Fox, P., 1977, Geological consequences of ridge subduction, in Pitman, W., and Talwani, M., eds., Island arcs, deep sea trenches, and back arc basins: American Geophysical Union, Maurice Ewing Series, v. 1, p. 221–228.

DeLong, S. E., Fox, P. J., and McDowell, F. W., 1978, Subduction of the Kula Ridge at the Aleutian Trench: Geological Society of America Bulletin, v. 89, p. 83–95.

DeLong, S. E., Schwarz, W. M., and Anderson, R. N., 1979, Thermal effects of ridge subduction: Earth and Planetary Science Letters, v. 44, p. 239–246.

Dickinson, W. R., and Snyder, W. S., 1979, Geometry of triple junctions related to San Andreas transform: Journal of Geophysical Research, v. 84, p. 561–572.

DiTullio, L. D., 1989, Evolution of the Eocene accretionary prism in SW Japan: Evidence from structural geology, thermal alteration, and plate reconstructions [Ph.D. thesis]: Providence, Rhode Island, Brown University, 161 p.

DiTullio, L. D., and Byrne, T., 1990, Deformation paths in the shallow levels of an accretionary prism: The Eocene Shimanto belt of southwest Japan: Geological Society of America Bulletin, v. 102, p. 1420–1438.

Dow, W. G., 1977, Kerogen studies and geological interpretations: Journal of Geochemical Exploration, v. 7, p. 79–99.

Dumitru, T., 1991, Effects of subduction parameters on geothermal gradients in forearcs, with an application to Franciscan subduction in California: Journal of Geophysical Research, v. 96, p. 621–641.

Engebretson, D. C., Cox, A., and Gordon, R. G., 1985, Relative motions between oceanic and continental plates in the Pacific Basin: Geological Society of America Special Paper 206, 59 p.

Ernst, W. G., 1974, Metamorphism and ancient continental margins, in Burk, C. A., and Drake, C. L., eds., Geology of continental margins: New York, Springer-Verlag, p. 907–919.

——, 1975, Systematics of large-scale tectonics and age progressions in Alpine and circum-Pacific blueschist belts: Tectonophysics, v. 26, p. 229–246.

——, 1977, Mineral parageneses and plate tectonic settings of relatively high-pressure metamorphic belts: Fortschritte der Mineralogie, v. 54, p. 192–222.

Ernst, W. G., Seki, Y., Onuki, H., and Gilbert, M. C., 1970, Comparative study of low-grade metamorphism in the California Coast Ranges and the Outer Metamorphic Belt of Japan: Geological Society of America Memoir 124, 276 p.

Faure, M., and Lalevee, F., 1987, Bent structural trends of Japan: Flexural-slip folding related to the Neogene opening of the Sea of Japan: Geology, v. 15,

p. 49–52.

Field, M. E., and Kvenvolden, K. A., 1985, Gas hydrates on the northern California continental margin: Geology, v. 13, p. 517–520.

Fisher, A. T., and Hounslow, M. W., 1990, Transient fluid flow through the toe of the Barbados accretionary complex: Constraints from Ocean Drilling Program Leg 110 heat flow studies and simple models: Journal of Geophysical Research, v. 95, p. 8845–8858.

Forsythe, R., and Nelson, E., 1985, Geological manifestations of ridge collision: Evidence from the Gulfo de Penas–Taitao Basin, southern Chile: Tectonics, v. 4, p. 477–495.

Forsythe, R., and 7 others, 1986, Pliocene near-trench magmatism in southern Chile: A possible manifestation of ridge collision: Geology, v. 14, p. 23–27.

Furlong, K. P., 1984, Lithospheric behavior with triple junction migration: An example based on the Mendocino triple junction: Physics of the Earth and Planetary Interiors, v. 36, p. 213–223.

Hamamoto, R., and Sakai, H., 1987, Rb-Sr age of granophyre associated with the Cape Muroto gabbroic complex: Kyushu University, Science Reports of the Department of Geology, Fukuoka Japan, v. 15, p. 1–5.

Hayashida, A., 1986, Timing of rotational motion of southwest Japan inferred from paleomagnetism of the Setouchi Miocene Series: Journal of Geomagnetism and Geoelectricity, v. 38, p. 295–310.

Hayashida, A., and Ito, Y., 1984, Paleoposition of southwest Japan at 16 Ma: Implications from paleomagnetism of the Miocene Ichishi Group: Earth and Planetary Science Letters, v. 68, p. 335–342.

Heasler, H. P., and Surdam, R. C., 1985, Thermal evolution of coastal California with application to hydrocarbon maturation: American Association of Petroleum Geologists Bulletin, v. 69, p. 1386–1400.

Henry, P., Lallemant, S. J., Le Pichon, X., and Lallemand, S. E., 1989, Fluid venting along Japanese trenches: Tectonic context and thermal modeling: Tectonophysics, v. 160, p. 277–291.

Hibbard, J. P., 1988, Evolution of anomalous structural fabrics in an accretionary prism; The Oligocene-Miocene portion of the Shimanto Belt at Cape Muroto, southwest Japan [Ph.D. thesis]: Ithaca, New York, Cornell University, 227 p.

Hibbard, J. P., and Karig, D. E., 1987, Sheath-like folds and progressive fold deformation in Tertiary sedimentary rocks of the Shimanto accretionary complex, Japan: Journal of Structural Geology, v. 9, p. 845–857.

—— , 1990a, An alternative plate model for the early Miocene evolution of the SW Japan margin: Geology, v. 18, p. 170–174.

—— , 1990b, Structural and magmatic responses to spreading ridge subduction: An example from southwest Japan: Tectonics, v. 9, p. 207–230.

Hibbard, J. P., Karig, D. E., and Taira, A., 1992, Anomalous structural evolution of the Shimanto accretionary prism at Murotomisaki, Shikoku Island, Japan: The Island Arc, v. 1, p. 133–147.

Hisatomi, K., 1988, The Miocene forearc basin of southwest Japan and the Kumano Group of the Kii Peninsula: Modern Geology, v. 12, p. 389–408.

Hyodo, H., and Niitsuma, N., 1986, Tectonic rotation of the Kanto Mountains, related with the opening of the Japan Sea and collision of the Tanzawa Block since middle Miocene: Journal of Geomagnetism and Geoelectricity, v. 38, p. 335–348.

Ishizaka, K., and Carlson, R. W., 1983, Nd-Sr systematics of the Setouchi volcanic rocks, southwest Japan: A clue to the origin of orogenic andesites: Earth and Planetary Science Letters, v. 64, p. 327–340.

Ishikawa, T., 1982, Radiolarians from the southern Shimanto Belt (Tertiary) in Kochi Prefecture, Japan: News Osaka Micropaleontology, Special Volume 5, p. 399–407.

Itoh, Y., 1986, Differential rotation of northeastern part of southwest Japan: Paleomagnetism of early to late Miocene rocks from Yatsuo area in Chubu district: Journal of Geomagnetism and Geoelectricity, v. 38, p. 325–334.

—— , 1988, Differential rotation of the eastern part of southwest Japan inferred from paleomagnetism of Cretaceous and Neogene rocks: Journal of Geophysical Research, v. 93, p. 3401–3411.

James, T. S., Hollister, L. S., and Morgan, W. J., 1989, Thermal modeling of the Chugach Metamorphic Complex: Journal of Geophysical Research, v. 94,

p. 4411–4423.

Jayko, A. S., Blake, M. C., Jr., and Brothers, R. N., 1986, Blueschist metamorphism of the eastern Franciscan Belt, northern California, *in* Evans, B. W., and Brown, E. H., eds., Blueschists and eclogites: Geological Society of America Memoir 164, p. 107–123.

Johnston, D. C., and White, S. H., 1983, Shear heating associated with movement along the Alpine fault, New Zealand: Tectonophysics, v. 92, p. 241–252.

Jolivet, L., 1987, America–Eurasia plate boundary in eastern Asia and the opening of marginal basins: Earth and Planetary Science Letters, v. 81, p. 282–288.

Jolivet, L., Huchon, P., and Rangin, C., 1989, Tectonic setting of Western Pacific marginal basins: Tectonophysics, v. 160, p. 23–47.

Kagami, H., Karig, D. E., Coulbourn, W. T., and the Shipboard Scientific Party, 1986, Initial reports of the Deep Sea Drilling Project, v. 87: Washington, D.C., U.S. Government Printing Office, 985 p.

Kano, K., Kosaka, K., Murata, A., and Yanai, S., 1990, Intra-arc deformations with vertical rotation axes: The case of the pre–middle Miocene terranes of southwest Japan: Tectonophysics, v. 176, p. 333–354.

Kikawa, E., Koyama, M., and Kinoshita, H., 1989, Paleomagnetism of Quaternary volcanics in the Izu Peninsula and adjacent areas, Japan, and its tectonic significance: Journal of Geomagnetism and Geoelectricity, v. 41, p. 175–201.

Kimura, G., and Tamaki, K., 1986, Collision, rotation, and back-arc spreading in the region of the Okhotsk and Japan Seas: Tectonics, v. 5, p. 389–401.

Kinoshita, H., and Yamano, M., 1986, The heat flow anomaly in the Nankai Trough area, *in* Kagami, H., Karig, D. E., and Coulbourn, W. T., eds., Initial Reports of the Deep Sea Drilling Project, v. 87: Washington, D.C., U.S. Government Printing Office, p. 737–743.

Klein, G. deV., and Kobayashi, K., 1980, Geological summary of the North Philippine Sea, based on Deep Sea Drilling Project Leg 58 results, *in* Klein, G. deV., and Kobayashi, K., eds., Initial reports of the Deep Sea Drilling Project, v. 58: Washington, D.C., U.S. Government Printing Office, p. 951–962.

Korgen, B. J., Bodvarsson, G., and Mesecar, R. S., 1971, Heat flow through the floor of Cascadia Basin: Journal of Geophysical Research, v. 76, p. 4758–4774.

Kosaka, K., Itoga, H., and Yanai, S., 1988, Macroscopic and mesoscopic kink folds of the Kobotoke Group in the southern Kanto Mountains, central Japan: Journal of the Geological Society of Japan: v. 94, p. 221–224.

Kulm, L. D., and Fowler, G. A., 1974, Cenozoic sedimentary framework of the Gorda–Juan de Fuca Plate and adjacent continental margin—A review, *in* Dott, R. H., Jr., and Shaver, R. H., eds., Modern and ancient geosynclinal sedimentation: Society of Economic Paleontologists and Mineralogists, Special Publication 19, p. 212–229.

Kulm, L. D., and 13 others, 1986, Oregon subduction zone: Venting, fauna, and carbonates: Science, v. 231, p. 561–566.

Kumon, F., and 8 others, 1988, Shimanto Belt in the Kii Peninsula, southwest Japan: Modern Geology, v. 12, p. 71–96.

Kvenvolden, K. A., and von Huene, R., 1985, Natural gas generation in sediments of the convergent margin of the eastern Aleutian Trench area, *in* Howell, D. G., ed., Tectonostratigraphic terranes of the circum-Pacific region: Circum-Pacific Council for Energy and Mineral Resources, Earth Science Series, n. 1, p. 31–49.

Lallemand, S., and Jolivet, L., 1986, Japan Sea; A pull-apart basin?: Earth and Planetary Science Letters, v. 76, p. 375–389.

Langseth, M., and Burch, T., 1980, Geothermal observations on the Japan Trench transect, *in* Shipboard Scientists, eds., Initial reports of the Deep Sea Drilling Project, v. 56, part 2: Washington, D.C., U.S. Government Printing Office, p. 1207–1210.

Langseth, M. G., Westbrook, G. K., and Hobart, M., 1990, Contrasting geothermal regimes of the Barbados Ridge accretionary complex: Journal of Geophysical Research, v. 95, p. 8829–8843.

Law, B. E., Nuccio, V. F., and Barker, C. E., 1989, Kinky vitrinite reflectance well profiles: Evidence of paleopore pressure in low-permeability, gas-bearing sequences in Rocky Mountain foreland basins: American Association of Petroleum Geologists Bulletin, v. 73, p. 999–1010.

Le Pichon, X., and 16 others, 1987a, Nankai Trough and the fossil Shikoku Ridge: Results of Box 6 Kaiko survey: Earth and Planetary Science Letters, v. 83, p. 186–198.

Le Pichon, X., and 11 others, 1987b, Nankai Trough and Zenisu Ridge: A deep-sea submersible survey: Earth and Planetary Science Letters, v. 83, p. 285–299.

Le Pechon, X., and 28 others, 1991, Water budgets in accretionary wedges: A comparison: Philosophical Transactions of the Royal Society of London, Series A, v. 335, p. 315–330.

Liou, J. G., Maruyama, S., and Cho, M., 1985, Phase equilibria and mineral paragenesis of metabasites in low-grade metamorphism: Mineralogical Magazine, v. 49, p. 321–333.

——, 1987, Very low-grade metamorphism of volcanic and volcaniclastic rocks—Mineral assemblages and mineral facies, *in* Frey, M., ed., Low temperature metamorphism: New York, Chapman and Hall, p. 59–113.

Lister, C.R.B., 1977, Estimates for heat flow and deep rock properties based on boundary layer theory: Tectonophysics, v. 41, p. 157–171.

Lonsdale, P., 1988, Paleogene history of the Kula Plate: Offshore evidence and onshore implications: Geological Society of America Bulletin, v. 100, p. 733–754.

Mackenzie, J. S., Taguchi, S., and Itaya, T., 1990, Cleavage dating by K-Ar isotopic analysis in the Paleogene Shimanto Belt of eastern Kyushu, S.W. Japan: Journal of Mineralogy, Petrology, and Economic Geology, v. 85, p. 161–167.

Marshak, S., and Karig, D. E., 1977, Triple junctions as a cause for anomalously near-trench igneous activity between the trench and volcanic arc: Geology, v. 5, p. 233–236.

Maruyama, S., Liou, J. G., and Seno, T., 1989, Mesozoic and Cenozoic evolution of Asia, *in* Ben-Avraham, Z., ed., The evolution of the Pacific Ocean margins: New York, Oxford University Press, p. 75–99.

Matsuda, T., Torii, M., Tatsumi, Y., Ishizaka, K., and Yokoyama, T., 1986, Fission-track and K-Ar ages of the Muro volcanic rocks, southwest Japan: Journal of Geomagnetism and Geoelectricity, v. 38, p. 529–535.

McCarthy, J., Stevenson, A. J., Scholl, D. W., and Vallier, T. L., 1984, Speculation on the petroleum geology of the accretionary body: An example from the central Aleutians: Marine and Petroleum Geology, v. 1, p. 151–167.

McLaughlin, R. J., Sorg, D. H., Morgon, J. L., Theodore, R. G., and Meyer, C. E., 1985, Paragenesis and tectonic significance of base and precious metal occurrences along the San Andreas fault at Point Delgada, California: Economic Geology, v. 80, p. 344–359.

Miyake, Y., 1985, MORB-like tholeiites formed within the Miocene forearc basin, southwest Japan: Lithos, v. 18, p. 23–34.

——, 1988, Petrology of the Shionomisaki igneous complex, southwest Japan and its implication for the ophiolite generation: Modern Geology, v. 12, p. 283–302.

Miyake, Y., and Hisatomi, K., 1985, A Miocene forearc magmatism at Shionomisaki, southwest Japan, *in* Nasu, N., Uyeda, S., Kushiro, I., Kobayashi, K., and Kagami, H., eds., Formation of active ocean margins: Tokyo, Terra Scientific Publishing Company, p. 411–422.

Moore, G. F., and Karig, D. E., 1976, Development of sedimentary basins on the lower trench slope: Geology, v. 4, p. 693–697.

Moore, G. F., and 7 others, 1990, Structure of the Nankai Trough accretionary zone from multichannel seismic reflection data: Journal of Geophysical Research, v. 95, p. 8753–8765.

Moore, J. C., 1989, Tectonics and hydrogeology of accretionary prisms: Role of the décollement zone: Journal of Structural Geology, v. 11, p. 95–106.

Moore, J. C., Mascle, A., and the ODP Leg 110 Scientific Party, 1987, Expulsion of fluids from depth along a subduction zone décollement horizon: Nature, v. 326, p. 785–788.

Moore, J. C., Mascle, A., Taylor, E., and the ODP Leg 110 Scientific Party, 1988, Tectonics and hydrogeology of the northern Barbados Ridge: Results from Ocean Drilling Program Leg 110: Geological Society of America Bulletin,

v. 100, p. 1578–1593.

Moore, J. C., and 5 others, 1991, Plumbing accretionary prisms: Effects of permeability variations: Philosophical Transactions of the Royal Society of London, Series A, v. 335, p. 275–288.

Mori, K., and Taguchi, K., 1988, Examination of the low-grade metamorphism in the Shimanto Belt by vitrinite reflectance: Modern Geology, v. 12, p. 325–339.

Mukhopadhyay, P. K., Rullkotter, J., Schaefer, R. G., and Welte, D. H., 1986, Facies and diagenesis of organic matter in Nankai Trough sediments, Deep Sea Drilling Project Leg 87A, *in* Kagami, H., Karig, D. E., Coulbourn, W. T., eds., Initial Reports of the Deep Sea Drilling Project, v. 87: Washington, D.C., U.S. Government Printing Office, p. 877–889.

Murata, A., 1987, Conical folds in the Hitoyoshi Bending, south Kyushu, formed by the clockwise rotation of the southwest Japan Arc: Journal of the Geological Society of Japan, v. 93, p. 91–105.

Murata, M., 1984, Petrology of Miocene I-type and S-type granitic rocks in the Ohmine district, central Kii Peninsula: Journal of Japan Association of Mineralogy, Petrology, and Economic Geology, v. 79, p. 351–369.

Nagihara, S., Kinoshita, H., and Yamano, M., 1989, On the high heat flow in the Nankai Trough area—A simulation study on a heat rebound process: Tectonophysics, v. 161, p. 33–41.

Nakada, S., 1983, Zoned magma chamber of the Osuzuyama acid rocks, southwest Japan: Journal of Petrology, v. 24, p. 471–494.

Nakada, S., and Takahashi, M., 1979, Regional variation in chemistry of the Miocene intermediate to felsic magmas in the Outer Zone and the Setouchi Province of southwest Japan: Journal of the Geological Society of Japan, v. 85, p. 571–582.

Nakajima, T., Banno, S., and Suzuki, T., 1977, Reactions leading to the disappearance of pumpellyite in low-grade metamorphic rocks of the Sanbagawa belt in central Shikoku, Japan: Journal of Petrology, v. 18, p. 263–284.

Nakajima, T., and Hirooka, K., 1986, Clockwise rotation of southwest Japan inferred from paleomagnetism of Miocene rocks in Fukui Prefecture: Journal of Geomagnetism and Geoelectricity, v. 38, p. 513–522.

Nelson, E. P., and Forsythe, R. D., 1989, Ridge collision at convergent margins: Implications for Archean and post-Archean crustal growth: Tectonophysics, v. 161, p. 307–315.

Niitsuma, N., 1988, Neogene tectonic evolution of southwest Japan: Modern Geology, v. 12, p. 497–532.

Niitsuma, N., and Akiba, F., 1985, Neogene tectonic evolution and plate subduction in the Japanese Island Arcs, *in* Nasu, N., Uyeda, S., Kushiro, I., Kobayashi, K., and Kagami, H., eds., Formation of active ocean margins: Tokyo, Terra Publishing Company, p. 75–108.

Nishi, H., 1988, Structural analysis of part of the Shimanto accretionary complex, Kyushu, Japan, based on planktonic foraminiferal zonation: Tectonics, v. 7, p. 641–652.

Oba, N., 1977, Emplacement of granitic rocks in the Outer Zone of southwest Japan and geologic significance: Journal of Geology, v. 85, p. 383–393.

Okamura, M., and Taira, A., 1984, Distribution of fossil ages in the Tertiary Shimanto Belt, *in* Saito, T., and Okada, H., eds., Biostratigraphy and international correlation of the Paleogene System in Japan, p. 81–83.

Okuda, Y., and Honza, E., 1988, Tectonic evolution of the Seinan (SW) Japan fore-arc and accretion in the Nankai Trough: Modern Geology, v. 12, p. 411–434.

Otofuji, Y., and Matsuda, T., 1983, Paleomagnetic evidence for the clockwise rotation of southwest Japan: Earth and Planetary Science Letters, v. 62, p. 349–359.

——, 1984, Timing of rotational motion of southwest Japan inferred from paleomagnetism: Earth and Planetary Science Letters, v. 70, p. 373–382.

——, 1987, Amount of rotation of southwest Japan—Fan shaped opening of the southwestern part of the Japan Sea: Earth and Planetary Science Letters, v. 85, p. 289–301.

Otofuji, Y., Hayashida, A., and Torii, M., 1985a, When was the Japan Sea opened?: Paleomagnetic evidence from southwest Japan, *in* Nasu, N., Uyeda, S., Kushiro, I., Kobayashi, K., and Kogami, H., eds., Formation of active ocean margins: Tokyo, Terra Scientific Publishing Company, p. 551–566.

Otofuji, Y., Matsuda, T., and Nohda, S., 1985b, Paleomagnetic evidence for the Miocene counter-clockwise rotation of northeast Japan—Rifting process of the Japan Arc: Earth and Planetary Science Letters, v. 75, p. 265–277.

Otsuki, K., 1990, Westward migration of the Izu-Bonin Trench, northward motion of the Philippine Sea Plate, and their relationships to the Cenozoic tectonics of Japanese Island arcs: Tectonophysics, v. 180, p. 351–367.

Oxburgh, E. R., and Turcotte, D. L, 1974, Thermal gradients and regional metamorphism in overthrust terrains with special reference to the eastern Alps: Schweizerische Mineralogische und Petrographische Mitteilungen, v. 54, p. 641–662.

Parsons, B., and Sclater, J. G., 1977, An analysis of the variation of ocean floor bathymetry and heat flow with age: Journal of Geophysical Research, v. 82, p. 803–827.

Ranken, B., Cardwell, R. K., and Karig, D. E., 1984, Kinematics of the Philippine Sea Plate: Tectonics, v. 3, p. 555–576.

Reck, B. H., 1987, Implications of measured thermal gradients for water movement through the northeast Japan accretionary prism: Journal of Geophysical Research, v. 92, p. 3683–3690.

Sakai, H., 1981, Olistostrome and sedimentary mélange of the Shimanto terrane in the southern part of Muroto Peninsula, Shikoku: Fukuoka, Japan, Kyushu University, Science Reports of the Department of Geology, v. 14, p. 81–101.

——, 1987, Active faults in the Muroto Peninsula of "non-active fault province": Journal of the Geological Society of Japan, v. 93, p. 513–516.

——, 1988, Origin of the Misaki Olistostrome Belt and re-examination of the Takachiho orogeny: Journal of the Geological Society of Japan, v. 94, p. 945–961.

Sample, J., and Moore, J. C., 1987, Structural style and kinematics of an underplated slate belt, Kodiak and adjacent islands, Alaska: Geological Society of America Bulletin, v. 99, p. 7–20.

Scholz, C. H., 1980, Shear heating and the state of stress on faults: Journal of Geophysical Research, v. 85, p. 6174–6148.

Seno, T., 1985, Age of subducting lithosphere and back-arc basin formation in the Western Pacific since the middle Tertiary, *in* Nasu, N., Yueda, S., Kushiro, I., Kobayashi, K., and Kagami, H., eds., Formation of active ocean margins: Tokyo, Terra Scientific Publishing Company, p. 469–481.

Seno, T., and Maruyama, S., 1984, Paleogeographic reconstruction and origin of the Philippine Sea: Tectonophysics, v. 102, p. 53–84.

Shi, Y., and Wang, C. Y., 1987, Two-dimensional modeling of the P-T-t paths of regional metamorphism in simple overthrust terrains: Geology, v. 15, p. 1048–1051.

Shi, Y., Wang, C. Y., Langseth, M. G., Hobart, M., and von Huene, R., 1988, Heat flow and thermal structure of the Washington-Oregon accretionary prism—A study of the lower slope: Geophysical Research Letters, v. 15, p. 1113–1116.

Shibata, K., 1978, Contemporaneity of Tertiary granites in the Outer Zone of southwest Japan: Geological Society of Japan Bulletin, v. 29, p. 551–554.

Shibata, K., and Ishihara, S., 1979, Initial $^{87}Sr/^{86}Sr$ ratios of plutonic rocks from Japan: Contributions to Mineralogy and Petrology, v. 70, p. 381–390.

Shibata, K., and Nozawa, T., 1967, K-Ar ages of granitic rocks from the Outer Zone of southwest Japan: Geochemical Journal, v. 1, p. 131–137.

Shih, T. C., 1980, Magnetic lineations in the Shikoku Basin, *in* Klein, G. deV., and Kobayashi, K., eds., Initial reports of the Deep Sea Drilling Project, v. 58: Washington, D.C., U.S. Government Printing Office, p. 783–788.

Shipley, T. P., and Shepard, L. E., 1982, Temperature data from the Mexico drilling area: Report on logging and inhole temperature experiments, *in* Watkins, J., and Moore, J. C., eds., Initial reports of the Deep Sea Drilling Project, v. 66: Washington, D.C., U.S. Government Printing Office, p. 771–774.

Sisson, V. B., Hollister, L. S., and Onstott, T. C., 1989, Petrologic and age constraints on the origin of a low pressure/high temperature metamorphic complex, southern Alaska: Journal of Geophysical Research, v. 94, p. 4392–4410.

Stach, E., Mackowsky, M-Th., Teichmuller, M., Taylor, G. H., Chandra, D., and Teichmuller, R., 1982, Stach's textbook of coal petrology: Berlin, Gerbruder Borntraeger, 511 p.

Suess, E., von Huene, R., and the Shipboard Scientific Party, 1988, Proceedings of the Ocean Drilling Program, initial reports, v. 112: College Station, Ocean Drilling Program, 1015 p.

Sweeney, J. J., and Burnham, A. K., 1990, Evaluation of a simple model of vitrinite reflectance based on chemical kinetics: American Association of Petroleum Geologists Bulletin, v. 74, p. 1559–1570.

Tagami, T., 1982, Paleomagmatism and fission-track ages of the Kumano acidic rocks: News of Osaka Micropaleontologists, no. 9, p. 23–32.

Taira, A., 1985, Sedimentary evolution of Shikoku subduction zone: The Shimanto Belt and Nankai Trough, *in* Nasu, N., eds., Formation of active ocean margins: Tokyo, Terra Scientific Publishing Company, p. 835–851.

Taira, A., Tashiro, M., Okmura, M., and Katto, J., 1980, The geology of the Shimanto Belt in Kochi Prefecture, Shikoku, Japan, *in* Taira, A., and Tashiro, M., eds., Geology and paleontology of the Shimanto Belt, Kochi: Rinya-Kosakai Press, p. 319–389.

Taira, A., Okada, H., Whitaker, J.H.McD., and Smith, A. J., 1982, The Shimanto Belt of Japan: Cretaceous to lower Miocene active margin sedimentation, *in* Leggett, J. K., ed., Trench-forearc geology: Geological Society of London Special Publication 10, p. 5–26.

Taira, A., Katto, J., Tashiro, M., Okamura, M., and Kodama, K., 1988, The Shimanto Belt in Shikoku, Japan—Evolution of Cretaceous to Miocene accretionary prism: Modern Geology, v. 12, p. 5–46.

Taira, A., Tokuyama, H., and Soh, W., 1989, Accretion tectonics and evolution of Japan, *in* Ben-Avraham, Z., ed., The evolution of the Pacific Ocean margins: New York, Oxford University Press, p. 100–123.

Taira, A., Hill, I., Firth, J., and the Leg 131 Shipboard Scientific Party, 1991, Proceedings of the Ocean Drilling Program, initial results, v. 131: College Station, Ocean Drilling Program, 434 p.

Takeuchi, A., 1986, Oligocene/Miocene rotational block-movement and paleostress field of Japan: Journal of Geomagnetism and Geoelectricity, v. 38, p. 495–511.

Takahashi, M., and Nomura, S., 1989, Paleomagnetism of the Chichibu quartz diorite—Constraints on the time of lateral bending of the Kanto Syntaxis: Journal of Geomagnetism and Geoelectricity, v. 41, p. 479–489.

Tatsu, A., 1989, P-T histories of peridotite and amphibolite tectonic blocks in the Sambagawa metamorphic belt, Japan, *in* Daly, J. S., Cliff, R. A., and Yardley, B.W.D., eds., Evolution of metamorphic belts: Geological Society of London Special Publication 43, p. 535–538.

Tateishi, M., 1978, Sedimentology and basin analysis of the Paleogene Muro Group in the Kii Peninsula, southwest Japan: Kyoto University, Memoirs of the Faculty of Science, Series of Geology and Mineralogy, v. 45, p. 187–232.

Tatsumi, Y., 1982, Origin of high-magnesian andesites in the Setouchi volcanic belt, southwest Japan; II, Melting phase relations at high pressures: Earth and Planetary Science Letters, v. 60, p. 305–317.

Tatsumi, Y., and Ishizaka, K., 1982, Origin of high-magnesian andesites in the Setouchi volcanic belt, southwest Japan; I, Petrographical and chemical characteristics: Earth and Planetary Science Letters, v. 60, p. 293–304.

Terakado, Y., Shimizu, H., and Masuda, A., 1988, Nd and Sr isotopic variations in acidic rocks formed under a peculiar tectonic environment in Miocene southwest Japan: Contributions to Mineralogy and Petrology, v. 99, p. 1–10.

Torii, M., and Ishikawa, N., 1986, Age estimation of a high-magnesian andesite from the Kii Peninsula, southwest Japan: Possible interpretation of anomalous magnetic direction: Journal of Geomagnetism and Geoelectricity, v. 38, p. 523–528.

Toriumi, M., and Teruya, J., 1988, Tectono-metamorphism of the Shimanto Belt: Modern Geology, v. 12, p. 303–324.

Tsunakawa, H., 1986, Neogene stress field of the Japanese arcs and its relation to igneous activity: Tectonophysics, v. 124, p. 1–22.

Underwood, M. B., 1989, Temporal changes in geothermal gradient, Franciscan subduction complex, northern California: Journal of Geophysical Research, v. 94, p. 3111–3125.

Underwood, M. B., and Howell, D. G., 1987, Thermal maturity of the Cambria slab, an inferred trench-slope basin in central California: Geology, v. 15, p. 216–219.

Underwood, M. B., O'Leary, J. D., and Strong, R. H., 1988, Contrasts in thermal maturity within terranes and across terrane boundaries of the Franciscan Complex, northern California: Journal of Geology, v. 96, p. 399–416.

van den Beukel, J., and Wortel, R., 1988, Thermo-mechanical modelling of arc-trench regions: Tectonophysics, v. 154, p. 177–193.

Vrolijk, P., 1990, The mechanical role of smectite in subduction zones: Geology, v. 18, p. 703–707.

Vrolijk, P., Myers, G., and Moore, J. C., 1988, Warm fluid migration along tectonic mélanges in the Kodiak accretionary complex, Alaska: Journal of Geophysical Research, v. 93, p. 10313–10324.

Wallis, S., and Banno, S., 1990, The Sambagawa belt—Trends in research: Journal of Metamorphic Geology, v. 8, p. 393–399.

Wang, C. Y., and Shi, Y., 1984, On the thermal structure of subduction complexes: A preliminary study: Journal of Geophysical Research, v. 89, p. 7709–7718.

Waples, D. W., 1980, Time and temperature in petroleum formation: Application of Lopatin's method to petroleum exploration: American Association of Petroleum Geologists Bulletin, v. 64, p. 916–926.

Watts, A. B., and Weissel, J. K., 1975, Tectonic history of the Shikoku marginal basin: Earth and Planetary Science Letters, v. 25, p. 239–250.

Yamano, M., and Uyeda, S., 1988, Heat flow, in Nairn, A.E.M., Stehli, F. G., and Uyeda, S., eds., The ocean basins and margins, volume 7B: Plenum Publishing Corporation, p. 523–557.

Yamano, M., Uyeda, S., Aoki, Y., and Shipley, T. H., 1982, Estimates of heat flow derived from gas hydrates: Geology, v. 10, p. 339–343.

Yamano, M., Honda, S., and Uyeda, S., 1984, Nankai Trough: A hot trench?: Marine Geophysical Researches, v. 6, p. 187–203.

Yamano, M., and 29 others, 1992, Heat flow and fluid flow regime in the northern Nankai accretionary prism: Earth and Planetary Science Letters, v. 109, p. 451–462.

Yanai, S., 1986, Megakink bands and Miocene regional stress field in outer southwestern Japan: University of Tokyo, Scientific Papers of the College of Arts and Sciences, v. 36, p. 55–79.

——, 1989, A horizontal buckle model as a dynamic mechanism for back arc spreading of the Japan Sea: Journal of Geology, v. 97, p. 569–583.

Yoshikawa, T., Kaizuka, S., and Ota, Y., 1964, Mode of crustal movement in the late Quaternary on the southeast coast of Shikoku, southwestern Japan: Geographical Review of Japan, v. 37, p. 627–648.

Zandt, G., and Furlong, K. P., 1982, Evolution and thickness of the lithosphere between coastal California: Geology, v. 10, p. 376–381.

MANUSCRIPT ACCEPTED BY THE SOCIETY APRIL 24, 1992

Geological Society of America
Special Paper 273
1993

Vitrinite reflectance and estimates of paleotemperature within the Upper Shimanto Group, Muroto Peninsula, Shikoku, Japan

Matthew M. Laughland* and Michael B. Underwood
Department of Geological Sciences, University of Missouri, Columbia, Missouri 65211

ABSTRACT

The Shimanto Group records a Cretaceous through Miocene history of subduction/accretion along the southwest margin of Japan. We used vitrinite reflectance on over 200 samples of shale and slate to determine regional trends in diagenesis and low-grade metamorphism on the Muroto Peninsula, Shikoku Island. Thermal structure throughout the section overprints all but the latest stage of a complicated polyphase structural evolution. Eocene strata of the Murotohanto subbelt display the highest levels of thermal maturity. Reflectance values range from 1.4 to 5.0%R_m; thermal maturity increases from north to south, and the %R_m values correspond to estimates of paleotemperatures of 180 to 315°C. The Shiina-Narashi fault marks the boundary between Eocene rocks and Oligocene-Miocene strata of the Nabae subbelt. South of this fault, thermal maturity for the Nabae strata ranges from 0.9 to 3.7%R_m, and these values correspond to estimates of peak paleotemperature of 140 to 280°C. Reflectance values increase in proximity to the Maruyama intrusive suite, and the maximum rock temperatures adjacent to the intrusions may have been as high as 500 to 550°C based on comparisons with laboratory heating experiments. Paleotemperatures within the Upper Shimanto Group were unusually high compared to shallow levels of other accretionary complexes. The geothermal gradient was at least 40°C/km on a regional scale and much higher locally. Subduction of juvenile oceanic crust can account for the anomalous levels of thermal maturity documented across the Muroto Peninsula.

INTRODUCTION

Since the pioneering work of Miyashiro (1961, 1967), subduction complexes of the circum-Pacific have been characterized as paired metamorphic belts: an outer belt (near trench) composed of low-temperature/high-pressure metamorphic mineral assemblages and an inner belt (volcanic arc) containing of high-temperature/low-pressure metamorphic facies. Models of accretionary margins illustrate the depression of isotherms in the outer belt owing to subduction of relatively "cold" oceanic lithosphere

(Ernst and others, 1970; Ernst, 1974; Oxburgh and Turcotte, 1970, 1971; Toksoz and others, 1971). More recently, workers have added detail to the predicted temperature distributions within the interiors of accretionary prisms (Wang and Shi, 1984; Van den Beukel and Wortel, 1988). Low geothermal gradients (<15°C/km) within the accretionary prism are an underpinning of all of these models. However, not all calculated and measured geothermal gradients within modern accretionary prisms support the contention of low thermal gradients (see Underwood, Hibbard, and DiTullio, this volume, their Table 1).

Many factors influence the thermal structure of accretionary prisms: the angle of subduction; the convergence velocity; radiogenic heat production from the accretionary prism and oceanic lithosphere; thermal conductivity of the accretionary prism and

---
*Present address: Mobil Research and Development, Exploration, and Producing Technical Center, 3000 Pegasus Park Dr., Dallas, Texas 75265-0232.

Laughland, M. M., and Underwood, M. B., 1993, Vitrinite reflectance and estimates of paleotemperature within the Upper Shimanto Group, Muroto Peninsula, Shikoku, Japan, *in* Underwood, M. B., ed., Thermal Evolution of the Tertiary Shimanto Belt, Southwest Japan: An Example of Ridge-Trench Interaction: Boulder, Colorado, Geological Society of America Special Paper 273.

oceanic lithosphere; mantle temperature; specific heat of the accretionary prism and oceanic lithosphere; bulk density of the accreted strata and subducting oceanic lithosphere; and frictional heating between the overriding prism and descending slab (Oxburgh and Turcotte, 1970, 1971; Toksoz and others, 1971; Wang and Shi, 1984; Van den Beukel and Wortel, 1988; Shi and others, 1988; James and others, 1989). Advective heat transfer can also occur as pore fluids migrate diffusively through the prism or become focused along the décollement zone and imbricate thrust faults (Kulm and others, 1986; Moore and others, 1987, 1988; Vrojlik, 1987, Vrojlik and others, 1988). Discharge of warm water through vents located near the trench axis and on trench-slope walls may contribute to local thermal anomalies (Anderson and others, 1978; Reck, 1987). In addition, accretionary prism thermal structure is affected by the age of the subducting oceanic lithosphere, since heat flux from the downgoing slab is inversely related to the square-root of crustal age (Lister, 1977; Parsons and Sclater, 1977). Because many combinations of these variables are realistic, the thermal models still require rigorous evaluation by empirical data sets.

One objective of our study, in conjunction with detailed analysis of structural relations (DiTullio and others, this volume; Hibbard and others, this volume), was to provide a thermal-maturity data set to test models of accretionary-complex thermal structure. The Upper Shimanto Group on the Muroto Peninsula, Shikoku Island, Japan, was investigated because: (1) it has been a site of active subduction from Cretaceous through Holocene time (Moore and Karig, 1976; Taira and others, 1982, 1988); (2) exhumed portions of the accretionary prism have been studied in detail for sedimentology, paleontology, structure, and stratigraphy (Taira and others, 1980, 1982, 1988; Taira, 1985; Hibbard and Karig, 1987; DiTullio and Byrne, 1990; Hibbard and others, 1992); (3) analogous offshore regions of the Nankai accretionary prism and the Shikoku Basin (Phillipine Sea Plate) have undergone extensive marine geophysical and drilling investigations (Karig and others, 1975; de Vries Klein and others, 1980; Kagami and others, 1986); (4) preliminary analyses of vitrinite reflectance for samples collected from Muroto Peninsula indicated that rocks were of subgreenschist facies and contained organic matter of sufficient type and quantity for analysis of organic maturation (Mori and Taguchi, 1988); and (5) present-day heat flow offshore southwest Japan is anomalously high (Yamano and others, 1984; Kinoshita and Yamano, 1986; Nagihara and others, 1989). Thus, the Eocene-Miocene age Upper Shimanto Group may represent one end member of the spectrum of thermal regimes associated with processes of subduction/accretion.

MODELS OF ACCRETIONARY PRISM THERMAL STRUCTURE

Early models of accretionary prism thermal structure were generally "calibrated" against the pressure-temperature stabilities of metamorphic mineral assemblages (Ernst and others, 1970; Ernst, 1974; Oxburgh and Turcotte, 1970, 1971). These models typically span the entire forearc (150 to 200 km), and vertical thermal gradients are extrapolated to depths of 30 to 40 km. Such models provide only a crude estimate of the temperature-pressure regimes because (1) of the problem of scale, (2) direct observations of the prism are impossible below about 1 km, and (3) the pressure-temperature stability fields are relatively broad for the index metamorphic minerals (Liou, 1971; Frey, 1987; Liou and others, 1987).

Some studies have directed attention to both the geologic and thermal processes that affect the accretionary prism. For example, Cloos (1982, 1985) proposed models to account for the uplift of high-grade (blueschist) metamorphic blocks now dispersed within the mélange matrix of the Franciscan Complex, California. Downbowing isotherms are superimposed on the tectonic-mixing model; upward heat flow is reduced by the underflow of cold oceanic crust, and heat is conducted from the hanging wall into the descending plate and the wedge of accreted sediments (Fig. 1a). Subduction of "cold" oceanic crust is a prerequisite of this model in order to maintain temperature levels within the accretionary wedge that are conducive to the formation of blueschist-grade (low T/high P) blocks.

Exhumed subduction complexes should record changes in thermal regimes associated with changes in orientations of plate convergence or convergence rates. As a consequence, another critical objective of our study is to determine how far back in geologic time the present-day abnormal geothermal gradient in and surrounding Nankai Trough can be traced. A quantitative finite-element computer model of thermal structure (Wang and Shi, 1984), which is restricted to the region from the trench to a position approximately 50 km inboard of the trench, emphasizes the importance of conductive heat transfer across the lower boundary (i.e., between the prism and the subducting oceanic plate). Figure 1b shows that for radioactive heat production of 8.4×10^{-10} W/kg and a convergence rate of 10 mm/yr the maximum temperature reaches 200°C near the rear corner of the prism, while the majority of the prism lies within a temperature range of 50 to 200°C. In contrast, the temperature near the rear corner rises nearly 150°C when a heat flux of only 25 mW/m^2 is allowed across the lower boundary near the toe of the prism (Fig. 1c). Isotherms are depressed more with rapid convergence (e.g., 100 mm/yr), because most of the heat transmitted into the prism from below and generated within the prism by radioactive heating is effectively transferred to the surface by relatively fast internal convection (Wang and Shi, 1984). Models have also been generated to predict the thermal response of prism sediments to the subduction of youthful, hot oceanic crust or an active spreading ridge (DeLong and others, 1979; James and others, 1989).

VITRINITE REFLECTANCE AS AN INDICATOR OF THERMAL MATURITY

As mentioned previously, metamorphic mineral assemblages are of limited value for establishing small changes in pressure or temperature within accretionary wedges because the

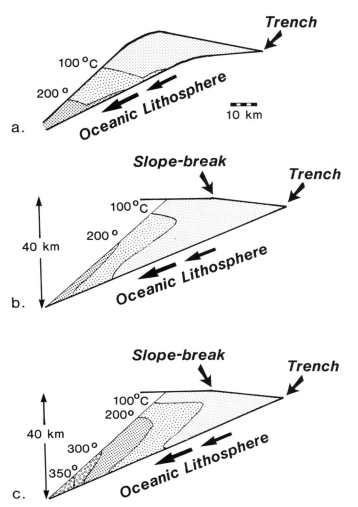

Figure 1. a, Schematic cross section of a mature convergent plate margin. Metamorphism occurs at depth as is shown by the depression of isotherms. From Cloos (1982). b, Cross section of an accretionary prism showing isotherm distribution. Model is based on 8.4×10^{-10} W/kg radiogenic heat production and a subduction rate of 10 mm/yr. No heat transfer is allowed across the lower boundary or the vertical boundary on the left of the model. Modified from Wang and Shi (1984). c, All boundary conditions are equal to those defined in Figure 1b, except that 25mW/m^2 heat flow is allowed across the lower boundary. Modified from Wang and Shi (1984).

phase equilibria (Dow, 1977; Bostick, 1979). Moreover, vitrinite reflectance is known to increase exponentially over a broad range of temperatures (50 to 400°C) (Barker, 1983; Barker and Pawlewicz, 1986) and is unaffected by retrograde metamorphism, thereby recording the highest temperatures to which enclosing strata have been exposed.

Although coals were not analyzed in this study, some common principles of coal petrology are important when interpreting vitrinite-reflectance data. Coal is largely composed of organic matter called "macerals," which are the individual components of the peat-forming environment (Stach and others, 1982). There are three fundamental maceral groups: exinite, vitrinite, and inertinite. Each maceral group possesses different chemical and physical properties, which are a function of their genesis. Petrographically, macerals are identified and classified on the basis of relative reflectance, color, morphology, size, and polishing hardness (Stach and others, 1982).

The rank of a coal refers to its position within the series of organic metamorphism from lignite through subbituminous, bituminous, anthracite, and meta-anthracite. Although rank can be assessed using a variety of chemical properties, the reflectivity of vitrinite (woody tissue of terrestrial plants) is commonly used because (1) vitrinite tends to be the most abundant maceral, (2) vitrinite reflectance increases with rank in a well-known manner, (3) vitrinite is generally easily recognized, and (4) vitrinite maintains a relatively homogeneous character (Davis, 1977).

Vitrinite is not a crystalline substance; however, vitrinite typically behaves as an optical uniaxial negative crystal. Therefore, vitrinite displays two reflectance minima and two maxima during a 360° rotation of the microscope stage. Whether these maxima and minima are "true" maxima and minima depends on whether the polished surface of the vitrinite particle is perpendicular to bedding (true maxima and true minima), parallel to bedding (true maximum reflectance in all directions), or oblique to bedding (true maxima and intermediate minima; Davis, 1977). A histogram of reflectance values from randomly oriented vitrinite particles in a sample theoretically should show a Gaussian distribution. The difference between the minimum and maximum values (anisotropy) is termed bireflectance. Mean vitrinite reflectance ($\%R_m$) is considered a more accurate indicator of thermal maturity than maximum reflectance at relatively low ranks ($<2.0\% R_m$). Because bireflectance increases as a function of coal rank, the standard deviation of randomly oriented particles increases as mean vitrinite reflectance increases (Fig. 2). Most coal characterization studies report a mean maximum reflectance value for a single coal (Davis, 1977). In this study, however, we report the mean random reflectance ($\%R_m$) obtained from nonrotated vitrinite particles. Random reflectance was measured because the size of particles dispersed in argillaceous rocks is generally small, thus making it difficult to keep a particle centered beneath an optical measuring spot during rotation of the microscope stage. Consequently, mean random reflectance is the standard method for evaluating dispersed organic matter (Bostick, 1979; Barker and Pawlewicz, 1986).

pressure-temperature stability fields are relatively wide, especially in subgreenschist-facies rocks (Coombs, 1961; Liou, 1971; Zen and Thompson, 1974; Frey, 1987; Liou and others, 1987). Moreover, inorganic reactions are dependent upon a number of parameters in addition to temperature, such as complex solid and fluid compositions, mixed component volatiles, metasomatism, slow kinetics, continuous solid-solution reactions, and retrograde reactions (Liou and others, 1985). Vitrinite reflectance, on the other hand, tends to be more sensitive to temperature and is relatively unaffected by cofactors that influence mineralogic

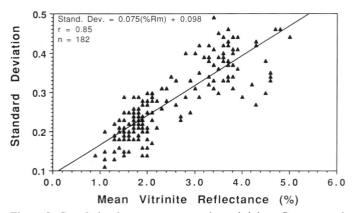

Figure 2. Correlation between mean random vitrinite reflectance and standard deviation. Standard deviation increases with coal rank owing to the concomitant increase in bireflectance.

Sample collection

Samples of shale and slate were collected from surface outcrop exposures of turbidites and mélange from the Upper Shimanto Group (see also, DiTullio and others, this volume; Hibbard and others, this volume). A total of 243 samples were analyzed for vitrinite reflectance. One hundred twenty-seven samples from the Eocene Murotohanto subbelt were collected along two separate north-south coastal transects (east and west coasts of the peninsula) and three separate transects within the interior of the peninsula (Fig. 3). One east-west traverse was located just north of the Shiina-Narashi fault. One hundred sixteen samples from the Oligocene-Miocene Nabae subbelt were collected along two separate north-south coastal transects that merged at the southern tip of the peninsula (Figs. 4 and 5). Interior sample control was emphasized adjacent to the Maruyama intrusive suite at the southern tip of the peninsula, and across the Shijujiyama Formation.

Sample preparation and petrographic procedure

Each sample was ground to powder-form using a mortar and pestel. Approximately 28 ml of sample powder were demineralized using hydrochloric and hydrofluoric acids in order to concentrate the organic matter (kerogen). The kerogen was washed with distilled water and centrifuged a minimum of four times to reduce acid residue. After drying (at temperatures less than 35°C), the kerogen was mixed with epoxy, mounted on a glass slide, and polished. The reflectivity of 50 different, randomly oriented vitrinite particles per sample was measured in oil (R.I. = 1.5180) using a digital photometer and a stable reflected light source mounted on a Leitz MPV I microscope. The microscope was fitted with 10X magnification occulars, a 50X magnification oil immersion objective, and 1.25 lense factor for a total magnification of 625X. A mean random reflectance value (%R_m) was calculated for each sample. If samples contained significantly

fewer than 50 different vitrinite particles of desirable quality a mean reflectance value was not determined.

One difficulty that may arise in the accurate assessment of vitrinite reflectance is the suppression of reflectance values by the presence of hydrogen-enriched macerals (e.g., exinites and alginites; Jones and Edison, 1978; Hutton and Cook, 1980; Hutton and others, 1980; Kalkreuth, 1982; Walker, 1982; Price and Barker, 1985). Reflectance suppression may affect samples with mean reflectance values as high as 4.0%R_m, despite the fact that hydrogen-enriched macerals are thermally destroyed at a maturity level of approximately 1.35% R_m (Price and Barker, 1985). No unusually large concentrations of exinite macerals were observed during reflected white light petrographic analysis. Moreover, Rock-Eval pyrolysis of shale samples from the study area indicates a predominance of Type III kerogen (terrestrial plant material) with extremely low H/C ratios (Hibbard and Larue, this volume). Exposure of organic matter to the atmosphere or oxygenated waters in the subsurface results in oxidation of organic matter and can reduce vitrinite reflectance (Chandra, 1962; Pearson and Kwong, 1979; Bustin, 1982; Bustin and others, 1989). Oxidized rims have a lower reflectance than the particle core (Crelling and others, 1979). When encountered, oxidized grains were completely avoided during reflectance measurement.

Data analysis

Even if a sample contains sufficient vitrinite for reflectance analysis, there can be problems in determining an accurate mean reflectance value. For example, a single sample may contain two or more vitrinite populations. Optical contrasts in reflectance for different vitrinite populations can be subtle and are more easily recognized by constructing a histogram of individual reflectance values. One philosophy is to select the lowest population as the "indigenous" population and compute a mean as the true maturation level (called the low-gray method). Higher populations are considered "recycled" vitrinite (reworked vitrinite that has been exposed earlier to greater levels of thermal maturity), and these data are eliminated (Fig. 6a; Bostick, 1979; Dow and O'Connor, 1982). We believe that such a technique will bias reflectance data toward low estimates of thermal maturity, especially for high-rank samples. With randomly oriented, high-rank (>2.0%R_m) vitrinite particles, the lowest mode alone does not represent the "indigenous" population. Owing to the increase in vitrinite bireflectance with an increase in coal rank, the statistical distribution is commonly polymodal (Fig. 6b). Accurate identification of maceral groups is relatively uncomplicated at lower ranks since vitrinite is typically more gray than the brightly reflecting inert macerals (<2.00%R_m). At higher ranks, however, vitrinite reflectance overlaps with the reflectance range of macerals from the inert group (e.g., semifusinite or fusinite). Moreover, maceral morphology is obscured with dispersed (transported) organic matter. As a consequence, inertinite reflectances may be measured inadvertently.

In our study, individual reflectance measurements (%R_o) for

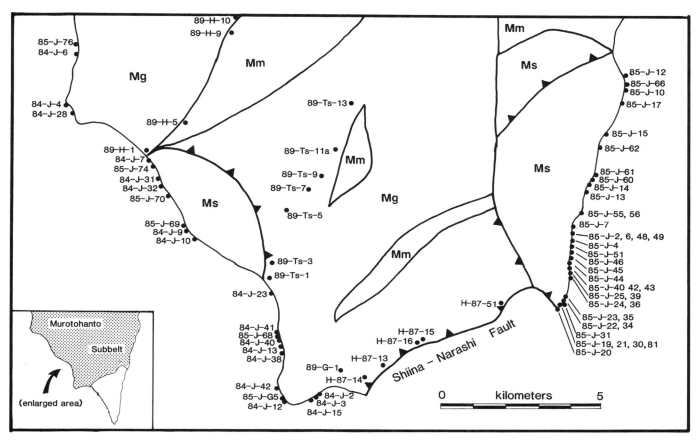

Figure 3. Simplified map of the Murotohanto subbelt showing the locations of all sampling stations. The Shiina-Narashi fault separates the Eocene Murotohanto subbelt to the north from the Oligocene-Miocene Nabae subbelt to the south. Refer to Figure 11 for symbol identification.

each sample were plotted in histogram form to test for truly abnormal statistical distributions. Since mean reflectance values greater than 2.0% are not unusual for samples from the Upper Shimanto Group, we adopted a rather conservative approach by discarding individual measurements only if values exceeded acceptable levels of vitrinite anisotropy based on the correlation between mean reflectance and maximum reflectance from whole-coal analyses (Fig. 7; Stach and others, 1982). Elimination of individual reflectance values was generally restricted to samples with mean reflectance values greater than 2.0%. Rarely were more than five individual reflectance values eliminated from a single population, and rarely did the elimination of those values reduce the initial mean reflectance by greater than $0.1\%R_O$.

Precision and accuracy

Analytical sensitivity extends to $0.01\%R_O$, and previous comparisons between UMC (University of Missouri-Columbia) vitrinite-reflectance data and results from two commercial laboratories[1] indicate the "accuracy" of the technique is better than $\pm0.1\%R_m$ over relatively low levels of thermal maturity

($<1.5\%R_m$). A reproducibility test that entailed two separate counts for each of 13 samples over a reflectance range of 0.9 to $4.0\%R_m$ yielded a correlation coefficient of 0.99. A higher degree of interlaboratory variability has been reported by Dembicki (1984), however, especially in higher-grade rocks. Thus, all mean reflectance values are reported herein to the nearest $0.1\%R_m$, even though analytical sensitivity extends to $0.01\%R_O$. Samples were also analyzed for illite crystallinity; as discussed in detail by Underwood and others (this volume). Crystallinity indices for the Upper Shimanto Belt confirm the levels of thermal maturity determined using vitrinite reflectance.

Estimates of paleotemperature

The procedure for calculating maximum paleotemperature based on vitrinite reflectance represents one of the major controversies in organic petrography and low-temperature metamor-

---

[1]GeoChem Laboratories, Inc., Houston, Texas; Clark Geological Services, Freemont, California.

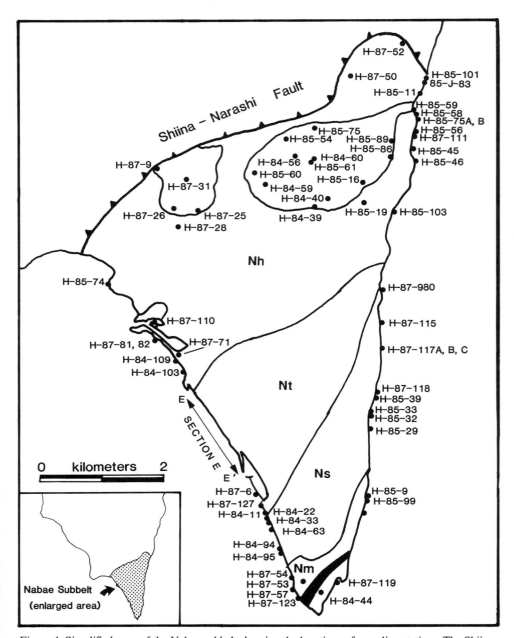

Figure 4. Simplified map of the Nabae subbelt showing the locations of sampling stations. The Shiina-Narashi fault separates the Eocene Murotohanto subbelt to the north from the Oligocene-Miocene Nabae subbelt to the south. Figure 5 shows the location of sampling stations along section E–E′. Refer to Figure 11 for symbol identification.

phism. The traditional contention is that vitrinite reflectance changes as a function of both peak burial temperature and effective heating time (Karweil, 1956; Hood and others, 1975; Bostick and others, 1978; Waples, 1980; Robert, 1988). Other scientists have argued, conversely, that vitrinite reflectance reaches equilibrium with maximum temperature within 1 m.y. or less; thus, for most situations involving spans of geologic time, heating duration can be ignored (Wright, 1980; Gretener and Curtis, 1982; Suggate, 1982; Barker, 1983, 1989; Price, 1983; Barker and Pawlewicz, 1986). In a recent review of the pertinent data, Kisch (1987) agreed that "stabilization" of coal rank is governed largely by temperature rather than heating time, and that once stabilization is achieved, continued heating at a constant temperature has negligible effect on coal rank. In addition, less time is necessary

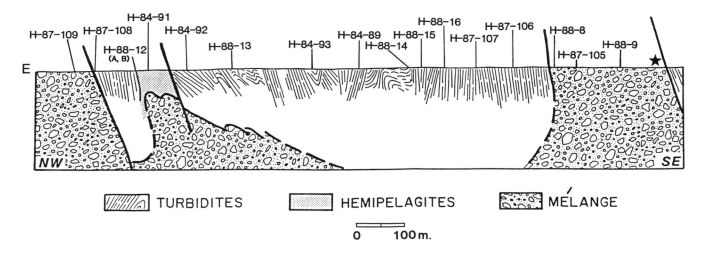

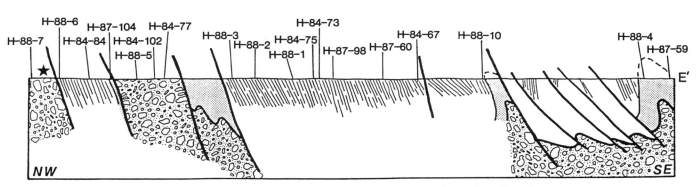

Figure 5. Geologic cross section (E–E′) showing the locations of sampling stations along the west coast transect of the Tsuro sequence (cross section modified from Hibbard and others, 1990). Upper and lower parts show a continuous cross section that joins at stars.

for stabilization of coal rank at higher temperatures (>200°C). The duration of heating seems to be critical only during relatively brief thermal events (see, e.g., the results of laboratory experiments by Bostick, 1979; and Ikehara and others, 1982).

Time-temperature models and sophisticated assessments of the kinetics of kerogen maturation have widespread appeal in petroleum exploration, particularly when applied to predictions of the timing and depths of oil and gas generation (e.g., Tissot and others, 1987; Wood, 1988). Most of the organic maturation models have been developed using data from first-cycle sedimentary basins where the stratigraphy, geochronology, and geothermal history are fairly well documented to begin with. The purpose of our study, in contrast, is to estimate peak burial temperatures within a geologic regime that has experienced a very complicated and poorly documented tectonic evolution; modeling under these circumstances is far from straightforward. To accommodate both the time-dependent and the time-independent schools of thought, we present paleotemperature estimates derived from most of the published correlations (Table 1). The resulting data, as expected, differ dramatically.

We recognize several problems with the use of time-dependent methods of paleotemperature calculations. The most fundamental problem involves the choice of a correct value for the effective heating time; this choice can be exceedingly difficult to make within a region of polyphase deformation, particularly if the thermal history is poorly known as well. Obviously, an erroneous value of effective heating time will lead to significant systematic errors in the resulting paleotemperature estimates. Even if the correct time factor is chosen, additional problems exist. The Karweil (1956) graph, for example, does not extend beyond temperatures of 240°C or heating durations less than 10 m.y. The empirical curves of Karweil (1956), Hood and others (1975), and Bostick and others (1978) do not include the original data used to produce the curves, so statistical analysis of the inherent error is impossible. The popular Lopatin method (Waples, 1980), if applied properly, requires a cumulative evaluation of incremental burial and heating history (i.e., changes in both depth and geothermal gradient through time). In the absence of these essential data, crude estimates of maximum temperature still can be made using the Lopatin method by assuming that the

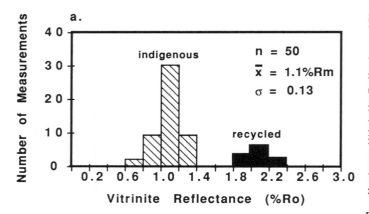

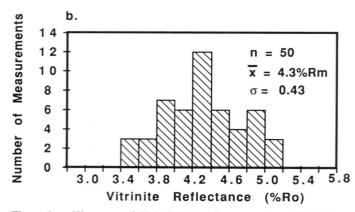

Figure 6. a, Histogram of 50 individual reflectance values (%R_o) for a single sample showing a population with a single mode and relatively low bireflectance. The recycled population would be eliminated based upon the correlation between mean vitrinite reflectance and maximum vitrinite reflectance (see Fig. 5). b, Histogram of 50 individual reflectance values (%R_o) for a single sample showing a polymodal distribution over a wide range of reflectance values. The degree of bireflectance shown by sample B does not exceed limits of maximum reflectance based on the mean reflectance versus maximum reflectance correlation of Stach and others (1982).

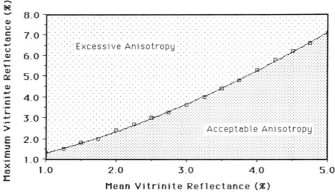

Figure 7. Plot of mean vitrinite reflectance (%R_m) versus maximum vitrinite reflectance (R_{max}), showing the effects of increasing bireflectance with coal rank. These data (from Stach and others, 1982, their Fig. 19A) were used to help identify nonindigenous vitrinite and macerals of the inert group. For a given mean reflectance value, individual reflectance values (%R_o) were eliminated from final calculations of mean reflectance (%R_m) if they exceeded acceptable limits of anisotropy.

TABLE 1. CORRELATION EQUATIONS FOR CONVERSION FROM %Rm TO MAXIMUM PALEOTEMPERATURE

Reference*	Heating Time	Equations
1	Independent	$T(°C) = 131 (\ln R_m) + 187$
2	Independent	$T(°C) = 146 (\ln R_m) + 122$
3	Independent	$T(°C) = 128 (\ln R_m) + 154$
4	Independent	$T(°C) = 104 (\ln R_m) + 148$
5	Independent	$T_h(°C) = 90 (\ln R_m) + 152$
6	Independent	$T_h(°C) = 123 (\ln R_m) + 155$
7	1 m.y.	$T(°C) = 192 + 85 (\ln R_m) - 13 (\ln [R_m])^2$
7	10 m.y.	$T(°C) = 153 + 85 (\ln R_m) - 13 (\ln [R_m])^2$
8	1 m.y.	$T(°C) = 62 (\ln R_m) + 169$
8	10 m.y.	$T(°C) = 63 (\ln R_m) + 135$
9	1 m.y.	$T(°C) = 93 (\ln R_m) + 174$
9	10 m.y.	$T(°C) = 90 (\ln R_m) + 158$

*1 = Price (1983); 2 = Barker (1983); 3 = Barker and Pawlewicz (1986); 4 = Barker (1988); 5 = Aizawa (1990); 6 = Barker and Goldstein (1990); 7 = Bostick and others (1978); 8 = Walples (1980); 9 = Sweeney and Burnham (1990).

strata were subjected to their maximum increment of burial temperature for the entire episode of heating (see, e.g., Mori and Taguchi, 1988).

The Lopatin method hinges on correlations between the so-called time-temperature index (TTI) and direct measures of thermal maturity, such as vitrinite-reflectance and thermal-alteration index (Waples, 1980). The cumulative TTI of a stratal horizon is equal to the sum of each incremental product of time and temperature factor. The common procedure is to plot TTI and %R_m values using a log-log format, but this type of graph tends to mask what amounts to a horrendous degree of scatter when data from around the world are plotted collectively (Waples, 1980). One might conclude, therefore, that a single calibration curve for TTI-%R_m is not applicable for all sedimentary

basins and burial conditions. This conclusion is supported by Issler (1984), who presented convincing evidence that the slopes and intercepts of the %R_m-TTI curves differ significantly when comparisons are made between individual basins in Canada. In the absence of vital data, selection of the "correct" %R_m-TTI curve would require intuitive a priori assumptions. Finally, other workers have proven that the "universal" %R_m-TTI correlation of Waples (1980) is faulty at levels of thermal maturity above about 2.0% R_m (Katz and others, 1982). Unfortunately for our purposes, much of the Shimanto data fall within this upper range of thermal maturities.

Another problem with the Lopatin method, and other models that describe vitrinite maturation as an Arrhenius first-order chemical reaction with a single activation energy, is that a single reaction does not adequately model complex reactions over a wide range of temperatures and heating rates. Larter (1989) concludes that vitrinite maturation is consistent with an activation energy distribution model using chemically meaningful parameters. Burnham and Sweeney (1989) used an integrated chemical kinetic model over a range of time and temperature in order to calculate vitrinite composition. Their model considers the separate and parallel reactions for elimination of water, carbon dioxide, methane, and higher hydrocarbons from the vitrinite structure. The chemical compositions thus are used to determine H/C and O/C ratios and carbon content, which were related to vitrinite reflectance through correlations between measured elemental composition and measured vitrinite reflectance from coal. Although the model is based largely on chemical analyses of Jurassic age coals with mean maximum reflectance values less than 2.0% (J. Sweeney, personal communication, 1990), the range of calculated values is extended to as high as $4.5\%R_m$.

The most recent version of the model, called Easy $\%R_o$, calculates values of vitrinite reflectance over a range of time and temperature, which are entered easily into a spreadsheet using a personal computer (Sweeney and Burnham, 1990). In order to evaluate the reflectance-temperature correlation, Easy $\%R_o$ was programmed to create time-temperature models in which the temperature instantaneously reaches a constant value and then remains there for a given time while vitrinite reacts isothermally. Using this approach, different values of vitrinite reflectance were calculated by entering different temperature values into the spreadsheet for a given effective heating time, 1 m.y. and 10 m.y., respectively. Although this model is based upon reaction-kinetics for whole-coal samples ($R_{max.} \leqslant 2.0\%$) rather than dispersed sedimentary organic matter, the temperature reflectance correlations for both t_{eff} = 1 m.y. and 10 m.y. produce a close-fit with the empirically derived, time-independent reflectance-temperature correlation of Barker (1988) (discussed below) to reflectance values as high as 4.5% (see Fig. 8a, b).

Several mathematical solutions now exist in which vitrinite reflectance is treated as an absolute measure of peak burial temperature (Fig. 9). One criticism of these curves is that errors are inherent in the measurements of both vitrinite reflectance and maximum burial temperature. The data of Barker and Pawlewicz (1986), nevertheless, display an impressive degree of statistical correlation, such that roughly 70% of the variations in mean reflectance can be attributed solely to changes in peak temperature; the remaining variance is probably due to differences in analytical procedure between different laboratories (Dembicki, 1984), erroneous corrections of borehole temperature, and perhaps difference in the duration of heating. Among the available curves, the major-axis regression of Barker (1988) yields the most conservative estimates of burial temperature over the range of $\%R_m$ values obtained in the Upper Shimanto Group, whereas the curve of Price (1983) yields the highest temperatures (Fig. 9). The

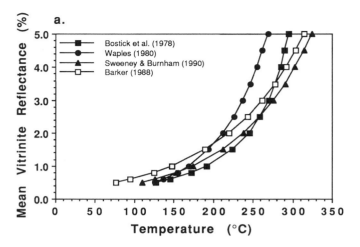

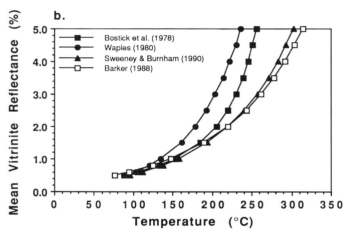

Figure 8. a, Comparison of the time-dependent correlations for calculating paleotemperature from values of vitrinite reflectance, using an effective heating time of 1 m.y., with Barker's (1988) time-independent correlation; b, Comparison of the time-dependent correlations for calculating paleotemperature from values of vitrinite reflectance, using an effective heating time of 10 m.y. with Barker's (1988) time-independent correlation. Correlation equations are listed in Table 3.

curve of Barker (1983) is based solely on data from liquid-dominated geothermal systems and, therefore, we believe that it should be used with caution during interpretations of the rock record unless there is evidence of hydrothermal alteration. With the exception of the curves (T_{eff} = 1 and 10 m.y.) from Easy $\%R_o$ (Sweeney and Burnham, 1990), the Barker (1988) correlation deviates markedly from the other time-dependent curves, particularly at R_m values above 1.5% (Fig. 8a, b).

Additional testing and calibration of the time-independent $\%R_m$-temperature curves is possible using other geothermometers, particularly fluid inclusions, which are not affected by heating time (Roedder, 1984; Burruss, 1989). Based upon fairly limited data sets, Barker and Halley (1986) and Pollastro and Barker (1986) demonstrated that excellent agreement can exist

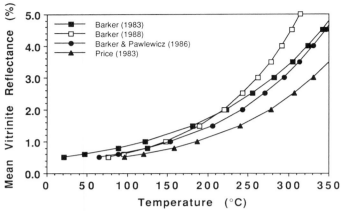

Figure 9. Graph shows the different time-independent correlations for calculating paleotemperatures from values of vitrinite reflectance. Correlations equations are listed in Table 3.

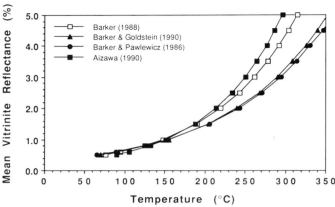

Figure 10. Reflectance-temperature correlations based upon direct comparisons of fluid inclusion homogenization temperatures and mean vitrinite reflectance. Barker's (1988) and Barker and Pawlewicz's (1986) time-independent correlations are plotted for comparison. Correlation equations are listed in Table 3.

between fluid-inclusion homogenization temperatures (T_h) and $\%R_m$-derived paleotemperatures calculated from the curve of Barker and Pawlewicz (1986). Similar results were obtained during studies by Tilley and others (1989). Tilley and others (1989), moreover, showed that (1) temperatures calculated using the Lopatin-Waples method are typically 20 to 50°C lower than the T_h values from methane-rich and hydrocarbon-rich inclusions in early phase (highest emplacement temperature) quartz veins; and (2) Lopatin-Waples temperatures are equal to or slightly lower than T_h from aqueous inclusions in later-phase calcite without coexisting methane; the calcite veins, moreover, precipitated after the quartz during a subsequent period of uplift and cooling. The T_h values (uncorrected for pressure) for this type of fluid inclusion should always be less than the original trapping temperature (Burruss, 1989). The incongruity between Lopatin-Waples temperatures and T_h is difficult to explain given the fact that most inclusion-hosting vein minerals (such as quartz and calcite) precipitate from migrating pore fluids as the fluids cool; in other words, even the fluid-inclusion trapping temperatures should be less than the maximum temperature of the host rock.

Two direct correlations between T_h and $\%R_m$ are shown in Figure 10 (Aizawa, 1990; Barker and Goldstein, 1990). The data of Barker and Goldstein (1990) is the more exhaustive of the two, and the resulting correlation curves is remarkably similar to the $\%R_m$-temperature curve of Barker and Pawlewicz (1986); conversely, the matches between $\%R_m$-T_h and all of the time-dependent $\%R_m$-temperature curves (Fig. 8a, b) are very poor with the exception of paleotemperatures derived from the model of Sweeney and Burnham (1990). The data compiled by Barker and Goldstein (1990) include abundant measurements from both liquid-dominated geothermal systems and "normal" sedimentary basins; thus, their correlation should serve effectively as a general standard. The correlation coefficient between T_h and $\%R_m$ is equal to 0.93. The results of Aizawa (1990) may be more germane to our investigation of the Upper Shimanto Group, how-

ever, because those data were obtained from coeval strata exposed to analogous thermal conditions in northwest Kyushu, Japan. The Aizawa (1990) data provide an impressive match with the $\%R_m$-temperature curve of Barker (1988), as T_h is equal to or slightly less than the peak rock temperature derived from $\%R_m$ (Fig. 10). Therefore, based largely on the congruity of both slope and intercept between these two curves, we conclude that the paleotemperatures derived from the Barker (1988) equation provide the most accurate measure of peak thermal conditions within the Upper Shimanto Group.

Sources of error

The errors associated with $\%R_m$-temperature correlations have been discussed by Barker and Pawlewicz (1986). The reflectance-temperature data of Price (1983) produces a correlation coefficient of 0.97, and visual inspection of his data points suggests a random error of approximately ±20°C. Correlation coefficients reported for the regression curves of Barker and Goldstein (1990) and Barker and Pawlewicz (1986) are 0.93 and 0.84, respectively. Estimates of paleotemperature are herein rounded to the nearest 10°C, and our estimated error is ±30°C.

THE UPPER SHIMANTO GROUP, MUROTO PENINSULA, SHIKOKU ISLAND

As summarized by Underwood, Hibbard, and DiTullio (this volume), the Upper Shimanto Group on Muroto Peninsula, Shikoku Island, is subdivided into the Eocene–early Oligocene (?) age Murotohanto subbelt and the late Oligocene–early middle Miocene age Nabae subbelt (Fig. 11). The Shiina-Narashi fault, a steeply dipping thrust fault (Hibbard and others, 1992), is the most fundamental structural break in the section. This fault sepa-

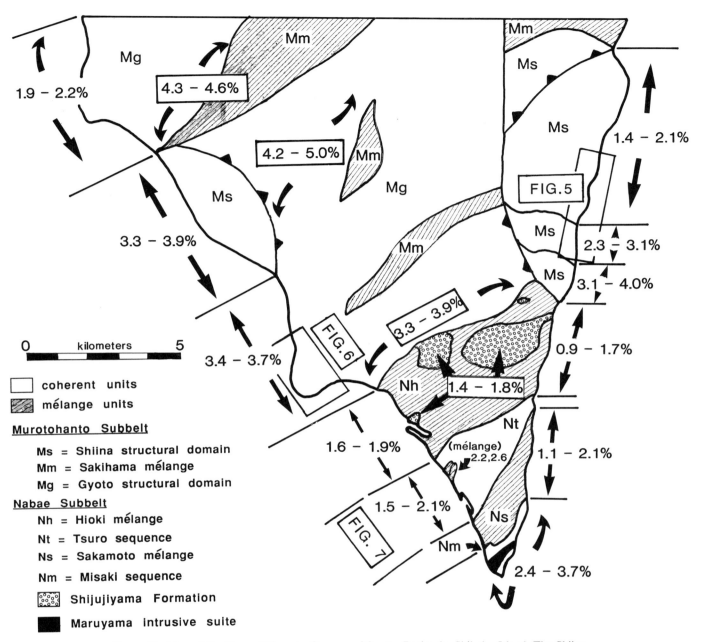

Figure 11. Map of the Upper Shimanto Group on Muroto, Peninsula, Shikoku Island. The Shiina-Narashi fault separates the Eocene age Murotohanto subbelt (M) (mapping by DiTullio and Byrne, 1990) from the Oligocene-Miocene age Nabae subbelt (N) (mapping by Hibbard and others, 1992). Map shows alternating belts of mélange and coherent units, as well as spatial trends in mean reflectance values. See DiTullio and others (this volume) and Hibbard and others (this volume) for additional details.

rates older strata of the Murotohanto subbelt to the north from the younger Nabae subbelt to the south and juxtaposes rocks of contrasting deformational histories (DiTullio and Byrne, 1990; DiTullio and others, this volume; Hibbard and Karig, 1987; Hibbard and others, 1992; Hibbard and others, this volume).

Both subbelts are subdivided into roughly northwest-striking, alternating lithostratigraphic units of coherent turbidites and mélange, and the Murotohanto subbelt is further subdivided into the Gyoto and Shiina structural domains (DiTullio and Byrne, 1990; DiTullio and others, this volume; Hibbard and others, this volume). Mafic dikes of the Maruyama intrusive suite intrude the Nabae subbelt along the east and west coasts and at the southern tip of the peninsula; radiometric dates indicate emplacement at approximately 14 Ma (Hamamoto and Sakai, 1987, referenced *in* Hibbard and others, 1992). Offshore, the Upper Shimanto Group merges with the active Nankai accretionary complex where the Shikoku back-arc basin (Phillipine Sea Plate) is subducting beneath the Eurasian Plate (Taira and others, 1988).

RESULTS

Based on detailed analyses of geologic structures on the Muroto Peninsula, the thermal structure overprints much of the Cenozoic polyphase deformational history (see DiTullio and others, this volume; Hibbard and others, this volume). The following %R_m data that define the thermal structure have been subdivided into two groups based upon the position of sampling stations with respect to the Shiina-Narashi fault. All relevant statistics are reported in Tables 2 and 3.

Murotohanto subbelt

Samples collected from the Murotohanto subbelt are typically not only high in rank, but also lean in vitrinite. Eighty out of 127 samples contain sufficient vitrinite for reflectance analysis. A histogram of mean reflectance values shows that thermal maturity varies significantly (Fig. 12). The majority of samples exhibit high levels of thermal maturity (3.0 to 4.8%R_m); a few samples maintain maturation levels as low as 1.4%R_m or as high as 5.0%R_m.

Mean reflectance values are shown in map distribution in Figure 11. Reflectance values increase generally from north to south along the east and west coast transects. Lower values (<2.5%R_m) are recorded along the northern limits of both transects. Reflectance values increase more abruptly with distance toward the south along the west coast of the peninsula, relative to the east coast. Samples collected from localities within the interior of the subbelt display the highest levels of thermal maturity for the Murotohanto subbelt, with values ranging from 3.3 to 5.0%R_m. According to the correlation of Barker (1988), reflectance values for the entire subbelt correspond to a paleotemperature range of 180 to 315°C (Fig. 12). Estimates of paleotemperature for the Gyoto and Shiina structural domains based upon on other reflectance-temperature conversion methods are shown in Table 4.

Nabae subbelt

One hundred two of 116 samples contained sufficient vitrinite for reflectance analysis. A histogram of mean reflectance values shows a wide range of thermal maturities for the Nabae subbelt (Fig. 13). In contrast to the Murotohanto subbelt, however, the majority of samples display thermal maturities on the order of 1.0 to 2.2%R_m; a few values are as high as 3.7%R_m.

Mean reflectance values are shown in map distribution in Figure 11. Albeit more subtle, a southward increase in reflectance

values for the Nabae subbelt is apparent and is similar to the southward increase in reflectance values recognized for the Murotohanto subbelt. With closer inspection, however, it is evident that vitrinite reflectance values are fairly uniform throughout most of the Nabae subbelt, but increase dramatically at the southern tip of the peninsula where the Maruyama intrusive suite crops out (see Hibbard and others, this volume). The highest reflectance values for the Nabae subbelt are for sample localities in the immediate vicinity of the mafic intrusions. According to the correlation of Barker (1988), reflectance values for the Nabae subbelt correspond to paleotemperatures of 100 to 280°C (Fig. 13). Estimates of paleotemperature based upon other reflectance-temperature conversion methods are shown in Table 4.

It is well documented that the thermal aureole surrounding an igneous dike or sill extends at least two times the width of the intrusion (Dow, 1977; Barker and Pawlewicz, 1986; Wang and others, 1989). The largest gabbroic dike of the Maruyama intrusive suite is nearly 200 m wide; therefore, thermal effects resulting from the igneous intrusion are superimposed on background levels of thermal maturity over an area that is about 800 m wide (1 km including the 200-m-wide intrusion). Heating related to intrusive emplacement at the southern tip of the peninsula can account for the higher levels of thermal maturity documented across the entire Misaki sequence, and to a large degree for the southward rise in reflectance values for the Sakamoto mélange to the north.

Shiina-Narashi fault

The Murotohanto subbelt, overall, yields an average mean reflectance value of 3.3%, which corresponds to an average paleotemperature of 270°C, whereas an average mean reflectance value of 1.8% for the Nabae subbelt corresponds to a paleotemperature of 215°C (Figs. 12 and 13). A sharp contrast in thermal maturity occurs across the Shiina-Narashi fault (Fig. 11). Mean reflectance values for sample localities immediately north of the fault are 3.3 to 3.9%, whereas those to the immediate south are 0.9 to 1.9%, and this increase in reflectance values translates to a difference of approximately 100°C in paleotemperature (using Barker's (1988 correlation).

DISCUSSION

Our estimates of thermal maturity are corroborated by other studies of organic metamorphism for the Upper Shimanto Group, Muroto Peninsula. Mori and Taguchi (1988) reported mean maximum vitrinite reflectance values for the east coast of the Muroto Peninsula that range from approximately 1.7 to 4.1% for the Murotohanta subbelt, whereas mean maximum reflectance values range from less than 1.0 to nearly 5.0% along the east coast of the Nabae subbelt. Caution is warranted in making direct comparison of the two study's data, however, because Mori and Taguchi (1988) report histograms of maximum reflectance values whereas we report mean random reflectance. Also, Mori and

TABLE 2. VITRINITE REFLECTANCE DATA FOR THE MUROTOHANTO SUBBELT

Sample	Structural Domain	Min. %Ro*	Max. %Ro[†]	n[§]	S.D.**	%Rm[‡]	Sample	Structural Domain	Min. %Ro*	Max. %Ro[†]	n[§]	S.D.**	%Rm[‡]
84-J-2	Gyoto	2.82	4.30	50	0.38	3.7	85-J-43	Shiina	2.72	4.24	49	0.39	3.5
84-J-3	Gyoto	2.40	4.20	47	0.49	3.4	85-J-44	Shiina	2.84	4.18	50	0.36	3.5
84-J-4	Gyoto	1.60	2.44	50	0.20	2.0	85-J-45	Shiina	2.36	3.73	50	0.29	3.1
84-J-6	Gyoto	1.39	2.72	48	0.28	2.2	85-J-46	Shiina	2.29	3.75	50	0.35	3.1
84-J-7	Shiina	3.73	4.61	49	0.42	3.7	85-J-48	Shiina	2.24	3.18	50	0.23	2.7
84-J-9	Shiina	2.55	3.95	55	0.34	3.3	85-J-49	Shiina	2.01	3.59	48	0.40	2.8
84-J-10	Shiina	2.52	4.09	49	0.42	3.3	85-J-51	Shiina	2.22	3.21	50	0.24	2.7
84-J-12	Gyoto	3.01	4.32	50	0.33	3.7	85-J-55	Shiina	1.43	2.30	50	0.21	1.8
84-J-13	Gyoto	3.01	4.58	47	0.46	3.7	85-J-56	Shiina	1.12	1.98	50	0.23	1.5
84-J-15	Gyoto	3.02	4.42	50	0.37	3.7	85-J-60	Shiina	1.20	2.04	49	0.19	1.6
84-J-23	Gyoto	2.95	4.26	50	0.30	3.5	85-J-61	Shiina	1.49	2.59	50	0.28	2.0
84-J-28	Gyoto	1.85	3.01	50	0.28	2.4	85-J-62	Shiina	1.35	2.25	44	0.22	1.8
84-J-31	Shiina	2.72	4.09	50	0.33	3.4	85-J-66	Shiina	1.13	2.02	50	0.20	1.6
84-J-32	Shiina	2.71	4.39	48	0.41	3.4	85-J-68	Gyoto	3.03	4.62	47	0.44	3.8
84-J-38	Gyoto	2.70	4.43	50	0.44	3.6	85-J-69	Shiina	2.90	4.67	54	0.43	3.7
84-J-40	Gyoto	2.99	4.35	50	0.34	3.7	85-J-70	Shiina	2.85	4.41	50	0.38	3.6
84-J-41	Gyoto	2.86	4.02	50	0.31	3.4	85-J-74	Shiina	2.69	4.53	52	0.40	3.6
84-J-42	Gyoto	2.72	4.38	50	0.45	3.5	85-J-76	Gyoto	1.46	2.54	50	0.28	1.9
85-J-2	Shiina	1.89	3.17	50	0.33	2.6	85-J-81	Shiina	2.88	4.85	50	0.44	3.9
85-J-4	Shiina	2.10	3.51	47	0.33	2.9	89-G-1	Gyoto	3.44	5.12	50	0.43	4.3
85-J-G5	Gyoto	3.01	4.64	50	0.41	3.8	89-H-1	Shiina	3.91	5.29	50	0.35	4.6
85-J-6	Shiina	2.27	3.24	52	0.25	2.9	89-H-5	Mélange	3.92	5.21	50	0.30	4.5
85-J-7	Shiina	1.81	2.98	47	0.31	2.3	89-H-9	Mélange	3.61	5.14	51	0.42	4.3
85-J-10	Shiina	1.40	2.73	50	0.30	2.1	89-H-10	Gyoto	3.73	5.19	50	0.38	4.5
85-J-12	Shiina	1.42	2.34	48	0.25	2.0	89-Ts-1	Shiina	4.21	5.81	50	0.44	5.0
85-J-13	Shiina	1.62	2.69	50	0.28	2.1	89-Ts-3	Shiina	3.91	5.21	51	0.33	4.6
85-J-14	Shiina	1.60	2.64	50	0.25	2.1	89-Ts-5	Gyoto	4.01	5.71	50	0.46	4.8
85-J-15	Shiina	0.94	1.94	50	0.20	1.4	89-Ts-7	Gyoto	3.92	5.18	50	0.34	4.6
85-J-17	Shiina	0.99	2.01	38	0.25	1.5	89-Ts-9	Gyoto	3.80	5.56	51	0.46	4.7
85-J-19	Shiina	3.18	4.51	50	0.34	3.8	89-Ts-11a	Gyoto	3.57	4.78	50	0.32	4.2
85-J-20	Shiina	3.42	4.67	50	0.29	4.0	89-Ts-13	Gyoto	3.80	4.85	52	0.27	4.3
85-J-21	Shiina	3.06	4.54	50	0.34	3.7	H-87-13	Gyoto	2.81	4.36	40	0.42	3.6
85-J-22	Shiina	3.11	4.50	50	0.31	3.8	H-87-14	Gyoto	2.86	4.45	50	0.42	3.6
85-J-23	Shiina	2.35	3.95	56	0.39	3.3	H-87-15	Gyoto	2.88	4.48	50	0.41	3.8
85-J-24	Shiina	2.36	3.99	48	0.42	3.3	H-87-16	Gyoto	2.91	5.60	49	0.41	3.9
85-J-25	Shiina	3.09	4.67	50	0.40	3.8	H-87-51	Gyoto	2.51	3.81	50	0.32	3.3
85-J-30	Shiina	3.10	4.56	42	0.36	3.9							
85-J-31	Shiina	2.99	4.45	37	0.37	3.8							
85-J-34	Shiina	2.84	4.45	50	0.33	3.6							
85-J-35	Shiina	2.92	4.49	50	0.33	3.6							
85-J-36	Shiina	3.00	4.34	49	0.39	3.6							
85-J-39	Shiina	2.69	4.10	50	0.37	3.4							
85-J-40	Shiina	2.25	3.95	49	0.42	3.1							
85-J-42	Shiina	2.30	3.86	50	0.32	3.3							

*Minimum reflectance value.
[†]Maximum reflectance value.
[§]Number of individual reflectance measurements used to calculate %Rm.
**Standard deviation.
[‡]Mean vitrinite reflectance.

Taguchi (1988) noted that no attempt was made to eliminate inert reflectance values during either petrographic analysis or statistical treatment. Analyses from both studies demonstrate that thermal maturity levels documented for the Upper Shimanto Group are anomalously high compared to other accretionary complexes (such as Barbados, Larue and others, 1985; the Eocene of Kodiak Island, Moore and Allwardt, 1980; the Olympic Peninsula of Washington, Snavely, 1987; and the Franciscan Complex of California, Underwood and Howell, 1987; Underwood and others, 1988; Laughland, 1991).

The highest calculated paleotemperatures are approximately 300°C for both the Murotohanto and Nabae subbelts. Traditional models of accretionary prism thermal structure illustrate that comparable levels of thermal maturity are attainable at depths of 20 to 40 km, depending upon the boundary conditions of the model (e.g., Wang and Shi, 1984). Direct determination of absolute burial depths for the Murotohanto and Nabae subbelts is impossible because stratigraphic relations have been destroyed by the polyphase deformation history, uplift, and erosion. However, field relations (Byrne and Fisher, 1990; Hibbard and others, this

M. M. Laughland and M. B. Underwood

TABLE 3. VITRINITE REFLECTANCE DATA FOR THE NABAE SUBBELT

Sample	Rock Unit	Min. %Ro*	Max. %Ro[†]	n[§]	S.D.**	%Rm[‡]	Sample	Rock Unit	Min. %Ro*	Max. %Ro[†]	n[§]	S.D.**	%Rm[‡]
85-J-83	Hioki	0.82	1.92	50	0.21	1.4	H-87-28	Hioki	1.15	1.81	50	0.15	1.5
H-84-11	Tsuro	1.51	2.50	45	0.30	2.0	H-87-31	Shijujiyama	0.96	1.66	50	0.16	1.4
H-84-22	Sakamoto	1.54	2.74	50	0.28	2.1	H-87-50	Hioki	1.13	1.98	50	0.19	1.5
H-84-33	Sakamoto	1.63	2.53	50	0.23	2.1	H-87-52	Hioki	0.82	1.90	50	0.23	1.4
H-84-39	Shijujiyama	1.09	2.02	50	0.20	1.6	H-87-53	Misaki	2.44	3.60	50	0.31	3.0
H-84-40	Shijujiyama	1.24	2.34	50	0.22	1.7	H-87-54	Misaki	2.26	3.41	50	0.29	2.8
H-84-44	Misaki	2.74	4.37	49	0.37	3.7	H-87-57	Misaki	1.85	3.22	50	0.34	2.5
H-84-56	Shijujiyama	1.14	2.21	50	0.25	1.7	H-87-59	Tsuro	1.11	2.08	50	0.22	1.6
H-84-59	Shijujiyama	1.36	2.23	50	0.21	1.8	H-87-60	Tsuro	1.37	2.31	50	0.23	1.8
H-84-60	Shijujiyama	1.66	2.22	50	0.23	1.7	H-87-71	Hioki	1.46	2.29	50	0.22	1.9
H-84-63	Sakamoto	1.63	2.92	50	0.26	2.3	H-87-81	Hioki	1.13	2.18	50	0.19	1.6
H-84-67	Tsuro	1.24	2.21	50	0.23	1.7	H-87-82	Hioki	1.10	2.04	50	0.22	1.6
H-84-73	Tsuro	1.55	2.54	50	0.27	2.0	H-87-98	Tsuro	1.44	2.35	50	0.19	1.8
H-84-75	Tsuro	1.46	2.59	50	0.29	2.1	H-87-102	Hioki	1.05	1.85	43	0.23	1.4
H-84-77	Hioki	1.33	2.06	50	0.18	1.7	H-87-104	Hilki	1.62	2.68	50	0.21	2.2
H-84-84	Tsuro	1.45	2.25	50	0.18	1.8	H-87-105	Hioki	1.93	3.22	50	0.33	2.6
H-84-89	Tsuro	1.33	2.12	42	0.20	1.7	H-87-106	Tsuro	1.13	2.09	50	0.19	1.6
H-84-91	Tsuro	1.63	2.30	50	0.14	2.0	H-87-107	Tsuro	1.55	2.38	50	0.14	1.9
H-84-92	Tsuro	1.37	2.34	50	0.24	1.8	H-87-108	Tsuro	1.08	2.25	50	0.24	1.7
H-84-93	Tsuro	1.30	2.16	50	0.24	1.8	H-87-109	Tsuro	1.15	2.04	50	0.20	1.6
H-84-94	Sakamoto	1.19	2.11	46	0.24	1.7	H-87-110	Shijujiyama	1.22	2.07	50	0.22	1.7
H-84-95	Sakamoto	1.42	2.36	50	0.21	1.9	H-87-111	Hioki	0.89	1.32	50	0.11	1.1
H-84-103	Hioki	1.07	1.85	50	0.29	1.5	H-87-115	Sakamoto	1.01	1.55	50	0.13	1.3
H-84-109	Hioki	1.56	2.22	50	0.16	1.9	H-87-117A	Sakamoto	0.98	1.82	50	0.18	1.4
H-85-6	Tsuro	1.70	2.92	51	0.31	2.4	H-87-117B	Sakamoto	0.83	1.78	50	0.22	1.3
H-85-7	Misaki	2.13	3.20	50	0.27	2.6	H-87-117C	Sakamoto	1.21	2.35	42	0.29	1.8
H-85-9	Misaki	2.70	4.25	50	0.38	3.3	H-87-118	Sakamoto	1.10	2.38	50	0.26	1.7
H-85-11	Hioki	0.61	1.27	50	0.14	0.9	H-87-119	Misaki	2.52	4.04	48	0.39	3.2
H-85-16	Shijujiyama	1.28	2.17	50	0.20	1.7	H-87-123	Misaki	2.55	4.25	50	0.39	3.3
H-85-19	Hioki	1.32	2.15	50	0.23	1.7	H-87-127	Tsuro	2.33	2.37	50	0.21	1.8
H-85-29	Sakamoto	1.76	2.41	50	0.17	2.1	H-87-980	Sakamoto	0.84	1.38	50	0.11	1.1
H-85-32	Sakamoto	1.26	2.10	50	0.18	1.7	H-88-1	Tsuro	1.25	1.75	45	0.14	1.5
H-85-33	Sakamoto	1.37	2.29	50	0.24	1.9	H-88-2	Tsuro	1.25	2.37	50	0.26	1.8
H-85-39	Sakamoto	1.36	2.13	46	0.21	1.7	H-88-3	Tsuro	1.27	2.27	50	0.24	1.8
H-85-45	Hioki	0.87	1.69	46	0.19	1.3	H-88-4	Tsuro	1.31	2.42	50	0.25	1.8
H-85-46	Hioki	1.19	2.19	50	0.29	1.7	H-88-5	Hioki	1.45	2.53	50	0.27	1.9
H-85-54	Shijujiyama	1.13	1.99	50	0.19	1.6	H-88-6	Tsuro	1.56	2.32	50	0.21	2.0
H-85-56	Hioki	0.78	1.39	50	0.15	1.1	H-88-7	Hioki	2.13	2.67	50	0.25	2.1
H-85-57A	Hioki	0.89	1.74	50	0.18	1.3	H-88-8	Hioki	1.42	2.12	50	0.18	1.8
H-85-57B	Hioki	1.02	1.66	48	0.15	1.3	H-88-9	Hioki	1.35	2.22	50	0.21	1.8
H-85-58	Hioki	1.11	1.85	50	0.18	1.5	H-88-10	Tsuro	1.65	2.40	50	0.18	2.0
H-85-59	Shijujiyama	1.13	1.99	47	0.21	1.5	H-88-12A	Tsuro	1.32	2.33	50	0.20	1.9
H-85-60	Shijujiyama	1.15	1.91	50	0.18	1.6	H-88-12B	Tsuro	1.62	2.27	50	0.16	2.0
H-85-61	Shijujiyama	1.17	1.93	50	0.18	1.5	H-88-13	Tsuro	1.35	1.90	49	0.15	1.6
H-85-74	Hioki	1.40	2.44	50	0.23	1.9	H-88-14	Tsuro	1.30	2.17	50	0.29	1.8
H-85-75	Shijujiyama	1.06	2.34	50	0.24	1.7	H-88-15	Tsuro	1.35	2.31	49	0.23	1.8
H-85-86	Shijujiyama	1.32	2.30	50	0.24	1.8	H-88-16	Tsuro	1.18	2.28	50	0.25	1.7
H-85-89	Shijujiyama	1.01	2.98	50	0.25	1.6							
H-85-99	Misaki	2.10	3.64	48	0.40	3.0							
H-85-101	Hioki	0.67	1.55	50	0.20	1.1							
H-85-103	Hioki	0.72	1.39	50	0.13	1.1							
H-87-6	Sakamoto	1.29	2.47	50	0.25	1.8							
H-87-9	Shijujiyama	1.16	2.10	50	0.22	1.6							
H-87-25	Shijujiyama	1.05	1.95	50	0.20	1.4							
H-87-26	Shijujiyama	0.92	1.92	48	0.20	1.5							

*Minimum reflectance value.
[†]Maximum reflectance value.
[§]Number of individual reflectance measurements used to calculate %Rm.
**Standard deviation.
[‡]Mean vitrinite reflectance.

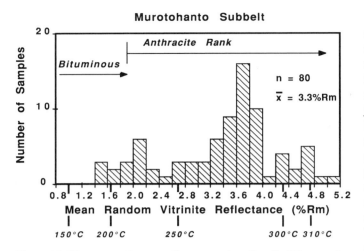

Figure 12. Histogram of mean reflectance values for all of the samples from the Murotohanto subbelt. Temperature scale based on Barker (1988).

volume), estimates of burial pressure obtained from mica b_0 lattice spacings (Underwood, Laughland, and Kang, this volume), and comparisons with analogous slope-basin sedimentary accumulations and porosity values (Hibbard and others, this volume), all indicate that burial depths were less than 10 km. Temperature levels for the Shimanto rocks, therefore, are incompatible with contentions of low heat flow associated with processes of shallow-level subduction/accretion.

A crude estimate of the geothermal gradient is obtained by dividing the maximum paleotemperature for the Murotohanto subbelt (320°C) by the maximum burial depth of 8 km determined from mica b_0 data (Underwood, Laughland, and Kang, this volume), which yields a minimum paleogeothermal gradient of 40°C/km. The maximum paleotemperature for the Shijujiyama Formation (210°C) divided by the estimate of maximum burial based on comparisons of porosity values for slope basin sediments (3 to 4 km; Hibbard and others, 1992) yields a paleogeothermal gradient of roughly 50 to 70°C/km. Temperature gradients must have varied from one portion of the prism to another through time, depending upon contrasts in thermal conductivity, proximity to igneous intrusions, and rates of uplift and erosion. Nonetheless, these estimates clearly demonstrate that the Upper Shimanto Belt experienced anomalously high geothermal gradients subsequent to early stage and intermediate-stage deformation. A model that allows for 25 mW/m^2 heat flux across the base of an accretionary wedge (e.g., Wang and Shi, 1984) cannot account for gradients this high. A critical question, then, is: "What geologic mechanism will allow for greater heat flux across the lower boundary of an accretionary prism?"

DeLong and others (1979) developed a finite-difference computer model to evaluate the temperature structure of a subduction zone during subduction of an active oceanic spreading ridge. Exact temperature gradients are a function of position within the accretionary prism, but the thermal state will alter significantly at least 2 m.y. in advance of ridge subduction, and isotherms remain elevated for longer than 10 m.y. after ridge

TABLE 4. RANGES OF REFLECTANCE VALUES AND PALEOTEMPERATURES (°C) FOR STRUCTURAL AND LITHOSTRATIGRAPHIC UNITS

Rock Unit	%R$_m$ Range	Paleotemperature Conversion Method*											
		1	2	3	4	5	6	7	8	9	10	11	12
Murotohanto Subbelt													
Gyoto domain	1.9–4.8	270–380	210–350	240–350	210–310	210–290	230–350	240–290	200–250	210–270	180–230	230–320	220–300
Shiina domain	1.4–5.0	230–400	170–360	200–360	180–315	180–300	200–350	220–300	180–260	190–270	160–240	210–320	190–300
Sakihama mélange	4.3–4.5	380	340	340–350	300		280–290	330–340	280–290	250	260	230	310	290
Nabae Subbelt													
Hioki mélange	0.9–2.6	170–310	110–260	140–280	140–250	140–240	140–270	180–260	140–220	160–230	130–200	160–260	150–240
Shijujiyama Formation	1.4–1.8	230–260	170–210	200–230	180–210	180–210	200–230	220–240	180–200	180–220	160–170	210–230	190–210
Tsuro sequence	1.4–2.4	230–310	170–260	200–280	180–250	180–230	200–260	220–260	180–220	190–220	160–190	210–260	190–240
Sakamoto mélange	1.1–2.1	200–280	140–280	170–250	160–230	160–220	170–250	200–250	160–210	180–220	140–180	180–240	170–230
Misaki sequence	2.4–3.7	300–360	250–310	270–320	240–290	230–270	260–320	260–280	220–240	220–250	190–220	260–300	240–280

*Sources: 1 = Price (1983); 2 = Barker (1983); 3 = Barker and Pawlewicz (1986); 4 = Barker (1988); 5 = Aizawa (1990); 6 = Barker and Goldstein (1990); 7 = Bostick and others (1978), T_{eff} = 1 m.y.; 8 = Bostick and others (1978), T_{eff} = 10 m.y.; 9 = Waples (1980), T_{eff} = 1 m.y.; 10 = Waples (1980), T_{eff} = 10 m.y.; 11 = Sweeney and Burnham (1990), T_{eff} = 1 m.y.; 12 = Sweeney and Burnham (1990), T_{eff} = 10 m.y.

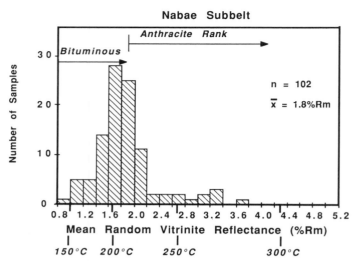

Figure 13. Histogram of mean reflectance values for all of the samples from the Nabae subbelt. Temperature scale based on Barker (1988).

subduction. In theory, temperatures can reach nearly 600°C at 20 km depth and 25 km inboard of the trench, whereas temperatures as high as ~350°C were calculated for a position within 10 km landward of the trench at 5 km depth. The geologic effects of ridge-trench collision will vary, but can include anomalous heat flow within the prism and overlying forearc, anomalous magmatism, underplating of the subducted slab; ophiolite obduction, tectonic erosion of the base of the forearc, and deformation of the prism and forearc (see also Marshak and Karig, 1977; Forsythe and Nelson, 1985; Cande and Leslie, 1986; Forsythe and others, 1986; Nelson and Forsythe, 1989).

Elevated reflectance values adjacent to the Maruyama intrusions prove that the Misaki sequence was affected by cooling intrusions. Estimates of paleotemperature derived for rocks adjacent to the Maruyama intrusive suite indicate that "equilibrium" temperatures were only as high as 285°C (Rm = 3.7%). However, peak temperatures in the vicinity of the igneous intrusions were probably higher than estimates based on organic maturation since a metamorphic mineral assemblage of garnet-biotite-quartz is found within centimeters of the intrusive bodies (Hibbard and others, this volume). Under conditions of rapid heating, such as with magmatic intrusion, vitrinite reflectance does not have enough time to equilibrate with peak temperature (Bostick, 1979; Barker and Pawlewicz, 1986). Laboratory heating experiments (Bostick, 1979) suggest that maximum emplacement temperatures for the intruded rock may have been on the order of 500 to 550°C.

Levels of organic metamorphism for coeval strata elsewhere in Japan are also anomalously high. Middle Miocene age coal fields of the Kumano Group, on the Kii Peninsula, Honshu Is-

land, display levels of thermal maturity ranging from 0.9%R_{max} to 6.1%R_{max} (Chijiwa, 1988). Intact stratigraphic gradients range from 0.15%/100 m to 0.66%/100 m, and borehole gradients are 0.6 to 0.32%/100 m. Paleogeothermal gradients for the Kumano Group were estimated to be 80 to 110°C/km based on the reflectance-temperature correlation of Bostick and others (1978; Suzuki and others, 1982); gradients could be as high as 160°C/km using the correlation of Barker (1988). Maturation of the Kumano Group has been attributed to intense hydrothermal activity associated with felsic igneous activity around 14 to 15 Ma, which was synchronous with Miocene igneous activity on Muroto Peninsula. Similarly, levels of organic alteration determined for late-middle Oligocene to late-middle Miocene age coal fields on southwest Kyushu correspond to paleogeothermal gradients of 65 to 70°C/km, according to Aihara (1989).

The occurrence of "near-field" effects (Nelson and Forsythe, 1989), such as anomalous magmatism and mechanical deformation of the accretionary prism, lend additional credence to the ridge-collision hypothesis for the Nabae subbelt. Hibbard and Karig (1990a, b) cite compelling evidence for subduction of the Shikoku Central Basin Ridge beginning at about 15 Ma. Whether this Miocene thermal event was also responsible for peak temperatures within the Murotohanta subbelt, however, remains a point of controversy. DiTullio (1989) modified a plate model that shows a change from convergence between the Kula and Eurasian Plates to Eurasian-Pacific subduction at about 43 Ma (see also Lonsdale, 1988). Based upon the DiTullio (1989) plate reconstruction, anomalous levels of thermal alteration within the Murotohanto may be related to an earlier collision between the extinct Kula-Pacific ridge and the Shimanto subduction front (DiTullio and others, this volume).

CONCLUSIONS

The Upper Shimanto Group records the effects of relatively high temperatures at shallow depths of burial for an accretionary prism. Mean vitrinite reflectance ranges from 0.9 to 5.0%, and this corresponds to temperatures of 140 to 315°C. Reconstructed geothermal gradients for the Eocene Murotohanto subbelt were approximately 40°C/km. Average gradients during middle Miocene time were perhaps 50°C/km, with higher values near igneous intrusions. Geothermal conditions within the Muroto study area, however, were subdued compared to the Kii Peninsula to the northeast. These thermal data are inconsistent with the low-temperature/high-pressure metamorphic conditions that promoted blueschist-facies metamorphism within other accretionary prisms. Thus, the Upper Shimanto Group represents one end member in a spectrum of possible thermal regimes. Anomalously high values of thermal maturity probably were caused by subduction of very young oceanic crust and perhaps collision with an active or recently extinct spreading ridge.

ACKNOWLEDGMENTS

Partial funding for this study was provided by a grant to Underwood from the National Science Foundation (Grant No. EAR-8706784). Acknowledgment is also made to the Donors of the Petroleum Research Fund, administered by the American Chemical Society, for partial support of this research (Grant No. 19187-AC2). Michael Batek, Robert Waddell, and Sin Mok Kang assisted in the laboratory. We thank R. M. Bustin and K. L. Shelton for their reviews of the manuscript.

REFERENCES CITED

Aihara, A., 1989, Paleogeothermal influence on organic metamorphism in the neotectonics of the Japanese Islands: Tectonophysics, v. 159, p. 291–305.

Aizawa, J., 1990, Paleotemperatures from fluid inclusions and coal rank of carbonaceous material of the Tertiary formations in northwest Kyushu, Japan: Journal of Mineralogy, Petrology, and Economic Geology, v. 85, p. 145–154.

Anderson, R. N., DeLong, S. E., and Schwarz, W. M., 1978, thermal model for subduction with dehydration in the downgoing slab: Journal of Geology, v. 86, p. 731–739.

Barker, C. E., 1983, Influence of time on metamorphism of sedimentary organic matter in liquid dominated geothermal systems, western North America: Geology, v. 11, p. 384–388.

——, 1988, Geothermics of petroleum systems: Implications of the stabilization of kerogen thermal maturation after a geologically brief heating duration at peak temperature, in Magoon, L., ed., Petroleum systems of the United States: U.S. Geological Survey Bulletin 1870, p. 26–29.

——, 1989, Temperature and time in the thermal maturation of sedimentary organic matter, in Naeser, N. D., and McCulloh, T. H., eds., Thermal history of sedimentary basins—Methods and case histories: New York, Springer-Verlag, p. 73–98.

Barker, C. E., and Goldstein, R. H., 1990, A fluid inclusion technique for determining maximum temperature in calcite and its comparison to the vitrinite reflectance geothermometer: Geology, v. 18, p. 1003–1006.

Barker, C. E., and Halley, R. B., 1986, Fluid inclusion, stable isotope, and vitrinite reflectance evidence for the thermal history of the Bone Spring limestone, southern Guadalupe Mountains, Texas, in Gautier, D. L., ed., Roles of organic matter in sediment diagenesis: Society of Economic Paleontologists and Mineralogists Special Publication, n. 38, p. 190–203.

Barker, C. E., and Pawlewicz, M. J., 1986, The correlation of vitrinite reflectance with maximum paleotemperature in humic organic matter, in Buntebarth, G., and Stegena, L., eds., Paleogeothermics: New York, Springer-Verlag, p. 79–83.

Bostick, N. H., 1979, Microscopic measurement of the level of catagenesis of solid organic matter in sedimentary rocks to aid exploration for petroleum and to determine former burial temperatures—A review, in Scholle, P. A., and Schluger, P. R., eds., Aspects of diagenesis: Society of Economic Paleontologists and Mineralogists Special Publication 26, p. 17–43.

Bostick, N. H., Cashman, S., McColloh, T. H., and Wadell, C. T., 1978, Gradients of vitrinite reflectance and present temperature in the Los Angeles and Ventura basins, California, in Oltz, D., ed., Symposium in geochemistry: Low temperature metamorphism of kerogen and clay minerals: Society of Economic Paleontologists and Mineralogists Pacific Section, p. 65–96.

Burnham, A. K., and Sweeney, J. J., 1989, A chemical kinetic model of vitrinite maturation and reflectance: Geochimica et Cosmochimica Acta, v. 53, p. 2649–2657.

Burruss, R. C., 1989, Paleotemperatures from fluid inclusions: Advances in theory and technique, in Naeser, N. D., and McCulloh, T. H., eds., Thermal history of sedimentary basins: Methods and case histories: New York, Springer-Verlag, p. 119–131.

Bustin, R. M., 1982, The effect of shearing on the quality of some coals in the southeastern Canadian Cordillera: Canadian Institute of Mining and Metallurgy Bulletin, v. 75, p. 76–83.

Bustin, R. M., Cameron, A. R., Grieve, D. A., and Kalkreuth, W. D., 1989, Coal petrology: Its principles, methods, and applications: Geological Association of Canada Short Course Notes, v. 3, 230 p.

Byrne, T., and Fisher, D., 1990, Evidence for a weak and overpressured décollement beneath sediment-dominated accretionary prisms: Journal of Geophysical Research, v. 95, p. 9081–9097.

Cande, S., and Leslie, R., 1986, Late Cenozoic tectonics of the southern Chile trench: Journal of Geophysical Research, v. 91, p. 471–496.

Chandra, D., 1962, Reflectance and microstructure of weathered coals: Fuel, v. 41, p. 185–193.

Chijiwa, K., 1988, Post-Shimanto sedimentation and organic metamorphism: An example of the Miocene Kumana Group, Kii Peninsula: Modern Geology, v. 12, p. 363–387.

Cloos, M., 1982, Flow mélanges: Numerical modeling and geologic constraints on their origin in the Franciscan subduction complex, California: Geological Society of America Bulletin, v. 93, p. 330–345.

——, 1985, Thermal evolution of convergent plate margins: Thermal modeling and reevaluation of isotopic Ar-ages for blueschists in the Franciscan Complex of California: Tectonophysics, v. 4, n. 5, p. 421–434.

Coombs, D. S., 1961, Some recent work on the lower grades of metamorphism: Australian Journal of Science, v. 24, p. 203–315.

Crelling, J. C., Schrader, R. H., and Benedict, L. G., 1979, Effects of weathered coal on coking properties and coke quality: Fuel, v. 58, p. 542–546.

Davis, A., 1977, The reflectance of coal, in Karr, C., Jr., ed., Analytical methods for coal and coal products, v. 1, chapter 2: New York, Academic Press, p. 27–81.

DeLong, S., Schwarz, W. M., and Anderson, R. N., 1979, Thermal effects of ridge subduction: Earth and Planetary Science Letters, v. 44, p. 239–246.

Dembicki, H., Jr., 1984, An interlaboratory comparison of source rock data: Geochimica et Cosmochimica Acta, v. 48, p. 2641–2649.

deVries, Klein, G., and 14 others, Initial Reports of the Deep Sea Drilling Project, v. 58: Washington D.C., U.S. Government Printing Office, 1022 p.

DiTullio, L. D., 1989, Evolution of the Eocene accretionary prism in SW Japan: Evidence from structural geology, thermal alteration, and plate reconstructions [Ph.D. thesis]: Providence, Rhode Island, Department of Geological Sciences, Brown University, 161 p.

DiTullio, L. D., and Byrne, T., 1990, Deformation paths in the shallow levels of an accretionary prism: The Eocene Shimanto Belt of southwest Japan: Geological Society of America Bulletin, v. 102, p. 1420–1438.

Dow, W. G., 1977, Kerogen studies and geological interpretations: Journal of Geochemical Exploration, v. 7, p. 79–99.

Dow, W. G., and O'Connor, D. E., 1982, Kerogen maturity and type by reflected light microscopy applied to petroleum generation, in How to assess maturation paleotemperatures: Society of Economic Paleontology and Mineralogy Short Course 7, p. 133–157.

Ernst, W. G., 1974, Metamorphism and ancient continental margins, in Burke, C. A., and Drake, C. L., eds., Geology of continental margins: New York, Springer-Verlag, p. 907–919.

Ernst, W. G., Seki, Y., Onuki, H., and Gilbert, M. C., 1970, Comparative study of low-grade metamorphism in the California Coast Ranges and the Outer Metamorphic Belt of Japan: Geological Society of America Memoir 124, 276 p.

Forsythe, R., and Nelson, E., 1985, Geological manifestations of ridge collision: Evidence from the Golfo de Penas–Taitao Basin, southern Chile: Tectonics, v. 4, p. 477–495.

Forsythe, R., and 7 others, 1986, Pliocene near-trench magmatism in southern Chile: A possible manifestation of ridge collision: Geology, v. 14, p. 23–27.

Frey, M., 1987, Very low-grade metamorphism of clastic sedimentary rocks, in Frey, M., ed., Low temperature metamorphism: New York, Chapman and Hall, p. 9–58.

Gretener, P. E., and Curtis, C. D., 1982, Role of temperature and time on organic metamorphism: American Association of Petroleum Geologists Bulletin, v. 66, p. 1124–1149.

Hamototo, R., and Sakai, H., 1987, Rb-Sr age of granophyre associated with the Cape Muroto gabbroic complex: Kyushu University, Science Reports of the Department of Geology, v. 15, p. 1–15.

Hibbard, J. P., and Karig, D., 1987, Sheath-like folds and progressive fold deformation in Tertiary sedimentary rocks of the Shimanto accretionary complex, Japan, Journal of Structural Geology, v. 9, p. 845–857.

—— , 1990a, Structural and magmatic responses to spreading ridge subduction; An example from southwest Japan: Tectonics, v. 9, p. 207–230.

—— , 1990b, An alternative plate model for the early Miocene evolution of the southwest Japan margin: Geology, v. 18, n. 2, p. 170–174.

Hibbard, J. P., Karig, D., and Taira, A., 1992, Oligocene-Miocene geology of the Shimanto accretionary prism, Murotomisaki, Shikoku, Japan: Journal of the Geological Society of Japan, v. 96 (in press).

Hood, A., Gutjahr, C.C.M., and Heacock, R. L., 1975, Organic metamorphism and the generation of petroleum: American Association of Petroleum Geologists Bulletin, v. 59, p. 986–996.

Hutton, A. C., and Cook, A. C., 1980, Influence of alginite on the reflectance of vitrinite from Joadja, NSW, and some other coals and oil shales containing alginite: Fuel, v. 59, p. 711–714.

Hutton, A. C., Kanstler, A. J., Cook, A. C., and McKirdy, D. M., 1980, Organic matter in oil shales: Journal of Australian Petroleum Exploration Association, v. 20, p. 44–63.

Ikehara, Y., Kano, K., and Taguchi, K., 1982, Heating effect on vitrinite reflectance: An experimental result: Science Reports of the Tohoku University, Mineralogy, Petrology, and Economic Geology, v. 15, p. 141–148.

Issler, D. R., 1984, Calculation of organic maturation levels for offshore eastern Canada—Implications for general application of Lopatin's method: Canadian Journal of Earth Sciences, v. 21, p. 477–488.

James, T. S., Hollister, L. S., and Morgan, W. J., 1989, Thermal modeling of the Chugach metamorphic complex: Journal of Geophysical Research, v. 94, p. 4411–4423.

Jones, R., and Edison, T., 1978, Microscopic observations of kerogen related to geochemical parameters with emphasis on thermal maturation, in Symposium in geochemistry, low temperature metamorphism of kerogen and clay minerals: Society of Economic Paleontologists and Mineralogists, Pacific Section, p. 1–12.

Kagami, H., and 14 others, 1986, Site reports 582 and 583, in Kagami, H., and others, eds., Initial reports of the Deep Sea Drilling Project, v. 87: Washington D.C., U.S. Government Printing Office, p. 35–246.

Kalkreuth, W.D., 1982, Rank and petrographic composition of selected Jurassic-Lower Cretaceous coals of British Columbia, Canada: Canadian Petroleum Geology Bulletin, v. 30, p. 112–139.

Karig, D. E., and 12 others, 1975, Initial Reports of the Deep Sea Drilling Project, v. 31: Washington D.C., U.S. Government Printing Office, 927 p.

Karweil, J., 1956, Die metamorphose der kohlen vom standpunkt der physikalischem chemie [The metamorphism of coal from the standpoint of physical chemistry]: Deutsche Geologische Gesellschaft Zeitschrift, v. 107, p. 132–139.

Katz, B. J., Liro, L. M., Lacey, J. E., and White, H. W., 1982, Time and temperature in petroleum formation: Application of Lopatin's method to petroleum exploration: Discussion: American Association of Petroleum Geologists, v. 66, p. 1150–1152.

Kinoshita, H., and Yamano, M., 1986, The heat flow anomaly in the Nankai Trough area, in Kagami, H., and 14 others, eds., Initial reports of the Deep Sea Drilling Project, v. 87: Washington, D.C., U.S. Government Printing Office, p. 737–743.

Kisch, H. J., 1987, Correlation between indicators of very low-grade metamorphism, in Frey, M., ed., Low temperature metamorphism: New York, Chapman and Hall, p. 227–246.

Kulm, L. D., and 13 others, 1986, Oregon subduction zone: Venting, fauna, and carbonates: Science, v. 231, p. 561–566.

Larter, S. R., 1989, Chemical models of vitrinite reflectance evolution: Geologische Rundschau, v. 78, p. 349–359.

Larue, D. K., Schoonmaker, J., Torrini, R., Lucas-Clark, J., Clark, M., and Schneider, R., 1985, Barbados: Maturation, source rock potential and burial history within a Cenozoic accretionary complex: Marine and Petroleum Geology, v. 2, p. 95–110.

Laughland, M. M., 1991, Organic metamorphism and thermal history of selected portions of the Franciscan accretionary complex of coastal California [Ph.D. thesis]: Columbia, Missouri, University of Missouri, Department of Geological Sciences, 318 p.

Liou, J. G., 1971, P-T stabilities of laumontite, wairakite, lawsonite, and related minerals in the system $CaAl_2Si_2O_8$-SiO_2-H_2O: Journal of Petrology, v. 12, p. 379–411.

Liou, J. G., Maruyama, S., and Cho, M., 1985, Phase equilibria and mineral parageneses of metabasites in low-grade metamorphism: Mineralogical Magazine, v. 49, p. 321–333.

Liou, J. G., Maruyama, S., and Cho, M., 1987, Very low-grade metamorphism of volcanic rocks—Mineral assemblages and facies, in Frey, M., ed., Low temperature metamorphism: New York, Chapman and Hall, p. 59–113.

Lister, C.R.B., 1977, Estimators for heat flow and deep rock properties based on boundary layer theory: Tectonophysics, v. 41, p. 157–171.

Lonsdale, P., 1988, Paleogene history of the Kula Plate: Offshore evidence and onshore implications: Geological Society of America Bulletin, v. 100, p. 733–754.

Marshak, S., and Karig, D., 1977, Triple junctions as a cause for anomalously near-trench igneous activity between the trench and volcanic arc: Geology, v. 5, p. 233–236.

Miyashiro, A., 1961, Evolution of metamorphic belts: Journal of Petrology, v. 2, p. 277–311.

—— , 1967, Aspects of metamorphism in the circum-Pacific region: Tectonophysics, v. 4, p. 519–521.

Moore, J. C., and Allwardt, A., 1980, Progressive deformationn of a Tertiary trench slope, Kodiak Islands, Alaska: Journal of Geophysical Research, v. 85, p. 4741–4756.

Moore, J. C., and Karig, D., 1976, Sedimentology, structural geology, and tectonics of the Shikoku subduction zone, southwestern Japan: Geological Society of America Bulletin, v. 87, p. 1259–1268.

Moore, J. C., and 22 others, 1987, Expulsion of fluids from depths along a subduction-zone décollement horizon: Nature, v. 326, p. 785–788.

Moore, J. C., and 22 others, 1988, Tectonics and hydrogeology of the northern Barbados Ridge: Results from Ocean Drilling Program Leg 110: Geological Society of America Bulletin, v. 100, p. 1578–1593.

Mori, K., and Taguchi, K., 1988, Examination of the low-grade metamorphism in the Shimanto Belt by vitrinite reflectance: Modern Geology, v. 12, p. 325–339.

Nagihara, S., Kinoshita, H., and Yamano, M., 1989, On the high heat flow in the Nankai Trough area–A simulation study on a heat rebound process: Tectonophysics, v. 161, p. 33–41.

Nelson, E., and Forsythe, R., 1989, Ridge collision at convergent margins: Implications for Archean and post-Archean crustal growth: Tectonophysics, v. 161, p. 307–315.

Oxburgh, E. R., and Turcotte, D. L., 1970, Thermal structure of island arcs: Geological Society of America Bulletin, v. 81, p. 1665–1688.

—— , 1971, Origin of paired metamorphic belts and crustal dialtion in island arc regions: Journal of Geophysical Research, v. 76, n. 5, p. 1315–1327.

Parsons, B., and Sclater, J. G., 1977, An analysis of the variation of ocean floor bathymetry and heat flow with age: Journal of Geophysical Research, v. 82, p. 803–827.

Pearson, D. E., and Kwong, J., 1979, Mineral matter as a measure of oxidation in coking coal: Fuel, v. 58, p. 63–66.

Pollastro, R. M., and Barker, C. E., 1986, Application of clay-mineral, vitrinite reflectance, and fluid inclusion studies to thermal and burial history of the Pinedale anticline, Green River basin, Wyoming, in Gautier, D. L., ed., Roles of organic matter in sediment diagenesis: Society of Economic Paleontologists and Mineralogists Special Publication, n. 38, p. 73–83.

Price, L. C., 1983, Geologic time as a parameter in organic metamorphism and vitrinite reflectance as an absolute paleogeothermometer: Journal of Petroleum Geology, v. 6, p. 5–38.

Price, L. C., and Barker, C. E., 1985, Suppression of vitrinite reflectance in amorphous rich kerogen–A major unrecognized problem: Journal of Petroleum Geology, v. 8, p. 59–84.

Reck, B. H., 1987, Implications of measured thermal gradients for water movement through the northeast Japan accretionary prism: Journal of Geophysical Research, v. 92, p. 3683–3690.

Robert, P., 1988, Organic metamorphism and geothermal history: Dordrecht, Netherlands, Reidel, 311 p.

Roeddner, E., 1984, Fluid inclusions, *in* Ribbe, P. H., ed., Reviews in mineralogy: Mineralogical Society of America, v. 12, 644 p.

Shi, Y., Wang, C. Y., Langseth, M. G., Hobart, M., and von Huene, R., 1988, Heat flow and thermal structure of the Washington-Oregon accretionary prism–A study of the lower slope: Geophysical Research Letters, v. 15, p. 1113–1116.

Snavely, P. D., Jr., 1987, Tertiary geologic framework, neotectonics, and petroleum potential of the Oregon-Washington continental margin, *in* Scholl, D. W., Grantz, A., Vedder, J. G., eds., Geology and resource potential of the continental margin of western North America and adjacent ocean basins—Beaufort Sea to Baja California: Circum-Pacific Council for Energy and Mineral Resources Earth Sciences Series, v. 6, p. 305–335.

Stach, E., Mackowsky, M-Th., Teichmuller, M., Taylor, G. H., Chandra, D., and Teichmuller, R., 1982, Stach's textbook of coal petrology: Berlin, Gerbruder Borntraeger, 511 p.

Suggate, R. P., 1982, Low-rank sequences and scales of organic metamorphism: Journal of Petroleum Geology, v. 4, p. 377–392.

Suzuki, S., Oda, Y., and Nambu, M., 1982, Thermal alteration of vitrinite in the Miocene sediments of the Kishu Mine area, Kii Peninsula: Kozau Chishitsu, v. 32, p. 55–65.

Sweeney, J. L., and Burnham, A. K., 1990, Evaluation of a simple model of vitrinite reflectance on chemical kinetics: American Association of Petroleum Geologists Bulletin, v. 74, p. 1559–1570.

Taira, A., 1985, Sedimentary evolution of Shikoku subduction zone: The Shimanto Belt and Nankai Trough, *in* Nasu, N., Kobayashi, K., Uyeda, S., Kushiro, I., Kagemi, H., eds., Formation of active ocean margins: Tokyo, Terra Scientific Publishing Company, p. 835–851.

Taira, A., Tashiro, M., Okmura, M., and Katto, J., 1980, The geology of the Shimanto Belt in Kochi Prefecture, Shikoku, Japan, *in* Taira, A., and Tashiro, M., eds., Geology and paleontology of the Shimanto Belt, Kochi: Rinya-Kosakai Press, p. 329–389.

Taira, A., Okada, H., Okada, H., Whitaker, J.H.McD., and Smith, A. J., 1982, The Shimanto Belt of Japan: Cretaceous to lower Miocene active margin sedimentation, *in* Leggett, J. K., ed., Trench-forearc geology: Geological Society of London, Special Publication 10, p. 5–26.

Taira, A., Katto, J., Tashiro, M., Okamura, M., and Kodama, K., 1988, The Shimanto Belt in Shikoku, Japan—Evolution of Cretaceous to Miocene accretionary prism: Modern Geology, v. 12, p. 5–46.

Tilley, B. J., Nesbitt, B. E., and Longstaffe, F. J., 1989, Thermal history of Alberta Deep basin: Comparative study of fluid inclusion and vitrinite reflectance data: American Association of Petroleum Geologists Bulletin, v. 73, p. 1206–1222.

Tissot, B. P., Pelet, R., and Ungerer, P. H., 1987, Thermal history of sedimentary basins, maturation indices, and kinetics of oil and gas generation: American Association of Petroleum Geologists Bulletin, v. 71, p. 1445–1466.

Toksoz, M. N., Minear, J. W., and Julian, B. R., 1971, Temperature field and geophysical effects of downgoing slab: Journal of Geophysical Research, v. 76, p. 1113–1138.

Underwood, M. B., and Howell, D. G., 1987, Thermal maturity of the Cambria slab, an inferred trench-slope basin in central California: Geology, v. 15, p. 216–219.

Underwood, M. B., O'Leary, J. D., and Strong, R. H., 1988, Contrasts in thermal maturity within terranes and across terrane boundaries of the Franciscan Complex, northern California: Journal of Geology, v. 96, p. 399–416.

Van den Beukel, J., and Wortel, R., 1988, Thermo-mechanical modeling of arc-trench regions: Tectonophysics, v. 154, p. 177–193.

Vrojlik, P., 1987, Tectonically driven fluid flow in the Kodiak accretionary complex, Alaska: Geology, v. 15, p. 466–469.

Vrojlik, P., Myers, G., and Moore, J. C., 1988, Warm fluid migration along tectonic mélanges in the Kodiak accretionary complex, Alaska: Journal of Geophysical Research, v. 93, p. 10313–10325.

Walker, A., 1982, Comparison of anomalously low vitrinite reflectance values with other thermal maturation indices at Playa del Rey oilfield, California [M.A. thesis]: University of Washington, 190 p.

Wang, C., and Shi, Y., 1984, On the thermal structure of subduction complexes: A preliminary study: Journal of Geophysical Research, v. 89, p. 7709–7718.

Wang, X., Lerche, I., and Walters, C., 1989, The effect of igneous intrusive bodies on sedimentary thermal maturity: Organic Geochemistry, v. 14, p. 571–584.

Waples, D. W., 1980, Time and temperature in petroleum formation: Application of Lopatin's method to petroleum exploration: American Association of Petroleum Geologists Bulletin, v. 64, p. 916–926.

Wood, D. A., 1988, Relationships between thermal maturity indices calculated using Arrhenius equation and Lopatin method: Implications for petroleum exploration: American Association of Petroleum Geologists Bulletin, v. 72, p. 115–134.

Wright, N.J.R., 1980, Time, temperature and organic maturation—The evolution of rank within a sedimentary pile: Journal of Petroleum Geology, v. 2, p. 411–425.

Yamano, M., Honda, S., and Uyeda, S., 1984, Nankai Trough: A hot wrench?: Marine Geophysical Research, v. 6, p. 187–203.

Zen, E., and Thompson, A. B., 1974, Low grade regional metamorphism: Mineral equilibrium relations: Annual Review of Earth and Planetary Science, v. 2, p. 179–212.

MANUSCRIPT ACCEPTED BY THE SOCIETY APRIL 24, 1992

Geological Society of America
Special Paper 273
1993

A comparison among organic and inorganic indicators of diagenesis and low-temperature metamorphism, Tertiary Shimanto Belt, Shikoku, Japan

Michael B. Underwood, Matthew M. Laughland,* and Sin Mok Kang
Department of Geological Sciences, University of Missouri, Columbia, Missouri 65211

ABSTRACT

Tertiary rocks of the Shimanto Belt represent the youngest subaerial part of the accretionary margin of southwest Japan. Measurements of mean vitrinite reflectance ($\%R_m$) from shales and cleaved metapelites show that the Eocene through early Miocene strata on the Muroto Peninsula of Shikoku Island were exposed to temperatures of approximately 140°C to 315°C. Analyses of inorganic phases corroborate these findings. Values of illite crystallinity index (CI) range from $0.87\Delta°2\theta$ to $0.21\Delta°2\theta$. Most of the CI data are consistent with conditions of advanced diagenesis and anchimetamorphism (transition into greenschist facies), and a few CI values fall within the zone of epimetamorphism (lowermost greenschist facies). A significant statistical correlation exists between $\%R_m$ and CI, with the best-fit curve corresponding to the following equation and correlation coefficient: $\%R_m = 0.57 - 5.99 \log(CI)$; $r = 0.84$. This curve conforms reasonably well with the boundaries of the anchizone, as established by independent compilations. Calibration of CI values with paleotemperature estimates (T, in °C), as derived from $R\%_m$ data, results in the following correlation: $CI = 1.197 - 0.0029(T)$. However, because of error propagation, uncertainties in the validity of extrapolation, and potential differences in the boundary conditions of low-grade metamorphism, this relation between CI and paleotemperature should be applied with caution to studies of other orogenic belts. Measured values of illite b_0 lattice spacings range from 9.001Å to 9.031Å; these data are consistent with moderate amounts of $(Mg+Fe_{total})$ in the illite unit cell. By analogy with other orogenic sequences, Shimanto metamorphism evidently was governed by intermediate pressure gradients. Maximum burial pressures were probably less than 2.5 kbar, and maximum burial depths were 9 km or less. Thus, at least when viewed within the blueschist-facies paradigm of subduction zones, the Tertiary Shimanto Belt must be regarded as somewhat unusual.

INTRODUCTION

The principal analytical method used to obtain estimates of paleotemperature during our investigation of the Tertiary Shimanto Belt (Fig. 1) is vitrinite reflectance (Laughland and Un-

derwood, this volume). Rather than relying solely on analyses of organic matter, however, we have also characterized diagnostic phyllosilicate minerals within the same sedimentary rocks. This approach follows a trend of the past decade in which studies of low-grade metamorphism have adopted a strong multidisciplinary flavor (e.g., Stalder, 1979; Frey and others, 1980; Cloos, 1983; Duba and Williams-Jones, 1983; Barker and Halley, 1986; Pollastro and Barker, 1986; Hesse and Dalton, 1991).

---

*Present address: Mobil Research and Development, Exploration, and Producing Technical Center, 3000 Pegasus Park Dr., Dallas, Texas 75265-0232.

Underwood, M. B., Laughland, M. M., and Kang, S. M., 1993, A comparison among organic and inorganic indicators of diagenesis and low-temperature metamorphism, Tertiary Shimanto Belt, Shikoku, Japan, *in* Underwood, M. B., ed., Thermal Evolution of the Tertiary Shimanto Belt, Southwest Japan: An Example of Ridge-Trench Interaction: Boulder, Colorado, Geological Society of America Special Paper 273.

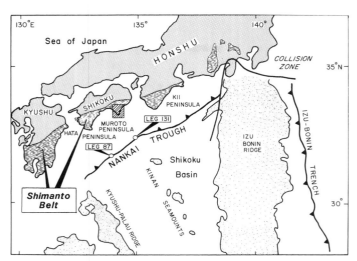

Figure 1. Index map of southwest Japan showing the principal outcrop distributions of the Shimanto Belt. Numbers in the offshore region (Nankai Trough) refer to sites associated with the Deep Sea Drilling Project and the Ocean Drilling Program.

The general metamorphic conditions of the Shimanto Belt have been outlined by Toriumi and Teruya (1988). Prehnite-pumpellyite facies minerals have been documented in basaltic rocks within the Cretaceous Shimanto Belt (Suzuki and Hada, 1983), and the Tertiary Shimanto strata are generally regarded as zeolite and/or prehnite-pumpellyite facies (Toriumi and Teruya, 1988). Traditional mineralogic studies of this type have suffered, however, from a lack of precision with respect to estimates of maximum burial temperature and burial pressure for individual specimens. This is largely because the common index minerals within the zeolite and prehnite-pumpellyite metamorphic facies are stable over wide ranges of burial conditions (Zen and Thompson, 1974; Kisch, 1983; Liou and others, 1985, 1987; Frey, 1987). For example, the mineral laumontite, which is quite common in low-grade arc-derived sedimentary successions, has been discovered at shallow burial depths in both geothermal fields (Boles, 1977; McCulloh and others, 1981) and on the sea floor (Sands and Drever, 1977). Laboratory experiments show that the upper pressure (P) and temperature (T) limits for laumontite equilibrium are approximately 3 kbar and 300°C, respectively (Liou, 1971).

As another measure of thermal maturation, the overall trends in the diagenesis of clay minerals are well documented. For example, detrital and authigenic smectite is replaced by illite as burial depth and temperature increase; in addition, intermediate phases of mixed-layer clay minerals display improved degrees of "ordering," plus progressive increases in the ratio of illite to smectite, as diagenesis proceeds (Burst, 1969; Dunoyer de Segonzac, 1970; Perry and Hower, 1970; Hower and others, 1976). As

conditions reach the stage of low-grade metamorphism, changes also occur in the dominant polytypes of white mica, as the disordered $1Md$ form is converted to the $2M_1$ polytype (Maxwell and Hower, 1970; Merriman and Roberts, 1985; Hunziker and others, 1986). Unfortunately, each one of these mineralogic changes is influenced by a large number of variables, including the composition and movement of fluid phases, lithostatic and fluid pressures, and variations in the original clay-mineral and bulk-rock chemistry (Nadeau and Reynolds, 1981; Roberson and Lahann, 1981; Kisch, 1983; Frey, 1987; Freed and Peacor, 1989; Pytte and Reynolds, 1989). Studies of illite/smectite diagenesis, therefore, are inadequate if one's goal is to reconstruct precisely the deformation history and the thermal evolution of complicated orogenic successions.

Perhaps the most widely used method for characterizing relative levels of inorganic diagenesis and metamorphism is the measurement of illite "crystallinity" by X-ray diffraction (Frey, 1987; Kisch, 1987). The procedure is rapid, simple, inexpensive, and applicable over a wide variety of geologic conditions. Another method that shows promise is the measurement of b_o lattice spacings of potassic white micas, also using X-ray diffraction (Sassi and Scolari, 1974; Guidotti and Sassi, 1976). These data are particularly useful for determining the general pressure/temperature gradient or baric type of metamorphism (Padan and others, 1982). Compilations of data from low-grade greenschist-facies and blueschist-facies terranes suggest that estimates of absolute pressure can be made for bona fide metamorphic micas if associated paleotemperatures are constrained independently (Guidotti and Sassi, 1986). The accuracy of the b_o method, especially under conditions of zeolite-facies alteration, remains suspect, in part because of the likelihood of mixing between multiple populations of detrital and authigenic minerals.

The purpose of this paper is to provide a direct and detailed comparison among organic and inorganic indicators of diagenesis and low-grade metamorphism in the Tertiary Shimanto Belt, specifically rocks exposed on the Muroto Peninsula of Shikoku Island, Japan (Fig. 1). As explained by Underwood, Hibbard, and DiTullio (this volume), the Shimanto Belt has experienced a complicated history of polyphase deformation within an accretionary-prism environment. Values of mean vitrinite reflectance vary from 0.9 to 5.0%, and this level of organic metamorphism corresponds to a range in paleotemperature of approximately 135 to 315°C (Laughland and Underwood, this volume). Geothermal gradients at the time of peak heating were relatively high for an accretionary prism (at least when compared with blueschist-facies terranes), and the thermal conditions probably were influenced by the subduction of very young oceanic crust.

Our analyses of inorganic phases are, by no means, exhaustive, but the correlations between organic and inorganic data allow comparisons to be made with general trends established elsewhere (e.g., Kisch, 1987). We also emphasize that our conclusions and statistical correlations may not apply beyond the specific geologic circumstances that affected southwest Japan during middle to late Cenozoic time.

BASIC PRINCIPLES

Illite crystallinity

The alteration of smectite and the monotonic transformation of intermediate mixed-layer illite-smectite phases to pure illite has been documented by many workers (e.g., Dunoyer de Segonzac, 1970; Perry and Hower, 1970). As summarized by Kisch (1983, 1987), illitization reactions are complicated by many variables, and the temperature at which any specific illite/smectite mixed-layer composition might be reached will vary from one locality to another. For example, a pronounced lag in clay diagenesis (relative to organic metamorphism) has been noted where geothermal gradients are abnormally high (Pevear and others, 1980), in zones immediately adjacent to igneous intrusions (Smart and Clayton, 1985), and in hydrothermal systems (Barker and others, 1986).

Systematic changes in the shape of the 10Å (001) illite peak on X-ray diffractograms serve as an indirect measure of the lattice reorganization of illite (Weaver, 1960). In reality, however, this mineralogic transformation involves more than just increases in the size of crystallites; it also entails changes in chemical composition and a progression toward greater regularity of the structural layers (Kisch, 1983). Frey (1970) demonstrated that improvements in the "crystallinity" of illite are accompanied by alterations of several other parameters: the intensity ratio of the illite basal reflections [$I(002)/I(001)$], the color of the host rock, mean bulk density, and microscopic rock texture due to reactions between clastic quartz grains and the clay-mineral matrix. In addition, iron and magnesium may be liberated from the illite during advanced diagenesis. Hunziker and others (1986) identified four chemical modifications that occur as illite is transformed into more crystalline K-mica: (1) expandable layers decrease in relative abundance and eventually disappear; (2) the 1Md illite polymorph is replaced by the $2M_1$ form; (3) the total layer charge and the K-content in the illite interlayer positions both increase; and (4) the chemical variability of individual illite/mica grains decreases. Because of this complexity, other authors have pointed out that the term illite "crystallinity" is not perfectly appropriate as a means of describing the mineralogic reorganization (Kisch, 1983; Frey, 1987). Nevertheless, the term will be used in the remainder of this paper without quotation marks.

Several methods have been proposed to quantify the shape of the 10Å illite peak. The so-called Weaver index (also known as the sharpness ratio) is defined as the ratio of the maximum 10Å peak height (minus background) to the peak-height (minus background) at 10.5Å (Weaver, 1960). The Weaver index in-

creases with increasing degrees of thermal alteration, but measurements by hand of analog X-ray diffraction (XRD) data become increasingly inaccurate as the peak narrows. The Kubler index (or crystallinity index) is equal to the width of the 10Å peak at one-half the peak height (Kubler, 1967); this parameter decreases as the diagenetic/metamorphic grade increases, and values can be calculated very precisely using digital XRD data. As originally defined, the units of crystallinity index (CI) were millimeters, but most subsequent workers have reported their values in units of $\Delta°2\theta$ to minimize the effects of variable machine settings (e.g., Kisch, 1980). Finally, Weber (1972) employed an external quartz standard to control the effects of instrument drift and calculated a peak-width ratio of the illite (001) peak-width divided by the quartz (100) peak-width (known as Hb_{rel} from the German word Halbwersbreite). Direct comparisons among the three indices, together with analyses of error, were presented by Blenkinsop (1988), who concluded that the Kubler crystallinity index is marginally superior to the Weaver and Hb_{rel} indices at all grades of diagenesis and low-temperature metamorphism.

Many workers have integrated measurements of illite crystallinity into detailed studies of regional and local structural evolution (e.g., Frey and others, 1980; Kisch, 1980; Roberts and Merriman, 1985; Kemp and others, 1985; Primmer, 1985; Laughland and others, 1990; Awan and Woodcock, 1991; Hesse and Dalton, 1991). Some caution is warranted in this regard, however, particularly if the CI data serve as the only indicator of paleotemperature. One might assume that CI values respond in a linear fashion to increases in burial temperature as a function of depth, as suggested by some borehole data (e.g., Yang and Hesse, 1991). Nevertheless, the reactions that affect CI are complicated by many internal and external variables, including the grain size of the specimen, the duration or rate of heating, fluid pressure, fluid composition (K^+ must be available), rates of fluid migration, original composition of the host sediment (inhibiting effects of Na^+ and Mg^{2+}), the chemical make-up of illite and/or mixed layer precursors, and the presence of organic matter (e.g., Ogunyomi and others, 1980; Eberl and others, 1987; Frey, 1987; Kreutzberger and Peacor, 1988; Yang and Hesse, 1991). Particularly noteworthy, for the purposes of our study, is the fact that very rapid conductive heat transfer and anomalous zones of advective hydrogeology (e.g., fluid-dominated geothermal fields and magmatic aureoles) can cause CI to lag behind organic metamorphism because of comparatively sluggish reaction rate (e.g., Roberts and Merriman, 1985; Kisch, 1987). In addition, CI values for intermediate-level diagenetic conditions can be notoriously unreliable because of a "contamination" effect caused by mixtures of authigenic illite, illite-smectite mixed-layer phases, and higher-grade detrital micas eroded from metamorphic terranes.

The effects of lithostatic pressure or effective stress on illite crystallanity remain unclear, but values evidently do change in response to some types of rock deformation (Frey, 1987). For example, CI anomalies have been documented in tectonic shear zones (Aldahan and Morad, 1986), perhaps in response to shear heating. Similarly, the hinge zones of some regional-scale folds coincide with the development of strong cleavage and increases in crystallanity (Roberts and Mereriman, 1985); an identical spatial relation might occur, however, if deeper-seated strata were uplifted and exposed within the core of a post-metamorphic anticline. Anomalies also exist in the hinge zones of some synformal mesoscale folds (Nyk, 1985), and Kreutzberger and Peacor (1988) demonstrated that CI values can change in response to pressure solution, even under isothermal conditions. Merriman and Roberts (1985) and Kemp and others (1985) noted clear correspondences between illite crystallinity and regional-scale zonations of cleavage, and Kisch (1991) has compiled an extensive data set showing how CI data vary with specific types of cleavage. On the other hand, Robinson and Bevins (1986) were unable to recognize a regular association of cleavage and CI. One reason for this inconsistency is that cleavage generation is not necessarily time-correlative with the attainment of maximum temperature.

As part of their analysis of error and precision inherent in the CI technique, Robinson and others (1990) concluded that geologic interpretations of CI gradients and anomalies should be based on differences of at least $0.1\Delta°2\theta$. The most reliable application of the technique, therefore, is simply to define broad zones or relative stages of diagenesis and low-grade metamorphism. A five-fold subdivision was introduced by Weaver (1960), but most workers follow the system of Kubler (1967, 1968), who defined the zones as diagenesis, anchimetamorphism or anchizone (transition into greenschist-facies metamorphism), and epimetamorphism or epizone (lowermost greenschist facies). Pinpointing the zonal boundaries has been problematic because of a failure to maintain uniformity in sample preparation, analytical equipment, and analytical technique (see Kisch, 1987, 1990; Kisch and Frey, 1987; and Robinson and others, 1990; for comprehensive discussions of these problems). Blenkinsop (1988) advocated the following values of CI for the two principal boundaries of thermal alteration: diagenesis-anchizone = $0.42\Delta°2\theta$ and anchizone-epizone = $0.25\Delta°2\theta$.

Vitrinite reflectance and paleotemperature

Correlations among illite crystallinity and other indicators of low-temperature metamorphism (both inorganic and organic) have been discussed in exceptional detail by Kisch (1987). Most noteworthy, for the purposes of our study of the Shimanto Belt, is the relation between CI and coal rank, which is typically defined by values of vitrinite reflectance ($\%R_m$). Unfortunately, the goal of establishing a universal statistical correlation between CI and $\%R_m$ is, perhaps, unrealistic because the two indices quite clearly respond to different (though overlapping) sets of external varia-bles. Unlike CI, for example, $\%R_m$ increases primarily in response to burial temperatures, thereby serving as a more direct indicator of paleotemperature (Barker, 1989, 1991).

Time-dependent models of organic metamorphism (Hood and others, 1975; Bostick and others, 1978; Waples, 1980; Middleton, 1982; Ritter, 1984; Wood, 1988; Hunt and others, 1991) are useful in many types of geologic studies, particularly those involving simple, first-cycle sedimentary basins. However, serious problems are encountered when these models are used to calculate paleotemperatures within uplifted orogenic sequences, where the tectonic and thermal histories can be quite complicated and poorly constrained. In addition to the guesswork involved in the selection of an appropriate value of activation energy for a given rock unit (e.g., Antia, 1986; Wood, 1988), large uncertainties usually exist in the inferred reaction rates for time-temperature integrals (e.g., Issler, 1984; Ritter, 1984). In addition, the actual effective heating times are generally unknown, and erroneous assumptions will lead to serious systematic errors in the resulting paleotemperature estimates. Conversely, many studies over the past decade have demonstrated that vitrinite reflectance is affected most by maximum temperature and almost imperceptibly by the duration of heating, at least under geologic situations where time is measured in millions of years (Wright, 1980; Gretener and Curtis, 1982; Suggate, 1982; Barker, 1983, 1989, 1991; Price, 1983; Barker and Pawlewicz, 1986). These workers, in other words, have argued that the time required for stabilization of vitrinite maturation within a given thermal regime is usually on the order of 1 m.y. or less.

Several statistical correlations now exist between $\%R_m$ and absolute paleotemperature (Barker, 1983, 1988; Price, 1983; Barker and Pawlewicz, 1986). We believe that the Barker (1988) equation yields the most reliable paleotemperature estimates, for three reasons: (1) the curve matches the chemical kinetic models of Burnham and Sweeney (1989) and Sweeney and Burnham (1990) fairly well, provided the inferred heating time is roughly 1 m.y.; (2) when compared, for example, to results from the Lopatin/Waples (1980) method, the calculated temperatures are in much better agreement with independent measures of paleotemperature, such as fluid-inclusion homogenization temperatures (Tilley and others, 1989; Barker and Goldstein, 1990); and (3) this equation is the most "conservative" of the time-independent models, in that it yields the lowest paleotemperature estimate for a given value of $\%R_m$ (Fig. 2). The relevant equation is: T (°C) = $148 + 104[\ln (\%R_m)]$.

No universal agreement has been reached on the anchizone boundaries in terms of either vitrinite reflectance or paleotemperature values. The suggested ranges are: diagenesis-anchizone = 2.5 to 3.1 $\%R_m$ (summarized by Kisch, 1987), and anchizone-epizone = 3.7 to 5.5$\%R_m$ (Frey and others, 1980; Kisch, 1980; Ogunyomi and others, 1980; Duba and Williams-Jones, 1983). The minimum and maximum temperature limits of the anchizone have been approximated at 200 and 300°C, respectively (Kisch, 1987). Based on recent compilations of data from circum-Pacific

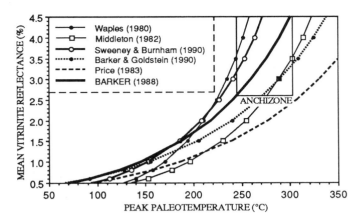

Figure 2. Curves used in calculations of peak paleotemperature based on values of mean vitrinite reflectance (%R_m). The Barker (1988) and Price (1983) curves are independent of heating time and based on data from both first-cycle sedimentary basins and geothermal systems. The Waples (1980) and Middleton (1982) models depend on the duration of heating; in both cases, time has been set at 1 m.y. The Barker and Goldstein (1990) curve follows a correlation between %R_m and homogenization temperatures of fluid inclusions. Also shown is one iteration of the Sweeney and Burnham (1990) EASY%R_o model, as illustrated in their Figure 5. See Laughland and Underwood (this volume) for additional discussion of these and other models.

orogenic belts, Underwood and others (1991) suggested the following limits for the anchizone: minimum boundary at 2.6%R_m and 245°C; epizone boundary at 4.5%R_m and 305°C (Fig. 2).

Mica b_o lattice spacing

As outlined by Velde (1965), phengitic micas are intermediate phases of a complex solid solution series between the end members muscovite (K+Al-rich mica) and celadonite (Fe+Mg-rich mica). The degree of cation substitution is a function of several variables, including pressure, temperature, oxygen fugacity, water pressure, and bulk-rock composition. Many studies have shown that phengitic micas are common in terranes that were affected by high-P, low-T metamorphic conditions (e.g., Ernst, 1963; Guidotti and Sassi, 1976). Estimates of relative burial pressure can be made if the other variables are held constant, and valid characterizations of baric type seem possible over a fairly wide range of metamorphic and geodynamic conditions (Sassi and Scolari, 1974; Fettes and others, 1976; Briand, 1980; Padan and others, 1982; Kemp and others, 1985; Offler and Prendergast, 1985; Guidotti and Sassi, 1986).

The most widely used method for determining the Fe+Mg content of white K-micas is through XRD measurements of the b_o lattice spacing (Sassi and Scolari, 1974; Guidotti and Sassi, 1976). The b_o spacing changes with metamorphic grade because larger Fe^{2+} and Mg^{2+} cations exchange with octahedral alumi-

num. The RM value for true metamorphic micas refers to the iron+magnesium content (i.e., the molar proportion of $2Fe_2O_3$ + FeO + MgO) in the unit cell. Naef and Stern (1982) demonstrated that the statistical correlation between b_o values (for single crystals) and phengite content (as quantified by chemical analysis) is not particularly good. Nevertheless, the use of large data sets plotted on histograms or cumulative frequency diagrams reduces some of the errors, and comparisons between different baric types are usually displayed in this fashion (Sassi and Scolari, 1974; Padan and others, 1982). Frey and others (1983) pointed out that the calculation of the b_o parameter from the d(060) value is technically incorrect, because the (060) mica reflection cannot be separated from the ($\bar{3}31$) reflection. In addition, Frey and others (1983) obtained a correlation, for well-crystallized white K-mica, between Si-content in the unit cell (based on chemical analyses) and b_o values. The low-pressure facies series (at least for true metamorphic micas crystallized at temperatures above 300°C) is associated with b_o values less than 9.000Å; the intermediate-pressure facies series (greenschist P-T gradient) ranges from 9.000Å to 9.040Å, and b_o values for the high-pressure facies series (blueschist P-T gradient) are greater than 9.040Å (Guidotti and Sassi, 1986).

Illite-b_o values also have been reported for regions of advanced diagenesis and anchimetamorphism (e.g., Padan and others, 1982; Cloos, 1983; Robinson and Bevins, 1986), but these data are much more difficult to interpret. For example, if the octahedral Fe+Mg content is held fixed in the illite, then the b_o value should increase during diagenesis because of increases in potassium content; this explains why the statistical relation between b_o and Fe+Mg abundance shifts for illitic clay minerals with respect to metamorphic muscovite (Hunziker and others, 1986). In addition, however, b_o lattice dimensions in nonmetamorphic illites may undergo a net reduction during anchimetamorphism because of a loss of Fe+Mg, even with a gain of potassium in the illite (Frey, 1970; Hunziker and others, 1986). Thus, it may not be possible to isolate the effects of cations inherited from illitic precursors, as opposed to the controls exerted by burial pressure on additional cation exchange during incipient metamorphism.

It is certainly tempting to extrapolate the P-T-b_o grid of Guidotti and Sassi (1986) into the field of zeolite-facies alteration, but several additional problems must be considered: (1) detrital micas may not recrystallize in the present host rock and, instead, may retain the b_o signature of their detrital source; (2) the effect of expandable smectite interlayers, which may survive even advanced stages of diagenesis, remains uncertain; (3) because the relevant dehydration reactions are dependent on the activity of water in the fluid phase, higher b_o values should be expected for very low-grade rocks containing abundant organic matter (Padan and others, 1982; Frey, 1987). Thus, estimates of absolute burial pressure based on illite b_o data (rather than data from well-crystallized mica) must be made with caution and viewed with some skepticism.

METHODS

Sample preparation

Specimens of shale and their low-grade metamorphic equivalents from the Shimanto Belt were collected from fresh outcrop exposures (see Laughland and Underwood, this volume; DiTullio and others, this volume; and Hibbard and others, this volume). The rocks were ground to a fine powder using a mortar and pestle. A split of the bulk powder was packed into round aluminum sample holders for XRD measurements of b_o lattice spacing. The remaining powder was washed in distilled water and disaggregated further using an Ultrasonics cell disrupter. A pinch of Calgon then was added to inhibit flocculation of suspended clay minerals, and the $<2 \mu m$ size fraction was segregated by centrifugation (at 770 rpm for 3.3 minutes; 500 rpm for 15 minutes). Oriented aggregates of the clay-sized material were placed on glass slides using the pipette-dropping technique (Moore and Reynolds, 1989) and allowed to air dry at room temperature. Prior to XRD analyses, the specimens were placed in a chamber containing ethylene glycol for at least 12 hours (at approximately 60°C) to saturate any existing expandable clay layers. This treatment was required to eliminate the effects of variable atmospheric humidity on clay lattice expansion.

Diffractometer settings and digital data processing

All measurements by X-ray diffraction were completed using a Scintag PAD V microprocessor-controlled diffractometer, interfaced with a Microvax 2000 microcomputer. X-ray scans were run under the following conditions: voltage = 30 Kv; amperage = 20 mA; radiation = $Cu_{K\alpha}$; filters = none; receiving slits = 0.3° (close to detector); scatter slits = 2° (close to beam emission); scan = continuous with 0.03° chopper increment; time constant = 1.8 sec/step; scan rate = $1°2\theta$/min; rate meter = automatic control. The (101) quartz peak (3.342Å) was used as an instrument standard for calibration of 2θ peak positions. Scans for illite-mica b_o lattice spacing ranged from $59.5°2\theta$ to $62.5°2\theta$ without spinning the sample holder, and the scan for illite crystallinity was run from $2°2\theta$ to $15°2\theta$ (with the sample holders spun). The resulting digital data were processed first for a background correction and $K_{\alpha2}$ stripping using Scintag software. The illite/mica (001) and (060) peaks were identified, and the peaks were deconvoluted using an interactive graphics subroutine. The deconvolution program is designed to fit XRD peaks to ideal models, based on a Split Pearson VII profile shape (Gaussian-Lorentzian hybrid). Automatic computations yield peak positions and d-spacings (in $°2\theta$ and Å), peak intensities (counts per second), integrated peak areas (total counts), and values of peak width at half maximum ($\Delta°2\theta$).

Illite crystallinity

As discussed previously, several methods are available to quantify illite crystallinity (Blenkinsop, 1988). Our data follow the parameter established by Kubler (1968), with CI values (glycolated) expressed in units of $\Delta°2\theta$. The analytical sensitivity for measurements of peak width using the Scintag diffractometer extends to $0.001°2\theta$. However, tests of reproducibility (using eleven splits of the same sample and eleven runs of the same split) resulted in standard deviations about the mean ranging from $0.009\Delta°2\theta$ to $0.015\Delta°2\theta$ (Kang, 1990); consequently, we report the data only to the nearest $0.01°2\theta$.

Illite b_o lattice spacing

The b_o lattice spacing of illitic mica was measured on randomly oriented bulk powders using the position of the $(060,\overline{3}31)$ reflection, which occurs at a scanning angle of approximately $61.5°2\theta$. Sample holders were not spun during these measurements because of problems with keeping the powder from falling out. To correct for the alignment of each position on the multiple sample holder, the (211) quartz reflection (approximately $60°2\theta$) was used as an internal standard, and the d-spacing for this stable peak is 1.5418Å. The analytical sensitivity of the Scintag diffractometer extends to 0.0001Å for measurement of d-spacing. However, tests of reproducibility (using ten splits of the same sample) resulted in a standard deviation about the mean b_o value of 0.002Å (Kang, 1990), so all reported b_o data have been rounded off to the nearest 0.001Å. In addition, because of the possibility of mixing among multiple populations of detrital and authigenic mica, particularly in lower-grade mudrocks, these b_o values should be regarded as bulk averages.

The following equation was used for estimates of the percentage of octahedral $Mg + Fe_{total}$ in bulk illite/mica assemblages, based upon the correlation of illite b_o and chemical data obtained by Hunziker and others (1986): $d(060,\overline{3}31) = 1.494 + 0.020 [(Mg + Fe_{total})/O_{11}]$. A separate equation for well-crystallized white K-micas from granitic rocks was derived by Frey and others (1983); that curve is roughly parallel in slope to the illite curve but yields a lower (Fe+Mg) number for a given $(060,\overline{3}31)$ d-spacing (intercept = 1.498Å, rather than 1.494Å).

RESULTS

Sampling

Samples were collected from outcrops of the following tectonostratigraphic units within the Shimanto Belt: the Murotohanto subbelt (Eocene–early Oligocene?), which itself includes tectonic mélange, the Gyoto domain, and the Shiina domain; and the Nabae subbelt (late Oligocene–early Miocene), which comprises the Shijujiyama Formation, the Hioki and Sakamoto mélanges, and the Tsuro and Misaki assemblages (Fig. 3). All of the sample localities are identified in papers by Laughland and Underwood (this volume), DiTullio and others (this volume), and Hibbard and others (this volume). Successful measurements of vitrinite reflectance were completed on a total of 182 samples (Laughland and Underwood, this volume), and values of mean random reflectance (%R_m) range from 0.9 to 5.0%. One hundred

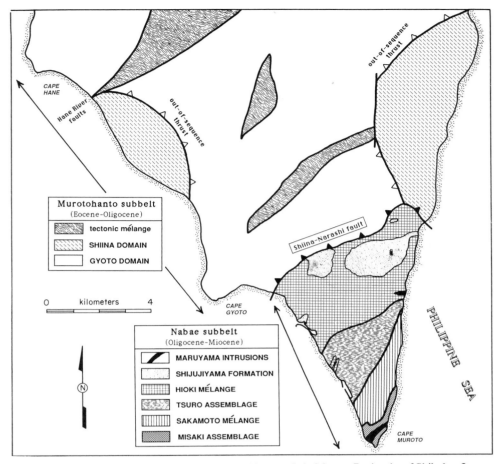

Figure 3. Simplified geologic map of the Tertiary Shimanto Belt, Muroto Peninsula, of Shikoku, Japan. See Underwood, Hibbard, and DiTullio (this volume) for descriptions of the individual rocks units assigned to the Murotohanto subbelt (Eocene–early Oligocene?) and the Nabae subbelt (late Oligocene–early Miocene), plus a summary of the structural-tectonic history of this region.

thirty samples were analyzed at the University of Missouri for illite crystallinity index (for additional CI data, see DiTullio and others, this volume; and DiTullio and Hada, this volume). From this suite, 105 samples yielded reliable data for both clay mineralogy and organic metamorphism (Table 1). A representative suite of 38 samples was chosen to characterize the illite-mica b_o lattice spacings (Table 2). Interpretations regarding the structural and stratigraphic context of all labodatory results are provided in companion papers (DiTullio and others, this volume; Hibbard and others, this volume; Underwood, Byrne and others, this volume).

Retention of expandable clay layers

Because of overlap between a strong discrete (001) illite peak and possible reflections caused by small amounts of ordered illite-smectite mixed-layer phases (001/002), we were not able to quantify the amount of mixed-layer clay or the percentage of smectite in the mixed-layer phase using the criteria of Reynolds and Hower (1970) or Moore and Reynolds (1989). Nevertheless, DiTullio and others (this volume) showed that the (001) illite peak produced by many air-dried specimens does change in response to treatment with ethylene glycol; this demonstrates that some expandable layers have been retained in the crystal lattice. Under typical circumstances of burial, conversion of the last 15 to 20% of the expandable layers seems to require temperatures in excess of 165 to 200°C (Perry and Hower, 1970; Hower and others, 1976; Hoffman and Hower, 1979). On the other hand, expandable mixed-layer phases are known to persist over much wider windows of temperature (135 to 255°C) in regions of unusually rapid heating (i.e., geothermal fields); this lag in clay-mineral diagenesis is probably a consequence of sluggish reaction rates (Barker and others, 1986). Thus, we do not believe that small amounts of lattice expansion are particularly diagnostic of the absolute thermal alteration in the Shimanto Belt.

TABLE 1. CORRELATIVE VALUES OF MEAN VITRINITE REFLECTANCE (%R_m), ILLITE CRYSTALLINITY INDEX (CI), AND PALEOTEMPERATURE IN THE TERTIARY SHIMANTO BELT, MUROTO PENINSULA, SHIKOKU, JAPAN

Sample	Rock Unit	%R_m	CI ($\Delta°2\Theta$)	Temp.* (°C)	Sample	Rock Unit	%R_m	CI ($\Delta°2\Theta$)	Temp.* (°C)
84j-2	Gyoto	3.65	0.35	280	h84-89	Tsuro	1.71	0.68	205
84j-4	Gyoto	2.03	0.55	220	h84-91	Tsuro	1.99	0.54	220
84j-6	Gyoto	2.15	0.53	225	h84-92	Tsuro	1.83	0.68	210
84j-7	Shiina	3.73	0.29	285	h84-95	Sakamoto	1.96	0.45	220
84j-9	Shiina	3.29	0.42	270	h85-6	Misaki	2.36	0.46	235
84j-10	Shiina	3.33	0.30	275	h85-7	Misaki	2.59	0.31	245
84j-12	Gyoto	3.71	0.50	285	h85-9	Misaki	3.32	0.39	275
84j-13	Gyoto	3.74	0.46	285	h85-11	Hioki	0.87	0.74	135
84j-15	Gyoto	3.72	0.34	285	h85-16	Shijujiyama	1.70	0.59	205
84j-23	Gyoto	3.51	0.45	280	h85-19	Hioki	1.71	0.59	205
84j-28	Gyoto	2.40	0.45	240	h85-32	Sakamoto	1.70	0.61	205
84j-31	Shiina	3.42	0.48	275	h85-39	Sakamoto	1.69	0.49	205
84j-38	Gyoto	3.56	0.47	280	h85-46	Hioki	1.65	0.75	200
84j-41	Gyoto	3.44	0.36	275	h85-54	Shijujiyama	1.57	0.65	195
84j-42	Gyoto	3.54	0.35	280	h85-57a	Hioki	1.27	0.70	175
85j-2	Shiina	2.61	0.46	245	h85-57b	Hioki	1.28	0.70	175
85j-4	Shiina	2.88	0.55	260	h85-59	Shijujiyama	1.53	0.45	190
85j-7	Shiina	2.34	0.52	235	h85-61	Shijujiyama	1.54	0.56	195
85j-12	Shiina	1.95	0.57	215	h85-74	Hioki	1.86	0.50	215
85j-13	Shiina	2.14	0.62	225	h85-75	Shijujjiyama	1.72	0.66	205
85j-17	Shiina	1.47	0.55	190	h85-86	Shijujiyama	1.79	0.68	210
85j-19	Shiina	3.83	0.30	285	h85-89	Shijujiyama	1.60	0.65	200
85j-20	Shiina	4.04	0.32	295	h85-99	Misaki	3.03	0.46	265
85j-21	Shiina	3.72	0.21	285	h85-101	Hioki	1.07	0.69	155
85j-22	Shiina	3.81	0.40	285	h87-13	Gyoto	3.61	0.30	280
85j-23	Shiina	3.25	0.34	270	h87-14	Gyoto	3.57	0.35	280
85j-25	Shiina	3.78	0.36	285	h87-15	Gyoto	3.79	0.32	290
85j-30	Shiina	3.86	0.33	290	h87-16	Gyoto	3.86	0.33	290
85j-31	Shiina	3.78	0.32	285	h87-26	Shijujiyama	1.50	0.62	190
85j-34	Shiina	3.62	0.34	280	h87-31	Shijujiyama	1.37	0.57	180
85j-35	Shiina	3.63	0.39	280	h87-50	Hioki	1.52	0.60	190
85j-36	Shiina	3.63	0.40	280	h87-52	Hioki	1.40	0.67	185
85j-39	Shiina	3.37	0.42	275	h87-53	Misaki	3.00	0.37	260
85j-40	Shiina	3.05	0.43	265	h87-54	Misaki	2.80	0.39	255
85j-44	Shiina	3.46	0.39	275	h87-59	Tsuro	1.57	0.54	195
85j-46	Shiina	3.05	0.51	265	h87-60	Tsuro	1.80	0.46	210
85j-49	Shiina	2.79	0.37	255	h87-81	Hioki	1.55	0.52	195
85j-55	Shiina	1.81	0.58	210	h87-98	Tsuro	1.83	0.53	210
85j-60	Shiina	1.62	0.68	200	h87-102	Hioki	1.40	0.59	185
85j-61	Shiina	2.01	0.60	220	h87-104	Hioki	2.15	0.61	230
85j-62	Shiina	1.82	0.63	210	h87-105	Hioki	2.55	0.57	245
85j-76	Gyoto	1.91	0.55	215	h87-106	Tsuro	1.63	0.78	200
85j-81	Shiina	3.85	0.31	290	h87-107	Tsuro	1.91	0.58	215
h84-103	Hioki	1.45	0.68	185	h87-108	Tsuro	1.71	0.86	205
h84-56	Shijujiyama	1.65	0.65	200	h87-110	Shijujiyama	1.65	0.65	200
h84-75	Tsuro	2.08	0.60	225	h87-111	Hioki	1.14	0.81	160
h84-11	Tsuro	2.03	0.44	220	h87-115	Sakamoto	1.29	0.58	175
h84-22	Sakamoto	2.12	0.47	225	h87-117c	Sakamoto	1.83	0.66	210
h84-33	Sakamoto	2.11	0.42	225	h87-119	Misaki	3.24	0.37	270
h84-44	Misaki	3.73	0.35	285	h87-123	Misaki	3.28	0.26	270
h84-59	Shijujiyama	1.80	0.56	210	h87-127	Tsuro	1.77	0.84	205
h84-60	Shijujiyama	1.66	0.63	200					
h84-63	Sakamoto	2.25	0.47	230	*Based on correlation of Barker (1988).				
h84-84	Tsuro	1.79	0.50	210					

TABLE 2. ILLITE/MICA b_o DATA, TERTIARY SHIMANTO GROUP, JAPAN

Sample	Rock Unit	b_o(Å)	d(060,$\overline{3}$31)	RM*
84j-4	Gyoto	9.017	1.5028	0.44
84j-6	Gyoto	9.013	1.5022	0.41
84j-7	Shiina	9.022	1.5036	0.48
84j-9	Shiina	9.022	1.5037	0.49
84j-12	Gyoto	9.010	1.5016	0.38
84j-15	Gyoto	9.010	1.5017	0.39
84j-41	Gyoto	9.020	1.5033	0.47
85j-4	Shiina	9.020	1.5033	0.47
85j-7	Shiina	9.013	1.5021	0.40
85j-12	Shiina	9.020	1.5033	0.47
85j-30	Shiina	9.014	1.5024	0.42
85j-34	Shiina	9.016	1.5026	0.43
85j-44	Shiina	9.023	1.5038	0.49
85j-54	Shiina	9.008	1.5014	0.37
85j-76	Gyoto	9.018	1.5030	0.45
h84-11	Tsuro	9.019	1.5032	0.46
h84-44	Misaki	9.014	1.5023	0.42
h84-63	Sakamoto	9.017	1.5029	0.44
h85-6	Misaki	9.020	1.5033	0.47
h85-11	Hioki	9.016	1.5027	0.44
h85-16	Shijujiyama	9.018	1.5030	0.45
h85-32	Sakamoto	9.003	1.5005	0.32
h85-61	Shijujiyama	9.011	1.5019	0.40
h85-74	Hioki	9.016	1.5027	0.44
h85-75	Shijujiyama	9.020	1.5033	0.47
h85-86	Shijujiyama	9.010	1.5017	0.39
h87-14	Gyoto	9.015	1.5025	0.42
h87-16	Gyoto	9.001	1.5002	0.31
h87-50	Hioki	9.031	1.5052	0.56
h87-54	Misaki	9.019	1.5031	0.46
h87-60	Tsuro	9.014	1.5024	0.42
h87-102	Hioki	9.015	1.5025	0.42
h87-106	Tsuro	9.016	1.5027	0.44
h87-108	Tsuro	9.016	1.5027	0.44
h87-110	Shijujiyama	9.022	1.5037	0.49
h87-111	Hioki	9.022	1.5036	0.48
h87-115	Sakamoto	9.003	1.5005	0.32
h87-123	Misaki	9.020	1.5033	0.47

*RM = Octahedral (Mg + Fe_{total})/O_{11}.

Illite crystallinity versus vitrinite reflectance

Histograms for each tectonostratigraphic unit on the Muroto Peninsula show a clear relation between vitrinite reflectance and illite crystallinity index (Figs. 4 and 5). As organic rank increases, the width of the 10Å illite peak decreases. In addition, pelitic rocks with lower CI values typically display well-developed pressure-solution cleavage (DiTullio and others, this volume). It is clear that most of the Eocene-Oligocene strata (Gyoto and Shiina domains) experienced burial conditions ranging from uppermost diagenesis through anchimetamorphism (Fig. 4). The

average CI value for Murotohanto strata is $0.43\Delta°2\theta$, and the lowest value ($0.21\Delta°2\theta$) extends into the epizone (Table 1).

Overall, the Oligocene-Miocene rock units (Nabae subbelt) display consistently lower levels of both inorganic and organic alteration (Fig. 5); most of these pelites are also weakly cleaved (Hibbard and others, this volume). CI values for the Hioki mélange, for example, average $0.64\Delta°2\theta$ (well within the realm of diagenesis), and the maximum value is $0.81\Delta°2\theta$ (Table 1). The Tsuro assemblage, which is less deformed than the surrounding mélange belts, yields a mean CI value of $0.62\Delta°2\theta$ and a maximum of $0.85\Delta°2\theta$. CI values decrease markedly toward the southern tip of the Muroto Peninsula; the average CI value for the Sakamoto mélange is $0.52\Delta°2\theta$, and the Misaki assemblage displays even higher levels of incipient metamorphism (Table 1). The Misaki strata have been intruded by mafic dikes and sills of middle Miocene age ($\approx$14 Ma), and the intrusive activity clearly affected both organic and inorganic constituents. The average $\%R_m$ value for the Misaki sequence increases to 3.0%, and the average CI decreases to $0.36\Delta°2\theta$, which places these pelites within the anchizone.

Figure 6 displays the results of regression analyses of our values of $\%R_m$ and CI. If the total data set is segregated according to stratal age (i.e., Murotohanto subbelt versus Nabae subbelt), two linear correlations can be made; the higher-grade Murotohanto rocks (Fig. 6A) show a steeper slope to the correlation curve, as compared to the Nabae strata (Fig. 6B). Note, however, that higher-rank samples from the Misaki sequence plot well above the regression line for the entire Nabae data set. This steepening of the curve with increasing metamorphic grade should be expected, given the well-documented logarithmic response of vitrinite reflectance to increases in temperature (e.g., Price, 1983; Barker and Pawlewicz, 1986). Conversely, we are not aware of any studies showing a logarithmic change in CI values as a function of increasing temperature and/or burial depth. When all of the data are plotted together, the best statistical fit is obtained using a logarithmic function, where $\%R_m = 0.57 - 5.99 \log (CI)$. The correlation coefficient for this curve is 0.84 (Fig. 6C). This correlation coefficient (r) is significantly larger than for comparable matches attempted elsewhere (Fig. 7), such as the Ouachita Mountains of Arkansas and Oklahoma, where r = 0.56 (Guthrie and others, 1986), or the Kandik region of Alaska, where r = 0.64 (Laughland and others, 1990; Underwood and others, 1991).

Temperature calibration of the anchizone

The relation outlined above between CI and $\%R_m$ agrees reasonably well with data from most other studies of this type (Fig. 7). In particular, if the accepted lower boundary of the anchizone is set at a CI value of $0.42\Delta°2\theta$, then the correlation curve for the Shimanto data clearly intersects the anchizone win-

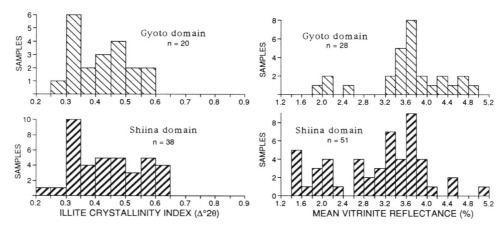

Figure 4. Histograms showing values of illite crystallinity index (CI) and mean random vitrinite reflectance (%R_m) for strata assigned to the Gyoto and Shiina domains of the Murotohanto subbelt, Muroto Peninsula, Shikoku, Japan. See Table 1 for a list of all data.

dow within an acceptable range of %R_m values (2.5 to 3.1%), as cited by Kisch (1987). Moreover, the point of intersection is just above the minimum anchizone limit of 2.6%R_m established by Underwood and others (1991). The Shimanto curve intersects the anchizone-epizone boundary of 0.25$\Delta°2\theta$ at an %R_m value of approximately 4.0%, which is also within the acceptable window defined by Kisch (1987). This intersection point, however, is lower than the 4.5%R_m value of Underwood and others (1991).

As discussed previously, it is not possible to make direct calibrations of maximum temperature over the full range of CI values using borehole data from active geothermal fields. This is because inorganic reactions are relatively sluggish and CI may not equilibrate at the present-day ambient temperature; in addition, there are complicating effects related to fluid chemistry and rates of fluid movement (e.g., Barker and others, 1986). One indirect way around this dilemma is first to convert the %R_m data to estimates of paleotemperature using the Barker (1988) equation (Fig. 2), then to plot the paleotemperature estimates against CI values measured from the same samples. We constructed a simple linear CI-temperature plot for the Shimanto Belt in this way, as shown in Figure 8. Naturally, the validity of this curve depends on the accuracy of Barker's (1988) equation. In addition, the CI-temperature curve suffers from the effects of error propagation. The error associated with the %R_m-temperature scale is roughly ±30°C because of inherent problems associated with the correction factors applied to borehole temperatures and the inconsistencies involved in measurements of vitrinite reflectance (see Barker and Pawlewicz, 1986; for a complete discussion). We suggest that the error associated with the CI-temperature curve propagates to at least ±50°C. With these limitations in mind, the Shimanto correlation places the diagenesis-anchizone boundary (0.42$\Delta°2\theta$) at a temperature of approximately 265°C. Linear extrapolation of the curve to the anchizone-epizone transition (0.25$\Delta°2\theta$) corresponds to a temperature of about 320°C (Fig.

8). Both of these limits are about 15 to 20°C higher than the limits suggested by Underwood and others (1991).

The linear-regression calibration of the CI scale (Fig. 8) results in higher paleotemperature estimates for any individual CI value than some of the other published temperature calibrations for the anchizone (Kisch, 1987). For example, within the diagenetic field, our paleotemperatures are higher (by approximately 20°C) than estimates made by Mitra and Yonkee (1985) within the Rocky Mountain thrust belt; however, their thermal reconstructions were based on the Lopatin model (Waples, 1980). Lopatin reconstructions yield systematically lower temperatures (compared to Barker, 1988), and some such calculations have been contradicted by fluid-inclusion data (e.g., Tilley and others, 1989).

As a second example, Frey and others (1980) showed that quartz veins in Alpine strata within the low-grade and medium-grade anchizone produce fluid inclusion homogenization temperatures between 200 and 270°C. One geologic explanation for this discrepancy (with respect to the Shimanto data) is that some of the Alpine veins (even first-generation veins) may have precipitated after the culmination of maximum heating, thereby lowering the paleotemperature estimates used to calibrate CI. A second possibility would be a modest retardation of the Shimanto CI values (with respect to %R_m) in response to rapid heating of the accreted strata. This effect should be most obvious in close proximity to the mafic dikes of the Maruyama intrusive suite (Misaki assemblage), but those data do not deviate much from the regional curve (Fig. 6). A third possibility worth considering is related to the effects of a strong pressure-solution cleavage, particularly within portions of the Murotohanto subbelt just north of the Shiina-Narashi fault (Fig. 3; DiTullio and others, this volume). However, we would expect phyllosilicate growth to be enhanced, rather than retarded, in zones of pronounced cleavage (e.g., Kisch, 1991). A final explanation is that the %R_m-derived

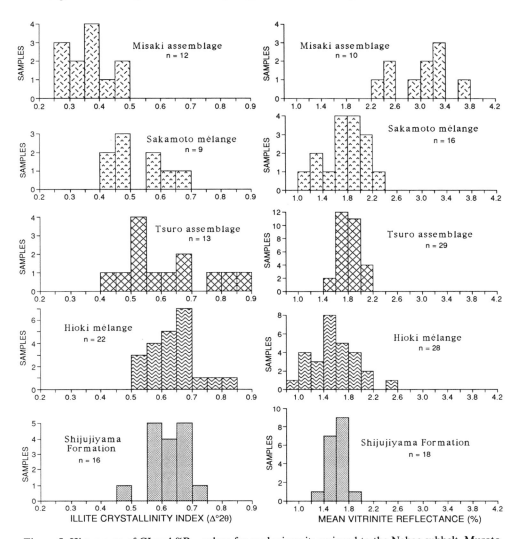

Figure 5. Histograms of CI and $\%R_m$ values for geologic units assigned to the Nabae subbelt, Muroto Peninsula, Shikoku, Japan. The degree of thermal alteration displayed by the Misaki assemblage is higher than for coeval rocks to the north, and this increase is due to the effects of igneous intrusions assigned to the Maruyama intrusive suite (see Fig. 3). See Table 1 for a list of all data.

temperature scale simply is not valid, especially as extrapolated beyond the limits of the Shimanto data set.

As a final case for comparison, we point out that Duba and Williams-Jones (1983) measured calcite fluid-inclusion homogenization temperatures that averaged about 300°C for rocks near the anchizone-epizone CI boundary, and some of the homogenization temperatures were as high as 340°C; moreover, the average organic-matter reflectance is about 4.6% for these rocks. Significantly, their study area in Quebec was reheated by buried igneous intrusions and was also affected by hydrothermal convection (Duba and Williams-Jones, 1983). Thus, it may be that the CI-temperature correlation for the Shimanto Belt is applicable only for orogenic regions exposed to moderately high geothermal gradients and relatively rapid heating events.

Illite b_o lattice spacing and burial depth

Most values of b_o lattice spacing for illites within the Tertiary Shimanto Belt range from 9.010Å to 9.024Å (Fig. 9A), and the mean value is 9.016Å. Only one sample yielded a value above 9.030Å. If we use the chemical data obtained from well-crystallized K-mica by Frey and others (1983), then the mean b_o value of 9.016Å corresponds to an Si-value of approximately 6.4, and the average content of octahedral $Mg+Fe_{total}$ would be about 25%. In addition, according to the criteria of Frey and others (1983), all but one of the Shimanto phyllosilicate assemblages would be classified chemically as muscovite (senso stricto); one sample would be classified as a "weak phengite."

Unfortunately, the chemical correlations cited above probably are not appropriate for the Shimanto samples because of differences in illite-mica crystallinity. Instead, our results should be compared with those of anchizone samples from the Swiss Alps analyzed by Hunziker and others (1986); following this analogy, the Shimanto illite b_o values of 9.012 to 9.18Å probably correspond to Si-values of 6.90 to 6.65. However, it is important to realize that chemical analyses of clay-sized fractions typically

produce inconsistent Si-values for a given d(060,$\overline{3}$31), and this effect is due largely to mixing among detrital and authigenic mineral populations (Hunziker and others, 1986). Similarly, the inferred percentages of octahedral iron and magnesium in the illitic clay-sized fraction are probably moderate, ranging from about 31 to 56% (Fig. 9B). For true metamorphic micas in the epizone, these same Mg+Fe values would be associated with larger b_o spacings of roughly 9.030Å to 9.048Å (Frey and others, 1983; Hunziker and others, 1986).

No significant differences exist in illite-b_o values between samples collected from the Murotohanto section and those from the Nabae section (Table 2). In addition, samples were tested over the full range of %R_m and CI values (Table 1). No statisti-

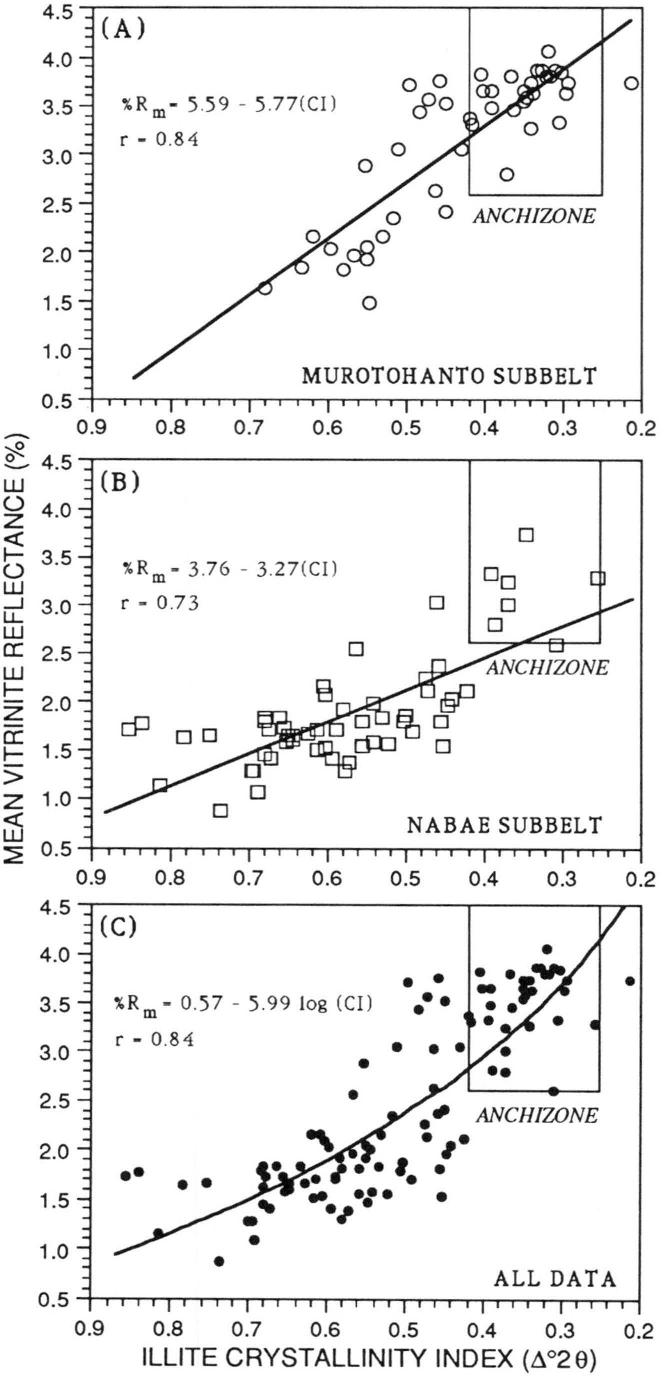

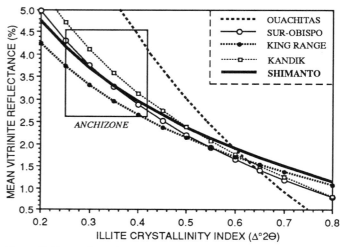

Figure 7. Comparisons between the %R_m-CI correlation for the Tertiary Shimanto Belt and curves derived from other orogenic belts. Also shown is the window of CI and %R_m values used to define the anchizone. The curve for the Ouachita Mountains (Arkansas and Oklahoma) is from Guthrie and others (1986) and Houseknecht and others (1987); curves for Kandik region (Alaska), the King Range (California), and the Sur-Obispo terrane (California) are from Underwood and others (1991).

Figure 6. Statistical correlations between values of mean vitrinite reflectance (%R_m) and values of illite crystallinity index (CI), Tertiary Shimanto Belt, Muroto Peninsula, Shikoku, Japan. Figure 6A displays a linear regression for the Murotohanto subbelt only, whereas the linear regression in Figure 6B is restricted to data from the younger Nabae subbelt. Correlation coefficients are given by the value of r. The slope for curve 6B is not as steep because most of these data correspond to conditions of advanced diagenesis rather than anchimetamorphism. Figure 6C shows the best-fit logarithmic curve for all of the data. Steepening of the curve at higher ranks of organic metamorphism should be expected because of the well-established logarithmic increase in vitrinite reflectance as a function of increasing temperature (e.g., Barker and Pawlewicz, 1986). Boundaries for the anchizone are based on Blenkinsop (1988) and Underwood and others (1991).

cally significant correlations exist either between illite-b_o and %R_m or between illite-b_o and CI. Thus, the b_o lattice dimensions seem to be independent of paleotemperature. A cumulative curve of the Shimanto illite-b_o data is consistent with baric conditions conforming to a metamorphic facies series of intermediate pressure (Fig. 10). Perhaps the best supporting evidence for our inferred intermediate P-T gradient is the similarity to data from anchizone samples of the Swiss Alps (Padam and others, 1982).

Any estimates of absolute burial pressure derived from b_o data must take into account the range of paleotemperatures encountered. Unfortunately, %R_m values for the Nabae subbelt are generally too low for reliable extrapolation of the P-T-b_o grid established by Guidotti and Sassi (1986). In contrast, the average paleotemperature value for the Murotohanto subbelt is approximately 270°C (derived from an average %R_m value of 3.3%), which requires only modest extension of the b_o curves into subgreenschist-facies P-T space. If this extrapolation into the anchizone is valid, then illite/mica-b_o values of less than 9.020Å indicate that burial pressures were less than about 2.5 kbar. This means that maximum burial depths were probably less than 9 km within most of the Murotohanto subbelt (assuming an average bulk density of 2.60 g/cm^3).

As one final analog for comparison, representative samples from the base of the Great Valley sequence in California, which are overlain by a measurable stratigraphic overburden of 8 to 10 km, typically yield illite/mica b_o values of 9.021Å to 9.026Å (Cloos, 1983). In response to a relatively low geothermal gradient, however, %R_m values for these Great Valley strata are typically less than 1.0%, so it is clear that burial conditions never passed beyond advanced diagenesis (Castano and Sparks, 1974; see also Dumitru, 1988). Consequently, our estimated depth

maximum of 9 km seems quite reasonable for the highest-grade rocks of the Tertiary Shimanto Belt. Paleotemperature estimates of about 300°C are typical of the Murotohanto subbelt, particularly for rocks near the Shiina-Narashi fault (Fig. 3; Laughland and Underwood, this volume). If these temperatures are divided by the maximum depth of 9 km, then a rough estimate for the background geothermal gradient at the time of peak heating would be approximately 33°C/km.

SUMMARY AND CONCLUSIONS

Inorganic phases from Tertiary shales and low-grade metapelites of the Shimanto Belt fully corroborate the relations shown by analyses of vitrinite reflectance. Illite crystallinity indices demonstrate that peak burial conditions ranged from the upper zone of diagenesis completely through the zone of anchimetamorphism, and the lowest CI values (0.21 Δ°2θ) place some of the Shimanto strata within the zone of epimetamorphism. A signifi-

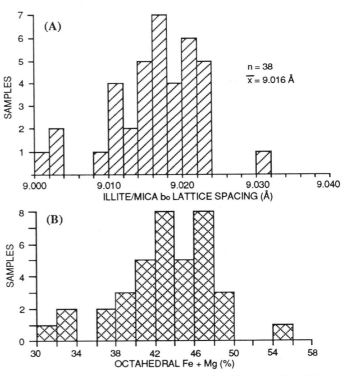

Figure 9. Illite/mica b_o data for the Tertiary Shimanto Belt, Muroto Peninsula, Shikoku, Japan. Figure 9A shows values of illite/mica b_o lattice spacing; n is the number of specimens and x̄ is the statistical mean. These values correspond to intermediate barometric conditions in anchizonal illites and metamorphic K-micas, according to Sassi and Scoleri (1974), Padan and others (1982), and Guidotti and Sassi (1986). Figure 9B shows comparable estimates of iron and magnesium in the illite/mica octahedral sites [Mg + Fe$_{total}$)/O$_{11}$], based on correlative chemical analyses completed by Hunziker and others (1986) on nonmetamorphic and anchimetamorphic illites. See Table 2 for a list of all data.

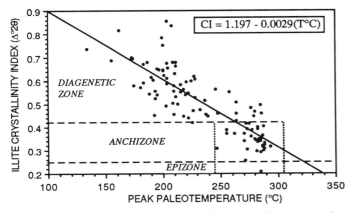

Figure 8. Temperature calibration of the illite crystallinity index using estimates derived from mean vitrinite reflectance and the Barker (1988) %R_m-temperature curve (see Fig. 2). The CI boundaries of the anchizone are from Blenkinsop (1988). This statistical relation applies only to results from the Tertiary Shimanto Belt, Muroto Peninsula, Shikoku, Japan, and should be used with caution elsewhere, particularly in regions where the boundary conditions of low-grade metamorphism might have been fundamentally different. Extrapolation of the linear regression beyond about 300°C is also dubious.

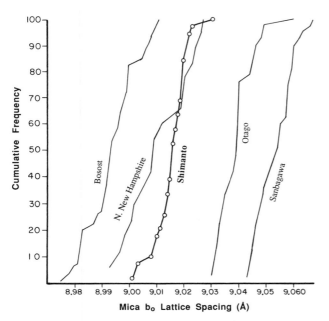

Figure 10. Cumulative frequency curve of illite/mica b_o data from the Tertiary Shimanto Belt, Muroto Peninsula, Shikoku, Japan (heavy line and open dots). Also shown are reference curves for crystalline K-mica (from Sassi and Scolari, 1974) for the Bosost facies series (high-temperature, low-pressure), northern New Hampshire (low-pressure, intermediate temperature), the Otago schist (Barrovian-type metamorphism), and the Sanbagawa belt of Japan (high-pressure, glaucophane-greenschist facies series). See Padan and others (1982) for additional reference curves.

cant statistical correlation exists between values of illite crystallinity index (CI) and values of mean vitrinite reflectance ($\%R_m$). The best statistical fit (r = 0.84) is provided by the following logarithmic regression: $\%R_m = 0.57 - 5.99 \log (CI)$.

The metamorphism of organic matter is sensitive mostly to temperature, at least over normal spans of geologic time. Conversely, many more variables need to be considered when analyzing corresponding changes in clay minerals and micas. We believe that $\%R_m$ values provide the most precise indications of diagenetic-metamorphic grade, as well as the most accurate estimates of absolute paleotemperature. The $\%R_m$ data from the Tertiary Shimanto Belt are consistent with maximum paleotemperatures ranging from 135 to 315°C.

The $\%R_m$-CI correlation curve for the Tertiary Shimanto Belt passes through the window of values from several comparable studies used to define the boundaries of the anchizone. Thus, there is no reason to believe that the processes of thermal alteration within southwest Japan were particularly unusual, as compared to the global "norm" of low-temperature metamorphism. In particular, there is no evidence of a widespread or pronounced lag in CI with respect to $\%R_m$, as might be expected under conditions of very rapid contact metamorphism or regional-scale hydrothermal alteration. Based upon our analyses of samples from the Misaki assemblage, for example, the thermal overprint

caused by mafic intrusions was sufficiently long in duration to allow equilibration of both $\%R_m$ and CI with the ambient temperature field.

If the CI data are calibrated against a scale of paleotemperature derived from measurements of $\%R_m$, then the transition from advanced diagenesis into the anchizone (CI = $0.42\Delta°2\theta$) occurs at a temperature of approximately 265°C. This temperature value is slightly higher (by about 20°C) than some of the other published calibrations, including those based on fluid-inclusion homogenization temperatures and oxygen isotopes. Thus, even though there is no obvious lag in CI values from the Shimanto Belt, with respect to $\%R_m$ data, the slope of the $\%R_m$-CI curve still may be sensitive to the rate of heating and the geothermal gradient, among other variables. We emphasize, therefore, that the temperature calibration of Shimanto CI data should be applied to other metamorphic belts with caution. The goal of establishing a universal correlation between these measures of thermal maturity seems impractical because of problems with error propagation, differences in instrumentation and analytical technique, and because of real physical and chemical changes in the many boundary conditions that affect low-temperature metamorphism.

We base our estimates of maximum burial pressure on measurements of illite/mica b_o lattice spacing. All but one of the b_o values from the Shimanto Belt fall within a range of 9.001Å to 9.023Å, which shows that the bulk Fe+Mg content of the illite is relatively modest. Assuming growth of true metamorphic micas, the illitic minerals conform to the chemical classification of muscovite (senso stricto). The limited amount of phengite substitution is consistent with a metamorphic facies series that is best regarded as the product of an intermediate pressure gradient. Maximum burial pressures were probably less than 2.5 kbar, and maximum burial depths were no greater than about 9 km. The quotient of maximum paleotemperature divided by maximum paleodepth yields a background geothermal gradient of approximately 33°C/km for much of the Eocene/Oligocene strata within Shimanto Belt. Local gradients must have been even higher at the time of peak heating, particularly near Miocene mafic intrusions. A subnormal geothermal gradient, which would be required to promote blueschist-facies metamorphism, obviously did not develop. Thus, the Tertiary Shimanto Belt fails to conform to most general models of subduction-zone metamorphism.

ACKNOWLEDGMENTS

The results of X-ray diffraction analyses provided the basis for a Masters thesis completed by Sin Mok Kang at the University of Missouri. Michael Batek and Robert Waddell assisted in the laboratory. Samples were collected during field-based studies by Jim Hibbard and Lee DiTullio. Partial funding for the laboratory work was provided by the National Science Foundation (Grant No. EAR-8706784). Acknowledgment is also made to the Donors of the Petroleum Research Fund, administered by the American Chemical Society, for partial support of this research

(Grant No. 19187-AC2). We thank J. G. Liou, Martin Frey, and Hannan Kisch for their insightful reviews of the manuscript. In addition, Martin Frey provided stimulating discussions regarding many aspects of very low-grade metamorphism.

REFERENCES CITED

Aldahan, A. A., and Morad, S., 1986, Mineralogy and chemistry of diagenetic clay minerals in Proterozoic sandstones from Sweden: American Journal of Science, v. 286, p. 29–80.

Antia, D.D.J., 1986, Kinetic method for modeling vitrinite reflectance: Geology, v. 14, p. 606–608.

Awan, M. A., and Woodcock, N. H., 1991, A white mica crystallinity study of the Berwyn Hills, North Wales: Journal of Metamorphic Geology, v. 9, p. 765–773.

Barker, C. E., 1983, The influence of time on metamorphism of sedimentary organic matter in selected geothermal systems, western North America: Geology, v. 11, p. 384–388.

—— , 1988, Geothermics of petroleum systems: Implications of the stabilization of kerogen thermal maturation after a geologically brief heating duration at peak temperature, *in* Magoon, L., ed., Petroleum systems of the United States: U.S. Geological Survey Bulletin 1870, p. 26–29.

—— , 1989, Temperature and time in the thermal maturation of sedimentary organic matter, *in* Naeser, N. D., and McCulloh, T. H., eds., Thermal history of sedimentary basins, Methods and case histories: New York, Springer-Verlag, p. 73–98.

—— , 1991, Implications for organic maturation studies of evidence for a geologically rapid increase and stabilization of vitrinite reflectance at peak temperature: Cerro Prieto geothermal field, Mexico: American Association of Petroleum Geologists Bulletin, v. 75, p. 1852–1863.

Barker, C. E., and Halley, R. B., 1986, Fluid inclusion, stable isotope, and vitrinite reflectance evidence for the thermal history of the Bone Spring Limestone, southern Guadalupe Mountains, Texas, *in* Gautier, D. L., ed., Roles of organic matter in sediment diagenesis: Society of Economic Paleontologists and Mineralogists, Special Publication 38, p. 189–203.

Barker, C. E., and Goldstein, R. H., 1990, Fluid-inclusion technique for determining maximum temperature in calcite and its comparison to the vitrinite reflectance geothermometer: Geology, v. 18, p. 1003–1006.

Barker, C. E., and Pawlewicz, M. J., 1986, The correlation of vitrinite reflectance with maximum temperature in humic organic matter, *in* Buntebarth, G., and Stegena, L., eds., Paleogeothermics: New York, Springer-Verlag, p. 79–93.

Barker, C. E., Crysdale, B. L., and Pawlewwicz, M. J., 1986, The relationship between vitrinite reflectance, metamorphic grade, and temperature in the Cerro Prieto, Salton Sea, and East Mesa geothermal systems, Salton Trough, United States and Mexico, *in* Mumpton, F. A., ed., Studies in diagenesis: U.S. Geological Survey Bulletin 1758, p. 83–95.

Blenkinsop, T. G., 1988, Definition of low-grade metamorphic zones using illite crystallinity: Journal of Metamorphic Geology, v. 6, p. 623–636.

Boles, J. R., 1977, Zeolites in low-grade metamorphic rocks, *in* Mumpton, F. A., ed., Mineralogy and geology of natural zeolites: Mineralogical Society of America, Short Course Notes, v. 4, p. 103–135.

Bostick, N. H., Cashman, S., McCulloh, T. H., and Wadell, C. T., 1978, Gradients of vitrinite reflectance and present temperature in the Los Angeles and Ventura basins, California, *in* Oltz, D., ed., Symposium in geochemistry: Low temperature metamorphism of kerogen and clay minerals: Society of Economic Paleontologists and Mineralogists, Pacific Section, p. 65–96.

Briand, B., 1980, Geobarometric application of the b_0 value of K-white mica to the Lot Valley and Middle Cevennes metapelites: Neues Jahrbuch fur Mineralogie Monatshefte, v. 1980, p. 529–542.

Burnham, A. K., and Sweeney, J. J., 1989, A chemical kinetic model of vitrinite maturation and reflectance: Geochimica Cosmochimica Acta, v. 53, p. 2649–2657.

Burst, J. F., 1969, Diagenesis of Gulf Coast clayey sediments and its possible relation to petroleum migration: American Association of Petroleum Geologists Bulletin, v. 53, p. 73–93.

Castano, J. R., and Sparks, D. M., 1974, Interpretation of vitrinite reflectance measurements in sedimentary rocks and determination of burial history using vitrinite reflectance and authigenic minerals: Geological Society of America Special Paper 153, p. 31–52.

Cloos, M., 1983, Comparative study of mélange matrix and metashales from the Franciscan subduction complex with the basal Great Valley sequence, California: Journal of Geology, v. 91, p. 291–306.

Duba, D., and Williams-Jones, A. E., 1983, The application of illite crystallinity, organic matter reflectance, and isotopic techniques to mineral exploration: A case study in southwestern Gaspé, Quebec: Economic Geology, v. 78, p. 1350–1363.

Dumitru, T. A., 1988, Subnormal geothermal gradients in the Great Valley forearc basin, California, during Franciscan subduction: A fission track study: Tectonics, v. 7, p. 1201–1222.

Dunoyer de Segonzac, G., 1970, The transformation of clay minerals during diagenesis and low-grade metamorphism: A review: Sedimentology, v. 15, p. 281–346.

Eberl, D. D., Srodon, J., Lee, M., Nadeau, P. H., and Northrop, R. H., 1987, Sericite from the Silverton Caldera, Colorado: Correlation among structure, composition, origin, and particle thickness: American Mineralogist, v. 72, p. 914–934.

Ernst, W. G., 1963, Significance of phengitic micas in low-grade schists: American Mineralogist, v. 48, p. 1357–1373.

Fettes, D. J., Graham, C. M., Sassi, F. P., and Scolari, A., 1976, The basal spacing of potassic white micas and facies series variation across the Caledonides: Scottish Journal of Geology, v. 12, p. 227–236.

Freed, R. L., and Peacor, D. R., 1989, Geopressured shale and sealing effect of smectite to illite transition: American Association of Petroleum Geologists Bulletin, v. 73, p. 1223–1232.

Frey, M., 1970, The step from diagenesis to metamorphism in pelitic rocks during Alpine orogenesis: Sedimentology, v. 15, p. 261–279.

—— , 1987, Very low-grade metamorphism of clastic sedimentary rocks, *in* Frey, M., ed., Low temperature metamorphism: New York, Chapman and Hall, p. 9–58.

Frey, M., and 7 others, 1980, Very low-grade metamorphism in external parts of the Central Alps: Illite crystallinity, coal rank and fluid inclusion data: Eclogae Geologica Helvetiae, v. 73, p. 173–203.

Frey, M., Hunziker, J. C., Jager, E., and Stern, W. B., 1983, Regional distribution of white K-mica polymorphs and their phengite content in the Central Alps: Contributions to Mineralogy and Petrology, v. 83, p. 185–197.

Gretener, P. E., and Curtis, C. D., 1982, Role of temperature and time on organic metamorphism: American Association of Petroleum Geologists Bulletin, v. 66, p. 1124–1129.

Guidotti, C. V., and Sassi, F. P., 1976, Muscovite as a petrogenetic indicator mineral in pelitic schists: Neues Jahrbuch fur Mineralogie Abhandlungen, v. 127, p. 97–142.

—— , 1986, Classification and correlation of metamorphic facies series by means of muscovite b_0 data from low-grade metapelites: Neues Jahrbuch fur Mineralogie Abhandlungen, v. 153, p. 363–380.

Guthrie, J. M., Houseknecht, D. W., and Johns, W. D., 1986, Relationships among vitrinite reflectance, illite crystallinity, and organic geochemistry in Carboniferous strata, Ouachita Mountains, Oklahoma and Arkansas: American Association of Petroleum Geologists Bulletin, v. 70, p. 26–33.

Hesse, R., and Dalton, E., 1991, Diagenetic and low-grade metamorphic terranes of Gaspé Peninsula related to the geological structure of the Taconian and Acadian orogenic belts, Quebec Appalachians: Journal of Metamorphic Geology, v. 9, p. 775–790.

Hoffman, J., and Hower, J., 1979, Clay mineral assemblages as low grade metamorphic geothermometers: Application to the thrust faulted Disturbed Belt of Montana, U.S.A., *in* Scholle, P. A., and Schluger, P. R., eds., Aspects of diagenesis: Society of Economic Paleontologists and Mineralogists, Special Publication 26, p. 55–80.

Hood, A., Gutjahr, C.C.C., and Heacock, R. L., 1975, Organic metamorphism and the generation of petroleum: American Association of Petroleum Geologists Bulletin, v. 59, p. 986–996.

Houseknecht, D. W., Johns, W. D., and Guthrie, J. M., 1987, Relationships among vitrinite reflectance, illite crystallinity, and organic geochemistry in Carboniferous strata, Ouachita Mountains, Oklahoma and Arkansas: Reply: American Association of Petroleum Geologists Bulletin, v. 71, p. 347.

Hower, J., Eslinger, E. V., Hower, M. E., and Perry, E. A., 1976, Mechanism of burial metamorphism of argillaceous sediment: Part I, Mineralogical and chemical evidence: Geological Society of America Bulletin, v. 87, p. 725–737.

Hunt, J. M., Lewan, M. D., and Hennet, R. J-C., 1991, Modeling oil generation with time-temperature index graphs based on the Arrhenius equation: American Association of Petroleum Geologists Bulletin, v. 75, p. 795–807.

Hunziker, J. C., and 8 others, 1986, The evolution of illite to muscovite: Mineralogical and isotopic data from the Glarus Alps, Switzerland: Contributions to Mineralogy and Petrology, v. 92, p. 157–180.

Issler, D. R., 1984, Calculation of organic maturation levels for offshore eastern Canada—Implications for general application of Lopatins method: Canadian Journal of Earth Science, v. 21, p. 283–304.

Kang, S. M., 1990, Clay mineralogy, diagenesis, and metamorphism of the Upper Shimanto Belt, Muroto Peninsula, Shikoku, Japan [M.S. thesis]: Columbia, University of Missouri, 93 p.

Kemp, A.E.S., Oliver, G.H.J., and Baldwin, J. R., 1985, Low-grade metamorphism and accretion tectonics: Southern Uplands terrain, Scotland: Mineralogical Magazine, v. 49, p. 335–344.

Kisch, H. J., 1980, Illite-crystallinity and coal rank associated with lowest-grade metamorphism of the Taveyanne graywacke in the Helvetic zone of the Swiss Alps: Eclogae Geologica Helvetiae, v. 73, p. 753–777.

——, 1983, Mineralogy and petrology of burial diagenesis (burial metamorphism) and incipient metamorphism in clastic rocks, in Larsen, G., and Chilingar, G. V., eds., Diagenesis in sediments and sedimentary rocks, Part 2; Developments in sedimentology, v. 25B: Amsterdam, Elsevier, p. 289–493.

——, 1987, Correlation between indicators of very low-grade metamorphism, in Frey, M., ed., Low temperature metamorphism: New York, Chapman and Hall, p. 227–300.

——, 1990, Calibration of the anchizone: A critical comparison of illite 'crystallinity' scales used for definition: Journal of Metamorphic Geology, v. 8, p. 31–46.

——, 1991, Development of slaty cleavage and degree of very-low grade metamorphism: A review: Journal of Metamorphic Geology, v. 9, p. 735–750.

Kisch, H. J., and Frey, M., 1987, Appendix: Effect of sample preparation on the measured 10Å peak width of illite (illite 'crystallinity'), in Frey, M., ed., Low temperature metamorphism: New York, Chapman and Hall, p. 301–304.

Kreutzberger, M. E., and Peacor, D. R., 1988, Behavior of illite and chlorite during pressure solution of shaly limestone of the Kalkberg Formation, Catskill, New York: Journal of Structural Geology, v. 10, p. 803–811.

Kubler, B., 1967, La cristallinite de l'illite et les zones tout a fait superieures de metamorphisme, in Etages tectoniques, Colloque de Neuchatel 1966: Neuchatel, Suisse, A la Baconniere, p. 105–121.

——, 1968, Evaluation quantitative du metamorphisme par la cristallinite de l'illite: Bulletin Centre de Recherches de Pau, SNPA, v. 2, p. 385–397.

Laughland, M. M., Underwood, M. B., and Wiley, T. J., 1990, Thermal maturity, tectonostratigraphic terranes, and regional tectonic history: An example from the Kandik area, east-central Alaska, in Nuccio, V. F., and Barker, C. E., eds., Applications of thermal maturity studies to energy exploration: Society of Economic Paleontologists and Mineralogists, Rocky Mountain Section, p. 97–111.

Liou, J. G., 1971, P-T stabilities of laumontite, wairakite, lawsonite, and related minerals in the system CaAl$_2$Si$_2$O$_8$-SiO$_2$-H$_2$O: Journal of Petrology, v. 12, p. 379–411.

Liou, J. G., Maruyama, S., and Cho, M., 1985, Phase equilibria and mineral parageneses of metabasites in low-grade metamorphism: Mineralogical Magazine, v. 49, p. 321–333.

——, 1987, Very low-grade metamorphism of volcanic and volcaniclastic rocks—Mineral assemblages and mineral facies, in Frey, M., ed., Low temperature metamorphism: New York, Chapman and Hall, p. 59–113.

Maxwell, D. T., and Hower, J., 1970, High-grade diagenesis and low-grade metamorphism of illite in the Precambrian Belt Series: American Mineralogist, v. 52, p. 843–857.

McCulloh, T. H., Frizzell, V. A., Jr., Stewart, R. J., and Barnes, I., 1981, Precipitation of laumontite with quartz, thenardite, and gypsum at Sespe Hot Springs, western Transverse Ranges, California: Clays and Clay Minerals, v. 29, p. 353–364.

Merriman, R. J., and Roberts, B., 1985, A survey of white mica crystallinity and polytypes in peletic rocks of Snowdonia and Llyn, North Wales: Mineralogical Magazine, v. 49, p. 305–319.

Middleton, M. F., 1982, Tectonic history from vitrinite reflectance: Geophysical Journal of the Royal Astronomical Society, v. 68, p. 121–132.

Mitra, G., and Yonkee, W. A., 1985, Relationship of spaced cleavage to folds and thrusts in the Idaho-Utah-Wyoming thrust belt: Journal of Structural Geology, v. 7, p. 361–373.

Moore, D. M., and Reynolds, R. C., Jr., 1989, X-ray diffraction and the identification and analysis of clay minerals: New York, Oxford University Press, 332 p.

Nadeau, P. H., and Reynolds, R. C., Jr., 1981, Burial and contact metamorphism in the Mancos Shale: Clays and Clay Minerals, v. 29, p. 249–259.

Naef, U., and Stern, W. B., 1982, Some critical remarks on the analysis of phengite and paragonite components in muscovite by X-ray diffractometry: Contributions to Mineralogy and Petrology, v. 79, p. 355–360.

Nyk, R., 1985, Illite crystallinity in Devonian slates of the Meggen Mine (Rhenish Massif): Neues Jahrbuch fur Mineralogie Monatshefte, v. 1985, p. 268–276.

Offler, R., and Prendergast, E., 1985, Significance of illite crystallinity and b$_o$ values of K-white mica in low-grade metamorphic rocks, North Hill End Synclinorium, New South Wales, Australia: Mineralogical Magazine, v. 49, p. 357–364.

Ogunyomi, O., Hesse, R., and Heroux, Y., 1980, Pre-orogenic and synorogenic diagenesis and anchimetamorphism in lower Paleozoic continental margin sequences of the northern Appalachians in and around Quebec City, Canada: Bulletin of Canadian Petroleum Geology, v. 28, p. 559–577.

Padan, A., Kirsch, H. J., and Shagam, R., 1982, Use of the lattice parameter b$_o$ of dioctahedral illite/muscovite for the characterization of P/T gradients of incipient metamorphism: Contributions to Mineralogy and Petrology, v. 79, p. 85–95.

Perry, E., and Hower, J., 1970, Burial diagenesis in Gulf Coast pelitic sediments: Clays and Clay Minerals, v. 18, p. 165–177.

Pevear, D. R., Williams, V. E., and Mustoe, G. E., 1980, Kaolinite, smectite, and K-rectorite in bentonites: Relation to coal rank at Tulameen, British Columbia: Clays and Clay Minerals, v. 28, p. 241–254.

Pollastro, R. M., and Barker, C. E., 1986, Application of clay-mineral, vitrinite reflectance, and fluid inclusion studies to the thermal and burial history of the Pinedale anticline, Green River Basin, Wyoming, in Gautier, D. L., ed., Roles of organic matter in sediment diagenesis: Society of Economic Paleontologists and Mineralogists, Special Publication 38, p. 73–83.

Price, L. C., 1983, Geologic time as a parameter in organic metamorphism and vitrinite reflectance as an absolute paleogeothermometer: Journal of Petroleum Geology, v. 6, p. 5–38.

Primmer, T. J., 1985, A transition from diagenesis to greenschist facies within a major Variscan fold/thrust complex in south-west England: Mineralogical Magazine, v. 49, p. 365–374.

Pytte, A. M., and Reynolds, R. C., 1989, The thermal transformation of smectite to illite, in Naeser, N. D., and McCulloh, T. H., eds., Thermal history of sedimentary basins: New York, Springer-Verlag, p. 133–140.

Reynolds, R. C., Jr., and Hower, J., 1970, The nature of interlayering in mixed-layer illite-montmorillonites: Clays and Clay Minerals, v. 18, p. 25–36.

Ritter, U., 1984, The influence of time and temperature on vitrinite reflectance: Organic Geochemistry, v. 6, p. 473–480.

Roberson, H. E., and Lahann, R. W., 1981, Smectite to illite conversion rates:

Effects of solution chemistry: Clays and Clay Minerals, v. 29, p. 129–135.

Roberts, B., and Merriman, R. J., 1985, The distinction between Caledonian burial and regional metamorphism in metapelites from North Wales: An analysis of isocryst patterns: Journal Geological Society of London, v. 142, p. 615–624.

Robinson, D., and Bevens, R. E., 1986, Incipient metamorphism in the lower Palaeozoic marginal basin of Wales: Journal of Metamorphic Geology, v. 4, p. 101–113.

Robinson, D., Warr, L. N., and Bevins, R. E., 1990, The illite 'crystallinity' technique: A critical appraisal of its precision: Journal of Metamorphic Geology, v. 8, p. 333–344.

Sands, C. D., and Drever, J. I., 1977, Authigenic laumontite in deep-sea sediments, *in* Sand, L. B., and Mumpton, F. A., eds., Natural zeolites, occurrence, properties, use: New York, Pergamon Press, p. 269–275.

Sassi, F. P., and Scolari, A., 1974, The b_0 value of the potassic white micas as a barometric indicator in low-grade metamorphism of pelitic schists: Contributions to Mineralogy and Petrology, v. 45, p. 143–152.

Smart, G., and Clayton, T., 1985, The progressive illitization of interstratified illite-smectite from Carboniferous sediments of northern England and its relationship to organic maturity indicators: Clay Minerals, v. 20, p. 455–466.

Stalder, P. J., 1979, Organic and inorganic metamorphism in the Taveyannaz Sandstone of the Swiss Alps and equivalent sandstones in France and Italy: Journal of Sedimentary Petrology, v. 49, p. 463–482.

Suggate, R. P., 1982, Low-rank sequences and scales of organic metamorphism: Journal of Petroleum Geology, v. 4, p. 377–392.

Suzuki, T., and Hada, S., 1983, Accretionary mélange of Cretaceous age in the Shimanto Belt in Japan, *in* Hashimoto, M., and Uyeda, S., eds., Accretion tectonics in circum-Pacific regions: Tokyo, Terra Scientific Publishing Company, p. 219–230.

Sweeney, J. J., and Burnham, A. K., 1990, Evaluation of a simple model of vitrinite reflectance based on chemical kinetics: American Association of Petroleum Geologists Bulletin, v. 74, p. 1559–1570.

Tilley, B. J., Nesbitt, B. E., and Longstaffe, F. J., 1989, Thermal history of Alberta deep basin: Comparative study of fluid inclusions and vitrinite reflectance data: American Association of Petroleum Geologists Bulletin, v. 73, p. 1206–1222.

Toriumi, M., and Teruya, J., 1988, Tectono-metamorphism of the Shimanto Belt: Modern Geology, v. 12, p. 303–324.

Underwood, M. B., and 8 others, 1991, Correlations among paleotemperature indicators within orogenic belts: Examples from pelitic rocks of the Franciscan Complex (California), the Shimanto Belt (Japan), and the Kandik Basin (Alaska): EOS Transactions of the American Geophysical Union, v. 72, p. 549.

Velde, B., 1965, Phengite micas: Synthesis, stability, and natural occurrence: American Journal of Science, v. 263, p. 886–913.

Waples, D. W., 1980, Time and temperature in petroleum exploration: Application of Lopatin's method to petroleum exploration: American Association of Petroleum Geologists Bulletin, v. 64, p. 916–926.

Weaver, C. E., 1960, Possible uses of clay minerals in search for oil: American Association of Petroleum Geologists Bulletin, v. 44, p. 1505–1518.

Weber, K., 1972, Notes on determination of illite crystallinity: Neues Jahrbuch fur Mineralogie Monatschefte, v. 1972, p. 267–276.

Wood, D. A., 1988, Relationships between thermal maturity indices calculated using Arrhenius equation and Lopatin method: Implication for petroleum exploration: American Association of Petroleum Geologists Bulletin, v. 72, p. 115–135.

Wright, N.J.R., 1980, Time, temperature, and organic maturation: The evolution of rank within a sedimentary pile: Journal of Petroleum Geology, v. 2, p. 411–425.

Yang, C., and Hesse, R., 1991, Clay minerals as indicators of diagenetic and anchimetamorphic grade in an overthrust belt, External Domain of southern Canadian Appalachians: Clay Minerals, v. 26, p. 211–231.

Zen, E.-An., and Thompson, A. B., 1974, Low grade regional metamorphism: Mineral equilibrium relations: Annual Review of Earth and Planetary Sciences, v. 2, p. 179–212.

MANUSCRIPT ACCEPTED BY THE SOCIETY APRIL 24, 1992

Geological Society of America
Special Paper 273
1993

Thermal maturity and constraints on deformation from illite crystallinity and vitrinite reflectance in the shallow levels of an accretionary prism: Eocene-Oligocene Shimanto Belt, southwest Japan

Lee DiTullio*
Department of Geological Sciences, Brown University, Providence, Rhode Island 02912
Matthew M. Laughland*
Department of Geological Sciences, University of Missouri, Columbia, Missouri 65211
Tim Byrne*
Ocean Research Institute, University of Tokyo, Tokyo, 164 Japan

ABSTRACT

Detailed vitrinite reflectance and illite crystallinity paleothermal data from a portion of the Eocene-Oligocene Shimanto accretionary prism in southwest Japan show that: (1) peak paleotemperatures range from ~200°C (diagenetic zone) in the northern part of the study area to ~300°C (anchizone) 10 km to the south. Preliminary b_o data from illite (Underwood and others, this volume) suggest that these rocks were buried no more than 10 km. This in turn implies that the Eocene-Oligocene Shimanto Belt experienced a relatively high paleogeothermal gradient of at least 30 to 40°C/km. (2) Shales with the best-developed pressure solution cleavages also record the highest temperatures suggesting that cleavage formation and peak heating were coeval. Structural differences with younger rocks and K-Ar dates on cleavage from correlative rocks in neighboring Ashizuri Peninsula that range from 43 to 18 Ma, with most falling in the range 34 to 26 Ma, suggest that major penetrative deformation of these rocks occurred in the middle to late Oligocene. Peak heating may have commenced in the late Eocene to earliest Oligocene when plate reconstructions suggest subduction of young crust (<10 m.y. old) of the fused Kula/Pacific Plate. An elevated paleogeothermal gradient may also have been maintained during the period of cleavage formation (34 to 26 Ma) by the decrease in subduction rates documented for the middle Oligocene. (3) Significant late Miocene or younger block-faulting, uplift, and flexure of the study area postdated the attainment of peak temperatures and substantially modified patterns of peak paleotemperatures. The most important displacement occurred along the subvertical contact fault between the Eocene-Oligocene ($4\%R_m$) and Oligocene-Miocene ($1.5\%R_m$) Shimanto Belts.

INTRODUCTION

The relation between metamorphism and deformation in accretionary prisms, particularly at shallow depths, is not well known. Classic studies of metamorphism within accretionary terranes have dealt with high-pressure phases formed at depths in excess of 10 km (e.g., Ernst, 1970; Cloos, 1983). Relatively few studies have analyzed rocks of subgreenschist facies such as would occur in the very shallow levels of a prism (e.g., Landis and Coombs, 1967; Moore and Allwardt, 1980; Kisch, 1980). In large part this is because many accreted sequences are typically

---

*Present addresses: DiTullio, 326 12th St., Apt. 2L, Brooklyn, New York 11215; Laughland, Mobil Research and Development, Exploration, and Producing Technical Center, 3000 Pegasus Park Dr., Dallas, Texas 75265-0232; Byrne, Department of Geology and Geophysics, University of Connecticut, Storrs, Connecticut 06269.

DiTullio, L., Laughland, M. M., and Byrne, T., 1993, Thermal maturity and constraints on deformation from illite crystallinity and vitrinite reflectance in the shallow levels of an accretionary prism: Eocene-Oligocene Shimanto Belt, southwest Japan, *in* Underwood, M. B., ed., Thermal Evolution of the Tertiary Shimanto Belt, Southwest Japan: An Example of Ridge-Trench Interaction: Boulder, Colorado, Geological Society of America Special Paper 273.

dominated by clastic turbidites that are compositionally inappropriate for the development of diagnostic metamorphic mineral assemblages (i.e., they lack zeolites) and/or are too fine-grained for the use of standard optical petrologic techniques. Consequently, the thermal history of most accreted rocks must be investigated by other techniques such as fluid inclusions, illite crystallinity, or vitrinite reflectance. Only recently have these techniques been used to link the thermal history with broad aspects of the structural evolution of these complex areas (Kubler, 1970; Kemp and others, 1985; Underwood and Howell, 1987; Vrolijk, 1987; Mori and Taguchi, 1988; Underwood and others, 1989).

In this study, we use detailed vitrinite reflectance and illite crystallinity data to (1) identify phases of deformation that predated, coincided with, and postdated peak heating; and (2) constrain the temperature conditions of the penetrative stage of deformation in a portion of the Eocene-Oligocene Shimanto accretionary prism in southwest Japan. The study area on the Muroto Peninsula of Shikoku Island (see Fig. 1) is an excellent place in which to carry out research of this type because the rocks are well exposed, their ages are relatively well controlled paleontologically, and they belong to one of the longest-lived and best-studied convergent margins anywhere (for an overview see Taira

and others, 1980, 1982, 1988, and references therein). In addition, the rocks at Muroto Peninsula have recently been the subject of a detailed structural examination (DiTullio and Byrne, 1990; DiTullio, 1989), which we now briefly summarize in the next few paragraphs.

The Eocene-Oligocene Shimanto Belt in the Muroto area is composed predominantly of coherent sandstone/shale turbidites that contain slump deposits, sandstone dikes, and trenchward-vergent structures, all of which are consistent with deformation and accretion at relatively shallow structural levels. A shallow burial ($\leqslant 10$ km) for these rocks is suggested by pressure estimates derived from the b_0 spacing of illite as reported in Underwood, Laughland, and Kang (this volume) and by the apparent absence of high-pressure minerals. Paleontologic ages of radiolaria and foraminifera in the coherent turbidites are Eocene with the youngest strata falling in the range middle Eocene to early Oligocene. For brevity, we refer to these rocks as the "Eocene Shimanto belt" or "Eocene section" in the rest of this paper. In addition to the coherent strata, a shale-matrix mélange containing blocks of hemipelagic shale, pillow basalt, and chert from the ocean floor is present. Fossils from both the blocks and matrix of this mélange are also Eocene in age, suggesting that young oceanic crust was being subducted during deposition and accretion of

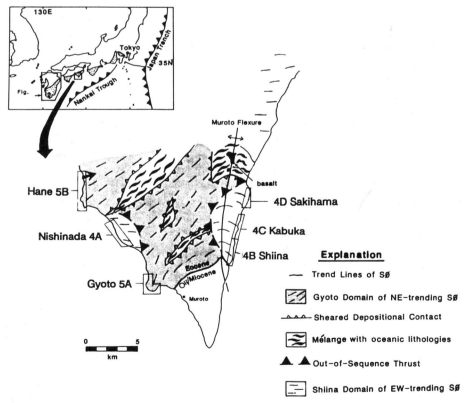

Figure 1. Location and geologic map of the Muroto Peninsula study area, Shikoku, Japan. Stippled area on inset shows the distribution of the Cretaceous through Oligocene/Miocene Shimanto Belt in southwest Japan. The locations of later figures and important place names are indicated.

the coherent trench-fill turbidites (Taira, 1985; Taira and others, 1988; and see discussion).

In the Muroto area, DiTullio and Byrne (1990) have divided the deformation history into three broad progressive stages: (1) accretion-related prelithification imbrication with associated clastic diking and meter-scale folding; (2) major intraprism shortening expressed by regional-scale folding and especially pressure solution cleavage; and (3) high-angle faulting and uplift. The first stage of deformation produced only mesoscale or smaller clastic dikes, faults, and folds that cannot be discriminated in the paleothermal data. These structures are therefore not discussed further in this paper; they do, however, clearly predate larger premetamorphic structures.

Stage 2 deformation produced regional-scale folds and penetrative pressure solution cleavages that together record a ~50° counterclockwise rotation in shortening direction. This rotation in shortening direction was recognized because the accreted rocks are distributed in two fault-bounded domains with differing strikes and structural histories (see Fig. 1). Crosscutting relations between clastic dikes, mesoscale faults, regional-scale folds, and cleavage indicate that the prerotation phase of shortening (D_1) was north-south directed, and the later phase (D_2) was northwest-southeast directed. The Shiina domain is characterized by a suite of clastic dikes and east-striking faults, regional-scale folds, and two cleavages, which together define D_1. In contrast, the Gyoto domain, which also contains D_1 faults and sandstone dikes, is characterized by northeast-striking regional-scale folds and two cleavages of D_2.

In order to explain the differing deformation histories of these domains, their relative structural positions, and the distribution of the tectonic mélange, DiTullio and Byrne (1990) proposed that since incorporation into the prism the two domains have always been separated by a major, originally subhorizontal, fault zone. DiTullio and Byrne theorized that the Gyoto domain was initially thrust beneath, and therefore isolated from, the Shiina domain as folds and cleavage developed in the latter during D_1. At the onset of D_2, a major out-of-sequence thrust (OST) carried the Gyoto domain to higher structural levels of the prism where it was folded and cleaved during northwest-directed shortening. This OST also carried the tectonic mélange at the base of the hanging-wall Gyoto domain and, because of the ocean floor nature of the blocks within the mélange, this suggests that the OST rooted in the décollement zone beneath the prism (see Fig. 8). The resultant relatively simple nappe geometry has unfortunately been significantly modified by further shortening and Stage 3 high-angle faulting.

We estimate the timing of Stage 2 penetrative deformation in these rocks to approximately span the Oligocene. The lower bound reflects the fact that some of the affected strata may be as young as early Oligocene, while the upper bound is constrained by the opposite (landward) vergence of penetrative deformation in the younger Oligocene-Miocene section to the south (Hibbard and Karig, 1990). K-Ar dates on cleavage from correlative rocks in neighboring Ashizuri Peninsula support this timing: the dates

range from 43 to 18 Ma with most falling in the range 34 to 26 Ma (Agar and others, 1989), suggesting that major penetrative deformation of these rocks occurred in the middle to late Oligocene. However, the relationship of cleavage formation to peak heating is not unequivocally demonstrated. Therefore, the timing of the peak thermal event in the Eocene rocks is controversial and may be as young as the mid-Miocene thermal event; our more detailed arguments in favor of an Oligocene syntectonic thermal event are presented below in the "Timing of peak heating" section of the discussion.

The third and final stage of deformation in the Eocene section included development of the east-trending Shiina-Narashi fault (which forms the contact with the Oligocene-Miocene rocks to the south) and the north-trending Muroto flexure (see Fig. 1). The flexure, which affects both the Eocene and younger rocks, occurs along the east side of the study area where it has rotated regional-scale fold axes from subhorizontal plunges on the west coast to subvertical plunges on the east coast. The flexure therefore appears to be a major, regional-scale structure and is similar to other cross folds in the Shimanto Belt that record Neogene, along-strike shortening. Relative uplift along these structures is inferred from the higher grade (coal rank) of rocks to the north of the Shiina-Narashi fault and locally to the west of the Muroto flexure.

The timing of Stage 3 deformation ranges from late Miocene to Present. Both the Shiina-Narashi fault and the Muroto flexure affect the Oligocene-Miocene section and therefore can be no older than late Miocene. However, the Shiina-Narashi fault appears to be folded by the Muroto flexure and thus is considered to be older than the flexure. The Muroto flexure may be active today because similar, parallel structures are presently forming offshore (Okamura, 1990; Sugiyama, 1989a). The east-west shortening accommodated by the Muroto flexure, and other similar cross folds along strike, may have begun in the middle Pliocene when the relative motion of the Philippine Sea Plate with respect to southwest Japan changed from north-northwest to west-northwest (Sugiyama, 1989b).

In the following sections we present a brief review of the illite crystallinity and vitrinite reflectance methods, and then discuss the paleothermal results that allow the peak paleotemperatures of the Eocene rocks to be determined. We then use these observations, together with structural relations, to constrain the timing of heating. Finally, we attempt to link that combined structural/metamorphic history to the regional plate-tectonic setting.

METHODS

Illite crystallinity and vitrinite reflectance methods were deemed most suitable for the Eocene section because it was shale-rich and contained sufficient organic matter to have measurable vitrinite as well as a measurable smectite/illite component. Reconnaissance search for fluid inclusions in Stage 2 quartz veins revealed that they were water-filled one-phase inclusions, almost all of which were too small for measurement. Note that in this paper "grade," which is inferred primarily from vitrinite rank, refers to changes in temperature not pressure and is used interchangeably with thermal maturity.

Illite crystallinity

Illite crystallinity is a technique that uses the width of the basal ~10Å (001) X-ray diffraction peak at half height as an indicator of diagenetic/metamorphic grade (Kubler, 1964). The "crystallinity index" (CI) is typically measured in units of $\Delta°2\theta$ and its value decreases with increasing crystallinity. At low grade, the material being measured is typically a mixed-layer clay containing illite and expandable smectite. Overlap from the basal reflection of smectite gives the illite 10Å peak an asymmetric low-angle tail resulting in a spuriously high CI values (low crystallinities). It is standard procedure therefore to glycolate samples before they are run in order to remove this interference effect (Kisch, 1987). This was done for all 88 samples used in this study. For comparison, untreated (air-dried) splits from ~60 of the samples were also measured. For the air-dried samples, general trends in crystallinity were the same as for the glycolated samples, but the broadening effect of the expandable smectite component was apparent in the lower grade samples (see Fig. 2).

Although the transformation of smectite to illite is affected by a variety of factors including temperature, time, the chemistry of the host rock and the circulating fluids, and possibly deformation, Frey (1987) concluded that temperature appears to be the dominant control. Nevertheless, it is clear from studies in hydrothermal systems and zones of contact metamorphism that more time is necessary for the illite to reach equilibrium with its surroundings than for vitrinite (Teichmuller, 1987). Deformation also apparently can be an important control and many workers have noted a strong correlation between higher crystallinity (lower CI values) and well-developed cleavage (e.g., Merriman and Roberts, 1985; Kreutzberger and Peacor, 1988; and this study). A genetic relationship has not been demonstrated (e.g., compare Kubler, 1967; and Roberts and Merriman, 1985), although we discuss evidence for such a link between the two below. In this study, the uniformity in composition and grain size of the samples (all samples are from shaly interbeds of arkosic turbidites) gives us confidence that the variations in CI values reflect real paleotemperature differences and not the effects of compositional variations. Locally, crystallinity appears to have been enhanced in some fault zones and retarded where cleavage formation was inhibited; these areas are discussed in the Results section below.

On the basis of their CI values from glycolated clays, low-

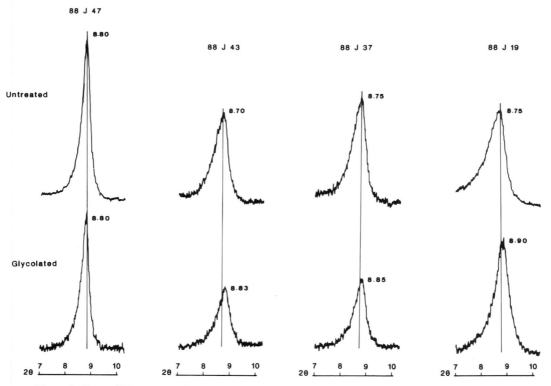

Figure 2. X-ray diffractograms from four selected samples covering the range of illite crystallinities in untreated and glycolated states. 88J47 = $0.37\Delta°2\theta$ unt. and $0.34\Delta°2\theta$ gly.; 88J43 = $0.60\Delta°2\theta$ unt. and $0.49\Delta°2\theta$ gly.; 88J37 = $0.58\Delta°2\theta$ unt. and $0.52\Delta°2\theta$ gly.; and 88J19 = $0.76\Delta°2\theta$ unt. and $0.62\Delta°2\theta$ gly. All but the highest crystallinity sample show the low-angle tail produced by expandable-layer clays in the untreated state and a narrowing and shift of the 10Å peak on glycolation.

grade metamorphic rocks are divided into three zones with increasing grade: diagenetic ($>0.42\Delta°2\theta$), anchizone (0.42 to $0.25\Delta°2\theta$), and epizone ($<0.25\Delta°2\theta$) (Kisch, 1987; Blenkinsop, 1988). The anchizone (transition into lowermost greenschist facies metamorphism) is defined only on the basis of CI values (Kubler, 1968). From coexisting mineral assemblages, coal rank, and fluid inclusion studies in many parts of the world, the diagenetic/anchizone boundary appears to correspond to temperatures of approximately 200 to 250°C and vitrinite reflectances of about 2.5 to $3.1\%R_m$ (Underwood, Laughland, and Kang, this volume). The anchizone/epizone boundary, although not nearly as well constrained, appears to correspond to temperatures of 300 to 350°C and reflectances of $>4.0\%R_m$ (Kisch, 1987). It is also found that the anchizone is typically associated with prehnite-pumpellyite facies mineral assemblages and may reflect similar metamorphic conditions.

For this study, CI was measured on glycolated <2 micron fractions of shale: 36 samples coming from the interior were run at Kochi University (see Fig. 3 for locations) and 52 samples from two coastal transects (see Figs. 4 and 5 below for locations) were run at the University of Missouri (see Underwood, Laughland, and Kang, this volume, for analytical techniques). The Kochi samples were run on an Rigaku X-ray diffractometer using CuKα radiation and a Ni filter. Settings were: 30kV and 15 mA; 1° divergence and scatter slits; and a 0.15-mm receiving slit. Scan speed was $0.5°2\theta$/min and chart speed was 1 cm/min. Each peak was scanned from ~6.5° to $10.5°2\theta$ four times and the values averaged. In tests to check reproducibility, it was found that the combined effects of measurement error, machine fluctuation, and differences in sample preparation produced errors of at most $\pm0.05\Delta°2\theta$. Thus, we only consider differences in CI of $0.1\Delta°2\theta$ as geologically significant.

We note here that CI values locally varied by more than $0.1\Delta°2\theta$ within small areas. In places this variation could be reasonably attributed to shearing or increased fluid flow in fault zones; in other places, however, there was no obvious explanation. Despite this unsatisfying variability, we emphasize that, on average, the pattern of increases and decreases in thermal maturity indicated by illite crystallinity is perfectly matched with that of vitrinite reflectance (see Fig. 6). Moreover, the thermal maturity indicated by the crystallinity values is in no case higher than that indicated by a corresponding vitrinite reflectance measurement, although it is commonly lower. Our results therefore suggest that illite crystallinity can be a reliable paleothermal tool if several data are averaged and that even single datum will never overestimate the thermal maturity of a sample.

Vitrinite reflectance

The vitrinite reflectance technique uses the increase in reflectivity of polished vitrinite macerals (organic matter) with temper-

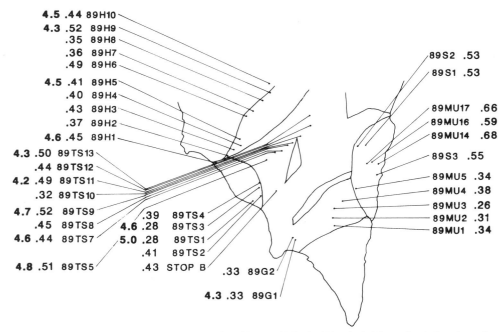

Figure 3. Location map and values for CI (in $\Delta°2\theta$) and VR (in $\%R_m$; in bold numbers) data from the interior of the study area. Note that coal ranks in the interior are consistently high and include the highest grade samples in the study area; however, many corresponding crystallinity values are surprisingly high (low grade)—this may be due to weathering of the samples. Even so, note that CI values east of the inferred trace of the Muroto flexure and north of the Flat Rock fault are still significantly lower grade, consistent with coastal thermal and structural data suggesting that this area exposes a shallower part of the prism than the rest of the Eocene section.

ature as a measure of thermal maturation (see the recent review by Teichmuller, 1987). Reflectance is measured in oil from 50 randomly oriented particles per sample and is reported as the average or mean reflectance ($\%R_m$); thus, in contrast to CI values, VR values increase with increasing temperature. For this part of the study, 76 splits from samples collected along the east and west coasts of the Muroto study area and an additional 15 samples from the interior were analyzed at the University of Missouri. See Laughland and Underwood (this volume) for all details regarding these data and the method and measurement techniques.

The increase in coal rank with temperature is not linear, and for many years it was assumed to follow first-order reaction kinetics with the reaction rate doubling for every 10°C increase in temperature (e.g., Waples, 1980). More recent work, however, indicates that the kinetics are far more complicated with the reaction rate varying with type of maceral and the temperature range being considered (e.g., Tissot and others, 1987; Sweeney and Burnham, 1990). In general, however, Price (1983) and Barker (1983, 1988, 1989) have argued, in part from comparisons of reflectance data from geothermally active areas and first-cycle sedimentary basins, that time plays a much less important role than implied by especially the early kinetic models and that 10^6 years, or in nongeothermal systems at most 10^7 years, is a reasonable upper limit on effective heating time necessary for vitrinite to accurately record ambient temperatures (see also Bostick, 1984; Kisch, 1987). In any case, we do not posssess a sufficiently detailed knowledge of the burial history of the rocks in the Muroto area to credibly use any kinetic model and therefore, like the other papers in this volume, we use Barker's (1988) time-independent correlation scheme for coal rank and temperature as the best with which to make our paleotemperature estimates. Laughland and Underwood (this volume) consider these estimates to have an error of ±30°C. We note that for the types of geothermal gradients we infer in the Eocene Shimanto Belt, Barker's model gives temperature estimates very similar to those of the more complex EASY $\%R_o$ model of Sweeny and Burnham for a burial time at maximum temperature of 10 m.y.

RESULTS

As our goal is to integrate the thermal and structural histories of the Eocene section, we present the paleothermal results in terms of the stages of deformation outlined above. Detailed maps and cross sections (Figs. 4 and 5) allow us to integrate the paleothermal and structural data and carefully discuss the relation of heating to the Stage 2 folds, cleavage, and the OST. The paleothermal data are summarized in two graphs along the coastal transects (Fig. 6) to give an overall view of the across-strike patterns. These transects clearly show how the "isograds" of thermal metamorphism crosscut Stage 2 structures but are offset by the Stage 3 Shiina-Narashi fault. The generally excellent correlation between the CI and VR methods as indicators of relative differences in grade is also apparent from Figure 6. The overall map distribution of grade illustrating the effects of the major Stage 3 faults (Fig. 9) is shown in Figure 10.

Stage 2 deformation

In this section, we first examine relations between thermal grade and folds in the Shiina and Gyoto domains, and then we discuss the relation of metamorphism to the OST where we believe it is exposed on the west coast between the Hane and Nishinada areas. In the last part of this section, we examine the relation of thermal metamorphism to cleavage in both domains. The increase in grade towards the south apparent in both coastal transects in Figure 6 is related to Stage 3 structures and will be discussed in the next section.

Shiina domain. The Shiina domain is represented on the west coast by the Nishinada section, shown in Figure 4a. These rocks were first imbricated and then deformed into an anticline /syncline pair, which DiTullio and Byrne (1990) interpreted to be an amplified fault-bend fold. The most important observation we make here is that across the over 2 km of predominantly steeply north dipping section the coal rank is fairly constant and high, ranging from $3.3\%R_m$ to $3.7\%R_m$. This crosscutting relationship of isorank contours to bedding and folds indicates that peak heating postdated folding. Illite crystallinity, although inherently more variable (ranging from $0.29\Delta°2\theta$ to $0.48\Delta°2\theta$ across the same section), also shows the same relationship to Stage 2 structures. Most local variations in CI with respect to VR (e.g., compare values for samples 84J10 and 84J7), are probably not significant, but the three CI data from the northern 500 m of the Nishinada section are all significantly higher grade than the section to the south. The sheared nature in the shale of this area, which we interpret as belonging to a mélange belt, may explain the better crystallinity of these samples.

The next area we discuss is the type locality for the Shiina domain shown in Figure 4b. (Bear in mind that this section and the neighboring Kabuka section to the north have been rotated by the Muroto flexure such that regional fold axes are steeply east plunging instead of subhorizontal as at Nishinada.) There are two things of note in this section. First, as with the fold at Nishinada, a remarkably similar high grade is seen across the Shiina anticline/ syncline (the low coal rank in sample 85J40 is anomalous and unexplained). Second, in the southernmost ~200 m of section, the coal rank steadily increases towards the Shiina-Narashi contact fault to some of the highest values in the study area; CI values are also consistently higher than in the rest of the section. The coal rank drops abruptly to $\sim1.4\%R_m$ south of this fault in the Oligocene-Miocene section (see Hibbard and others, this volume) indicating significant post-metamorphic uplift discussed in the next section.

In the Kabuka transect, contiguous with the Shiina section to the south (Fig. 4c), we see that crystallinity and coal rank both drop significantly in value across a narrow, poorly exposed zone where we infer the presence of the Shimizu fault. Grade continues to decrease steadily northwards until the Flat Rock fault is reached. Beyond this fault the grade is, on average, fairly constant and low (CI = $\sim0.60\Delta°2\theta$, VR = $\sim1.8\%R_m$) for the remaining 4 km of sandy section to Sakihama (Fig. 4d). Again, "isograds" from both paleothermal indicators crosscut all of the regional-scale folds. Only the Stage 3 high-angle structures, such as the Flat Rock or Ozaki faults, appear to affect the pattern of metamorphism. Other prominent drops in coal rank, such as at 85J66 in the Sakihama section and at 85J60 in the Kabuka section, are very local and their cause is unknown.

Gyoto domain. The type section of the Gyoto domain at Cape Gyoto (Fig. 5A) consists of a large (~1 km wavelength), moderately southwest plunging syncline. The CI and VR values across the entire syncline are essentially constant and the thermal grade is high, similar to that in the Nishinada section of the Shiina domain, with a VR of $3.7\%R_m$ and CI values of between $0.34\Delta°2\theta$ and $0.50\Delta°2\theta$. As noted in the vicinity of faults in other parts of the study area, there is a significant increase in illite crystallinity at the Moudodanigawa (MO) fault (not the Moudodani fault of Sakai, 1987) at the north end of the Gyoto section and at the South Gyoto (SG) fault at the south end of the section. These local increases in crystallinity may reflect shearing or enhanced fluid flow in the fault zones.

The other coastal exposure of the Gyoto domain, in the northern 2 km of the west coast transect at Cape Hane (Fig. 5B), also has a fairly constant grade (VR = ~$2.0\%R_m$ and CI = ~$0.55\Delta°2\theta$), but it is much lower than that of the rocks at Cape Gyoto. Although data are sparse, the distribution of grade appears to be unaffected by Stage 2 folds; however, both CI and VR values do show a small increase near the Hane Thrust (HA) (Sakai, 1987; see Figs. 5B and 6B). The Hane section is unique in the study area in that it consists predominantly of thick-bedded sandy turbidites that are isoclinally folded and overturned. On the basis of its distinctive lithology and structural style, it has been suggested that this section represents a slope basin (DiTullio and Byrne, 1990). Although we hoped to find supporting evidence for this interpretation from the thermal data, our results are equivocal. The Hane section is of much lower thermal maturity than other rocks to the south, but it is of equivalent or slightly higher rank than the Shiina domain rocks north of the Flat Rock fault on the east coast, which we consider to have been accreted. Thus, while we cannot rule out a slope basin origin for these rocks, they have experienced temperatures (and presumably burial) similar to at least some of the accreted strata.

Out-of-sequence thrust. The fact that coal rank and illite crystallinity maintain fairly constant values over wide areas and across the limbs of both D_1 and D_2 folds suggests that peak heating postdated Stage 2 deformation. If this is correct, then following the structural model of DiTullio and Byrne, the OST should also be crosscut by the thermal peak. In fact, no evidence of a major break in thermal maturity is seen when we consider data from the northern part of the Nishinada section where the OST appears to be exposed (see "fossil OST" in Fig. 4A). In this area, near Hirayama where two rivers discharge into the sea, there are northward changes in (1) strike (from 90 to 60°), (2) lithology (interbedded sand/shale turbidites to less typical shale-rich facies with carbonate concretions), and (3) cleavage style (spaced to anastomosing). The shale-rich facies to the north

of this structural break probably belongs to the mélange belt exposed in the interior to the east at the base of the Gyoto domain. VR and especially CI values indicate a slightly higher grade in the shale-rich facies versus the rest of the Nishnada section possibly reflecting a remnant of the OST thrust relationship, but more likely reflecting the effects of shearing and fluid circulation as in other fault zones. Regardless of its origin, this difference in grade is very slight when compared to the drop in thermal maturity in the Hane section to the north of the mélange belt across the Stage 3 Hane River fault(s).

More significant differences in grade are seen in the interior where coal rank data (shown in Fig. 3), from both the mélange belt and the structurally higher Gyoto domain rocks, are of significantly higher grade ($4.3\%R_m$ to $4.8\%R_m$) than footwall Shiina domain rocks at Nishinada. We note, however, that the highest rank measured in the entire study area ($5.0\%R_m$) comes from the interior of the Nishinada section. Thus, there is no evidence that movement on the OST postdated peak heating as recorded by vitrinite reflectance; although it appears that the interior of the Eocene section has been uplifted more than the coastal exposures—possibly by arching due to the Muroto flexure or on Stage 3 faults (see below).

Cleavage. The evidence discussed so far suggests that peak heating postdated Stage 2 deformation. However, the relation of cleavage to coal rank appears to require that peak heating was initiated during formation of the cleavages and persisted until after Stage 2 deformation ceased. As stated earlier, both domains contain two pressure solution cleavages: a weak axial planar cleavage, and a well-developed cleavage that transects the axial planes of the folds at a low angle. For the better-developed cleavages in both domains, there is a clear positive correlation between the degree of cleavage development and the coal rank of the shales (see Fig. 7). For example, the high-grade area just north of the contact with the Oligocene-Miocene section has the best-developed cleavages observed in the study area—this is true in both the Gyoto and Shiina domains. In contrast, areas with coal ranks of $\leq2.0\%R_m$ or illite crystallinities of $\geq0.50\Delta°2\theta$ have virtually no pressure solution cleavage (compare D and E in Fig. 7). These observations strongly suggest that the cleavages formed synchronously with peak heating.

To explain the apparently different timing and orientation of cleavage formation in the two domains, we note that pressure solution cleavage development requires both elevated temperatures and shortening. In the Eocene section, structural relations suggest that underthrust strata (which prior to development of the OST belonged to the Gyoto domain, and after development of the OST belonged to the presently exposed portion of the Shiina domain) escaped most horizontal shortening whereas rocks above the décollement were highly shortened (see DiTullio and Byrne, 1990, and references therein). Accordingly, we suggest that the internal lower levels (depths of ~7 to 10 km) of the Eocene prism experienced temperatures of ~250 to 300°C over the entire period of Stage 2 deformation, but that the Shiina and Gyoto domains occupied different positions with respect to the décolle-

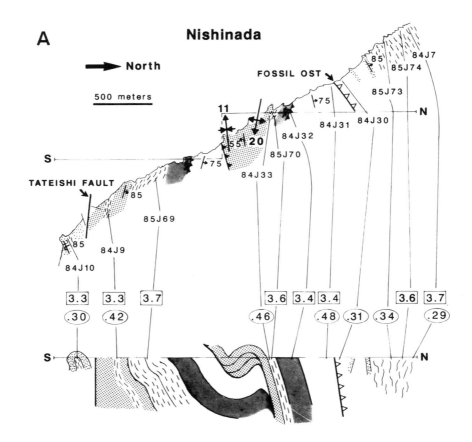

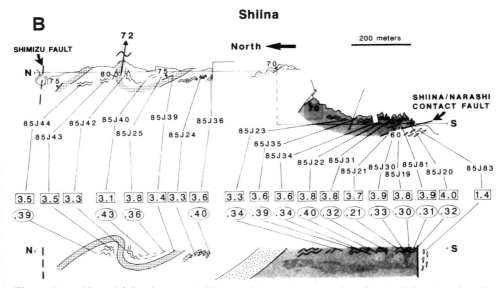

Figure 4 (on this and following pages). Maps and down-plunge sections for the Shiina domain with sample locations and VR (boxes) and CI (ovals) values shown. Note that the stippled patterns signify similar rock types but do not necessarily imply correlative units except in (A) the Nishinada section on the west coast, (B) the southernmost Shiina section, (C) the central Kabuka section, and (D) the northernmost Sakihama section. See text for discussion.

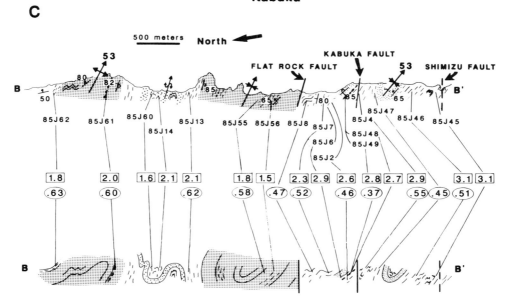

Kabuka

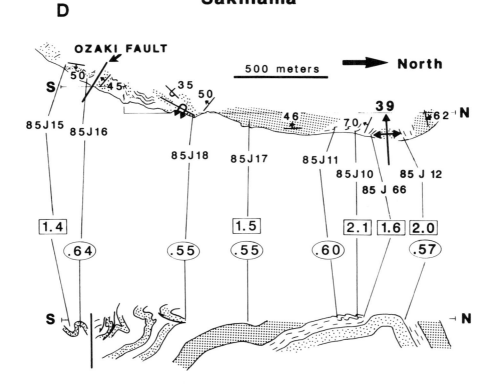

Sakihama

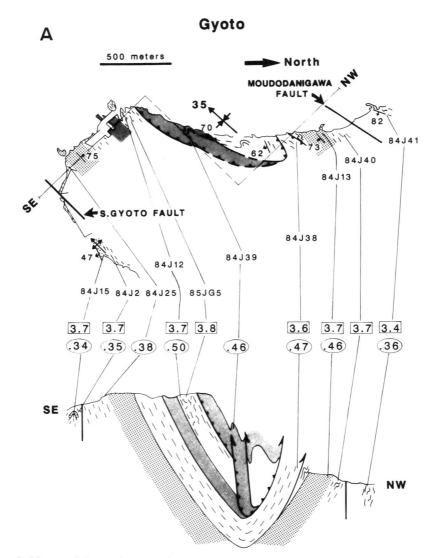

Figure 5. Maps and down-plunge sections for the Gyoto domain with sample locations and %R$_m$ (boxes) and CI (ovals) values shown. (A) is the type Gyoto section in the southwest; and (B) is the Hane section in the northwest, which is interpreted as a possible slope basin on structural and lithologic grounds. Note that the stippled patterns signify similar rock types and imply correlative units in (A), but not in (B). See text for discussion.

ment through time, thus recording different thermal/cleavage relationships (see Fig. 8 for a schematic drawing of this scenario).

Although the best-developed cleavage is always found in the highest grade rocks, it is not always true that the highest grade rocks display good cleavage. Where cleavage is absent, it appears to reflect the important role of preexisting anisotropies in the development of cleavage (see e.g., Knipe, 1981; Ishii, 1988). For example, an extensive slumped zone to the north of the Moudodanigawa fault exhibits the same high thermal maturity (as measured by coal rank) as the rest of the Gyoto section, but lacks cleavage. In this area, and also in the smaller slumped zone to the south represented by sample 84J12, we have noticed that the

level of crystallinity (and cleavage development) appears to be retarded with respect to neighboring well-bedded (and well-cleaved) strata of the same coal rank. Throughout the study area bedding and cleavage are typically subparallel and apparently the original bedding fissility was very important in promoting subsequent cleavage development.

Stage 3 deformation

The pattern of metamorphism, and in particular the increase in thermal maturity toward the Shiina-Narashi fault (Fig. 6), suggest that Stage 3 deformation has been significant. The pa-

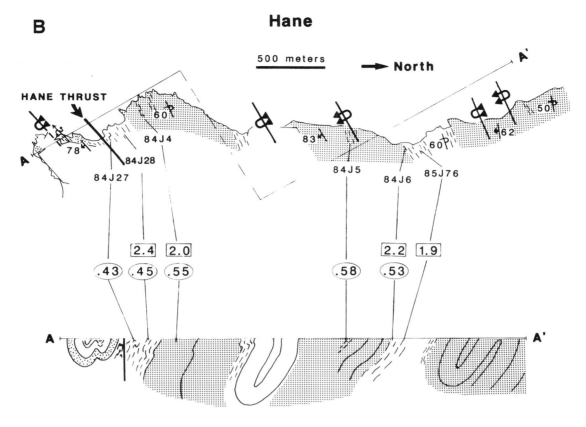

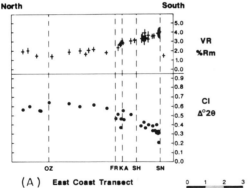

(A) **East Coast Transect**

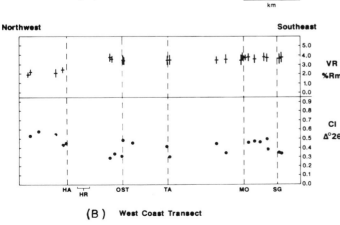

(B) **West Coast Transect**

Figure 6. Plots of CI (filled circles) and VR (crosses; length of vertical bar indicates error) data along the coastal transects. In the east coast transect (A), note the fairly constant grade north of the Flat Rock (FR) fault and the steady increase in grade to the south in both the VR and CI data. We believe, based on structural arguments outlined in the text, that the increase in grade is accommodated by a set of high-angle faults; four inferred to be the most important are illustrated by the dashed vertical lines (FR, Flat Rock; KA, Kabuka; SH, Shimizu; and SN, Shiina-Narashi). Also note that the Ozaki fault in the north part of the area (OZ, Ozaki), which separates zones of markedly different plunge of regional fold axes, appears to have at most only a slight reflection in the data. Also, see maps and cross sections in Figures 4B, 4C, and 4D. In the west coast transect (B), note that there are basically two plateaus in the data (especially in the VR data) indicating low grades to the north of the HR (Hane River) fault(s) and high grades to the south. Illite data indicate that the Hane Thrust (HA) does not separate zones of appreciably different grade; rather the increase in grade occurs over a zone ~1.5 km wide, which was not sampled. Note that the increase in CI values to the north of the OST is consistent with its interpretation as a thrust fault; but that CI values across other major faults along this transect (TA, Tateishi; MO, Moudodanigawa; and SG, South Gyoto), are more variable and difficult to interpret. VR data across all of these faults show no significant change. See Figures 4A, 5A, and 5B, and text for discussion.

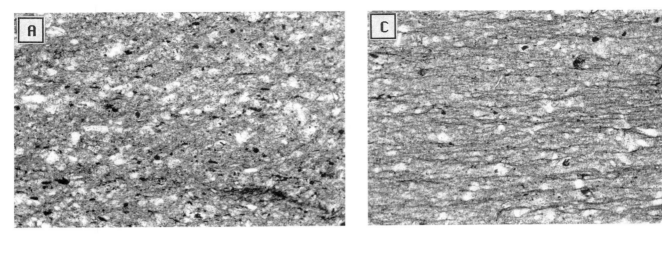

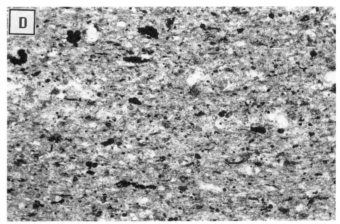

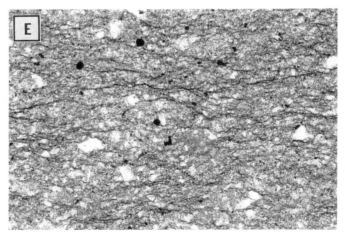

leothermal data indicate that the whole Eocene section was up-lifted with respect to the much cooler Oligocene-Miocene rocks to the south and that the northeastern part (north of the Shimizu fault and east of the Muroto flexure) experienced relatively less uplift than the rest of the Eocene section. Thus, post-metamorphic deformation appears to consist of relative uplift of fault-bounded blocks within the area on two sets of faults: an east-west system, dominated by the Shiina-Narashi fault; and a north-south system, dominated by the Muroto flexure. We discuss each of these aspects of post-metamorphic deformation below.

Shiina-Narashi and Hane River faults. The subvertical Shiina-Narashi fault is exposed on the east coast and locally in the interior where it has been mapped by Hibbard (1988). It separates relatively high rank ($\sim$4.0%R_m) Eocene rocks to the north from much less thermally mature ($\sim$1.5%R_m) Oligocene-Miocene rocks to the south (see Laughland and Underwood, this volume; Hibbard and others, this volume). It is thus the most significant post-metamorphic fault on the Muroto Peninsula. Assessment of the true offset on this fault, however, depends on the interpretation of the timing of peak heating in the Eocene rocks; this controversial point is discussed later.

Figure 7. Photomicrographs of samples from different parts of the study area illustrating the increasing cleavage development from north to south (or low to high grade). On the west coast: (A) Hane, 85J76, VR = 1.9 %R_m; (B) Nishinada, 85J74, VR = 3.6 %R_m; and (C) South Gyoto, 84J2, VR = 3.7 %R_m and CI = 0.35$\Delta°2\theta$. On the east coast: (D) Sakihama, 85J11, CI = 0.60$\Delta°2\theta$; and (E) Shiina, near the contact, 85J81, VR = 3.9 %R_m and CI = 0.39$\Delta°2\theta$. The width of view in each photograph is $\sim$0.8 mm.

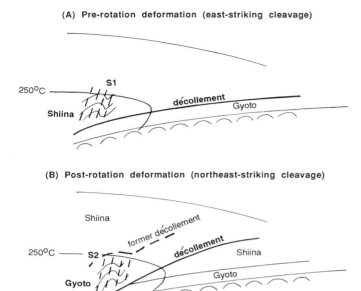

(A) **Pre-rotation deformation (east-striking cleavage)**

250°C

S1

Shiina

décollement

Gyoto

(B) **Post-rotation deformation (northeast-striking cleavage)**

Shiina

former décollement

250°C

S2

décollement

Shiina

Gyoto

Gyoto

Figure 8. Schematic diagram illustrating the development, during steady state thermal conditions, of the prerotation S_1 and post-rotation S_2 cleavages in the Shiina and Gyoto domains, respectively. (A) development of S_1 cleavage in the Shiina domain while the Gyoto domain is being underthrust beneath the décollement. (B) development of S_2 cleavage in the Gyoto domain after the décollement steps beneath it and it is transported to higher levels of the prism on the OST. Note that subhorizontal shortening (which is necessary for cleavage development) is inferred to act primarily above the décollement (in the post-rotation time this is the OST), and that cleavage is best developed at temperatures of 250 to 300°C. In the study area, only the portion of the Gyoto domain just above the OST and the part of the Shiina domain just below the OST are exposed.

The other major east-trending fault, not apparently related to the Muroto flexure, is inferred to exist within a 2-km-wide data gap spanning the change from low-grade rocks of the Hane section in the northwest to high-grade rocks of the Nishinada and Gyoto sections in the 7 km to the south. Paleothermal data from the mélange belt exposed in the interior, along strike from the data gap, are of very high coal rank suggesting that the break in thermal maturity occurs to the north of the mélange (perhaps along its northern boundary) at one or more of the high-angle faults mapped by Taira and others (1980). We have informally deisgnated this fault or fault zone the Hane River (HR) fault(s) (see Figs. 6 and 10).

Muroto flexure and related faults. The second important post-metamorphic deformation within the Eocene sequence involved the relative displacement of the low-grade Kabuki and Sakihama sections in the northeast with respect to rest of the high-grade Eocene section along a north-trending structure. This displacement appears to have been accommodated by the Muroto flexure and related east-trending splays. The flexure, which affects Cretaceous through Miocene Shimanto rocks, is defined

by the axis of major curvature in strike of bedding, cleavage, and fold axial surfaces (see Taira and others, 1980, Fig. 45; Hibbard and Karig, 1990; and Fig. 1). Previous work in the Eocene section (DiTullio and Byrne, 1990) has demonstrated that the flexure is a major cross fold that rotated regional fold axes from dominantly subhorizontal plunges on the west coast to subvertical plunges along the east coast. Taira and others (1980, their Fig. 45) indicate a major fault in its axial zone (see Fig. 9) and we consider the axial zone to be a high-angle, west-side-up reverse fault. This interpretation is supported by the CI and VR data from the interior which indicate that the area of low thermal maturity is confined to rocks east of the trace of the Muroto flexure (Fig. 10). Other north-trending faults in the interior of the peninsula shown by Taira and others (1980; and Fig. 9) may also have accommodated differential uplift as suggested by the very high thermal maturity of some of the samples in the interior (see Fig. 3).

Where exposed on the east coast, the southern boundary of the low-grade zone spans a 3-km-wide zone (from the Flat Rock to the Shiina-Narashi faults) with at least three and possibly more east-striking faults producing a gradient in thermal maturity (refer to Figs. 4 and 6). By their geometry and displacement sense, we infer that these faults are related to the Muroto flexure. A schematic diagram illustrating the major effects of the Muroto flexure and related faults is shown in Figure 11.

Another important subsidiary fault to the Muroto flexure is the Ozaki fault (OZ), located at the south end of the Sakihama section (Fig. 4D). This easterly striking high-angle fault is a structural boundary that separates east-striking bedding and moderately west-plunging folds to the north from northwest-striking bedding and steeply southeast-plunging folds to the south. Because the rotation of fold axes to the south of this fault was caused by the Muroto flexure, we consider that the Ozaki fault is a related, Stage 3 post-metamorphic fault. To the north and south of this fault, the VR values of samples 85J17 and 85J15 (1.5%R_m and 1.4%R_m, respectively) are of markedly lower rank than the average in the northern part of the east coast transect (north of the Flat Rock fault as discussed above). This decrease in grade is probably real and is also seen in the illite data at sample 85J16 (although not in the more sheared shale samples at 85J17 and 85J18). We hypothesize that it is related to drag on the Ozaki fault. The northern boundary of this area of lower grade, between samples 85J10 and 85J17, is probably also bounded by an unobserved late fault.

Summary

The paleothermal results just presented demonstrate that peak heating bears no relation to and apparently postdates all Stage 2 folds and the OST. However, we have also shown a positive correlation between coal rank and cleavage development which seems to require that peak heating was synchronous with much of the Stage 2 deformation. Together, these relations lead us to conclude that peak heating conditions were maintained for

Late High-angle Faults

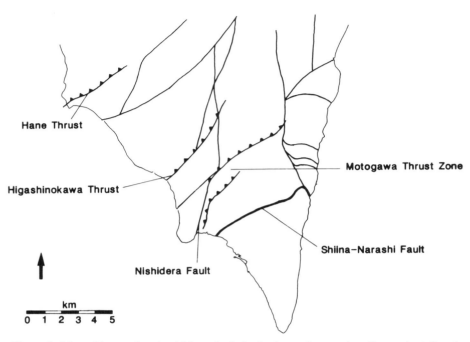

Figure 9. Map of late and active high-angle faults in the study area that disrupt the inferred original nappe distribution of the Gyoto and Shiina domains. Many of these faults also juxtapose rocks of different thermal maturity. The Shiina-Narashi fault is from Hibbard (1988); other labeled faults are taken from Sakai (1987). With the exception of the inferred Kabuka and Shimizu faults, the rest of the faults are from Taira and others (1980, Fig. 45).

several million years during Stage 2 and to some degree beyond; the age of this metamorphic event is discussed below.

Post-metamorphic deformation is seen to have been dominated by uplift on high-angle, north- and east-trending fault systems. The paleothermal data was instrumental in allowing us to identify faults other than the structurally obvious Shiina-Narashi and Muroto flexure as significant; faults such as the Hane River, Flat Rock, or Shimizu were either not exposed or not clearly important. In addition, the paleothermal data allowed us to gauge the vertical displacement on all of the Stage 3 faults.

Finally the paleothermal data have enabled us to identify the grade of the accreted rocks of the Eocene section. At Shiina, Nishinada, Gyoto, and in most of the interior, the grade is relatively high; in terms of illite crystallinity these areas fall in the anchizone and their coal rank ranges from on average ~3.6%R_m in the coastal exposures to ~4.5%R_m in the interior. Elsewhere, particularly in the northeast, the rocks are diagenetic grade with coal ranks ranging from 1.4 to 2.4%R_m. Paleotemperature estimates derived from these data are discussed in the next section. The distribution of grade and the effects of Stage 3 deformation are summarized in the map and cross sections in Figure 10 and the schematic block diagram in Figure 11.

DISCUSSION AND INTERPRETATION

In this section, we estimate the peak paleotemperatures of these rocks from the VR data. We then combine these paleotemperatures with geologic data to derive a paleogeothermal gradient. Next, we discuss the absolute age of the metamorphism and model its relation to Stage 2 deformation. Then, we take these results and integrate them with plate reconstructions to infer the nature of the subducting plate. Finally, we relate Stage 3 deformation, which controls the present distribution of grade, to later plate interactions.

Estimates of paleotemperature

Illite crystallinity values from the Eocene section fall primarily within the anchizone (0.42 to 0.25$\Delta°2\theta$) suggesting paleotemperatures of from ~250 to 300°C (Kisch, 1987). In these areas, the VR values are mainly ~3.0 to 4.0%R_m, which, according to Barker's (1988) correlation scheme, also correspond to temperatures of ~250 to 300°C. The highest crystallinity value in the study area of 0.21$\Delta°2\theta$ (sample 85J21 in the Shiina section) is within the epizone (300 to 350°C according to Kisch, 1987);

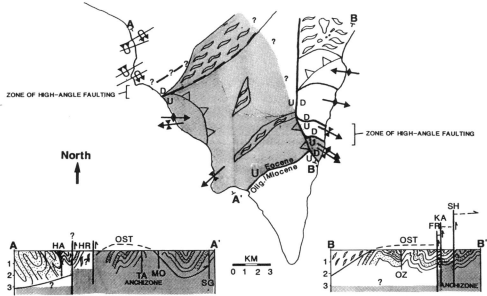

Figure 10. Map and cross sections of the study area summarizing the results of the thermal data and generalizing the distribution of anchizone grade rocks (stippled pattern) and diagenetic rocks (unpatterned) as determined from the CI data. The distribution in the interior is partly based on regional structural considerations discussed in the text. Major features illustrated by this map are the relatively high grade of most of the field area relative to the Oligocene-Miocene section to the south and the decrease in grade east of the Muroto flexure in the Kabuka and Sakihama sections. The increase in grade to the south is best seen in the two cross sections. The offsets of the diagenetic/anchizone boundary shown in the cross sections are approximate and were estimated assuming a paleogeothermal gradient of 30°C/km. The distribution of grade indicates that metamorphism postdated folding and steepening of strata in both the Gyoto and Shiina domains. See text for discussion and Figure 6 caption for definition of abbreviations.

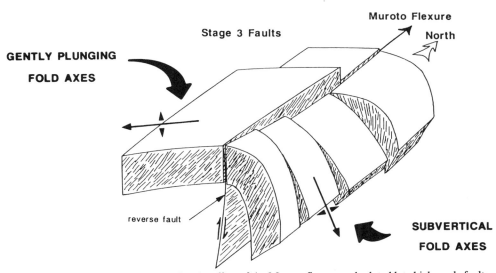

Figure 11. Schematic diagram showing the effect of the Muroto flexure and related late high-angle faults on the preexisting geometry of the Eocene prism. The southernmost boundary represents the Shiina-Narashi fault and the other prominent fault bounding the western uplifted block on the north is the Hane River fault(s). To the east of the flexure, the decrease in thermal grade from south to north is illustrated as accommodated by north-side-down throw on the Shiina-Narashi, Shimizu, and Flat Rock faults. The rotation of regional folds to steep plunges along the east coast of the field area, south of the Ozaki fault, is also shown.

the coal rank of $3.7\%R_m$ of this sample, however, indicates a somewhat lower temperature of 284°C. Interior areas, and particularly mélange, have VR values of ~4.5 to $5.0\%R_m$ suggesting temperatures of ~300 to 315°C. However, corresponding CI values are lower than the expected epizone values. This may indicate that the high temperatures suggested by the vitrinite data were relatively short-lived; or possibly that the illite values have been adversely affected by weathering of the samples. In the northern Hane and Sakihama sections, thermal maturity is markedly lower: many of the CI values are within the diagenetic zone suggesting to temperatures of from ~200 to 250°C and VR values are $\sim2\%R_m$ indicating temperatures of ~200°C. More precise estimates give a range of temperatures for the entire Eocene section of from 183 to 315°C.

Estimate of paleogeothermal gradient

In order to estimate the paleogeothermal gradient, we must know the thickness of the section over which it is being measured. At Shiina, the fairly smooth increase in grade from the Flat Rock to the Shiina-Narashi fault initially suggests that we may be observing a cross section through an upturned fossil paleogeothermal gradient. The increase in thermal maturity from an average of $\sim2.0\%R_m$ north of the Flat Rock fault to $4.0\%R_m$ at the Shiina-Narashi contact fault represents a temperature difference of 72°C (Barker, 1988). If the isotherms followed bedding, which dips approximately 40° north when rotation due to the Muroto flexure is removed, we would derive a paleogeothermal gradient of 37°C/km.

However, this way of deriving the paleogeothermal gradient is untenable because: (1) all available evidence indicates that peak heating postdated folding (paleoisotherms do not follow bedding), and (2) the enveloping surface of the folds along the east coast does not appear to have been backtilted from the horizontal (see cross sections Figs. 4B and 4C). Thus, as discussed in the preceding section and shown in Figure 10 (cross section B–B′), we have inferred that the southward increase in grade along the east coast is caused by stepped uplift of discrete blocks on subvertical faults.

The best estimate of the paleogeothermal gradient that we can make uses b_o data from Underwood, Laughland, and Kang (this volume) that suggest maximum burial for the Eocene rocks of 8 to 10 km. Because we know that peak heating postdated imbrication, regional-scale folding and the OST, we assume that the paleoisotherms were horizontal. Unfortunately, because of the structural complexity, it is not possible to accurately estimate the structural thickness of the Eocene section. However, the geometry of the regional-scale folds suggests that each domain is at least 500 to 1,000 m thick (see cross sections Figs. 4 and 5). Combining this information with our maximum paleotemperature estimate of ~300°C leads to a minimum paleogeothermal gradient estimate of ~30 to 40°C/km.

Such a gradient seems rather high when compared to some models of the thermal structure of accretionary prisms that indicate much lower geothermal gradients of 10 to 15°C/km (e.g., Oxburgh and Turcotte, 1970; Wang and Shi, 1984; van den Beukel and Wortel, 1988). However, recent studies have shown that geothermal gradients in active prisms, at least in the toe region, may be much higher than previously thought due to the important role of advective heat transfer by fluid flow (e.g., Davis and Langseth, 1986; Reck, 1987; Moore and others, 1988). Studies of heat flow at active margins using the gas hydrate method as well as direct temperature measurements in boreholes show that many accretionary prisms are characterized by geothermal gradients of from 30 to 60°C/km or more (e.g., Yamano and others, 1982; and see Table 1 in Underwood, Hibbard, and DiTullio, this volume). Another way of generating high geothermal gradients in prisms is through the subduction of either active spreading ridges or very young crust (DeLong and others, 1979; James and others, 1989; and see below).

Estimate of absolute age of the thermal peak

The absolute age of the thermal peak in the Eocene section is controversial and hinges on the interpretation of the age of the cleavages in these rocks. We have presented evidence in this paper that peak heating both postdates Stage 2 regional-scale folds and the OST and was synchronous with the development of D_1 and D_2 cleavages. The available data suggest two possible ways to satisfy both of these conditions. The first possibility is that the cleavages are young—middle Miocene in age—and correlate with emplacement of the igneous rocks exposed at Cape Muroto (Hibbard and others, this volume). The second possibility is that peak metamorphic conditions were related to accretion of the Eocene section and occurred throughout Stage 2 deformation—a period that approximately spanned the Oligocene (~10 m.y.).

The first possibility is suggested by the presence of a regionally extensive middle Miocene tectonothermal event that is recorded by local intrusions throughout the Shimanto Belt and that has been related to subduction of the actively spreading Shikoku basin (see Underwood, Hibbard, and DiTullio, this volume; Hibbard and others, this volume). At Cape Muroto, a 14-Ma gabbroic intrusion has substantially metamorphosed the Oligocene-Miocene accreted sediments and a north-northeast–striking pressure solution cleavage is well developed within the contact zone. Detrital zircons from the sediments near the intrusion appear to be completely reset and yield 14-Ma ages (Hasabe, this volume). It is therefore possible that the Eocene section was also heated and cleaved at this time. In fact, if this interpretation is correct, the 50° rotation in shortening direction recognized by DiTullio and Byrne (1990) may be related to the clockwise rotation of southwest Japan as the Japan Sea opened in the middle Miocene (Otofuji and Matsuda, 1987).

Another possibility, which we believe is more strongly supported by the available data, is that the cleavages in the Eocene section developed as part of its progressive deformation during accretion in the Oligocene. In support of this interpretation, we note that the deformation paths recorded in the Eocene and

Oligo-Miocene rocks are quite different. Detailed structural studies (Hibbard, 1988) indicate that the younger rocks are characterized primarily by landward-verging structures with only a locally developed pressure solution cleavage. In contrast, structures in the Eocene rocks are consistently seaward-verging and the pressure solution cleavages are regionally extensive—showing no spatial correlation with the Cape Muroto gabbro or any other igneous body. Also, there is no evidence for a rotation in shortening direction in the Oligocene-Miocene section; only one cleavage is developed and its orientation is different (north-northeast) from either the easterly or northeasterly trending cleavages in the Eocene section.

Finally, a total of nine K-Ar cleavage ages obtained by Agar and others (1989) and Mackenzie and others (1990) from correlative rocks along strike from the Muroto Peninsula, support the interpretation that the cleavages in the Eocene section are Eocene to Oligocene in age. Agar and other's (1989) data come from the eastern sections of the Eocene Hiromi and Kurosuno Formations on the Hata Peninsula of western Shikoku Island and range in age from 43.4 to 18.5 Ma; six of the dates form a tight cluster ranging from 33.7 to 25.7 Ma. Mackenzie and other's single cleavage age comes from the eastern part of the Eocene Kitagawa Formation on Kyushu Island and yields a 48.4-Ma age.

Agar and others (1989) and Mackenzie and others (1990) considered these dates to be representative of the time of cleavage formation because microstructural studies showed that very fine grained phyllosilicates have grown within, and parallel to, the cleavage domains and because only very fine grained fractions (<0.5 microns) were used for dating. The possibility of contamination from detrital or diagenetic phyllosilicates could not be completely eliminated, however, because of the fine grain size of the cleavage-forming phyllosilicates. The ages obtained by Agar and others and Mackenzie and others, therefore, may be maximum ages of the cleavages in these rocks; however, we believe that all the evidence taken together supports our interpretation of the timing of the first and second phases of cleavage development in the Eocene section of the Muroto area.

Nature of the subducting plate

Having tentatively concluded that the thermal peak in the Eocene prism occurred over at least a 10-m.y.-long period in the late Eocene through Oligocene, we turn our attention to how an apparently high paleogeothermal gradient of 30 to 40°C/km could have been maintained for such a period. Such long-lived heating must reflect a steady state condition rather than a short-lived thermal pulse related to a specific tectonic event. In particular, we examine the effect of the subduction of a very young oceanic plate. The model of James and others (1989) concentrated on the thermal structure in shallow levels of an accretionary prism (≤10 km) and they were able to show that paleogeothermal gradients similar to those estimated in the Eocene section can be generated by the subduction of very young crust. They demonstrated that the subduction of 1.5 to 2-m.y.-old crust (assuming a convergence rate of 7.5 cm/yr) can produce temperatures at 10 km depth of over 300°C at 60 km from the trench and 250°C at 100 km from the trench. Moreover, if subduction of such young crust ceases, temperatures of >300°C may be reached in a large part of the prism at 10-km depth within a few million years (see James and others, 1989, their Fig. 5). Lesser, but still significant, heating of the shallow levels of a prism may be expected if the rate of subduction of young crust (≤5 m.y. old) slows appreciably.

Using these model results, we propose that the relatively high paleogeothermal gradient in the Eocene Shimanto belt was caused by a combination of the subduction of young, hot crust and substantial slowing of the rate of convergence at 40 to 30 Ma. Taira and others (1988) have reviewed how subduction has occurred beneath southwest Japan from the Cretaceous to the present. This convergence was generally oriented north-northwest to northwest and occurred at an average rate of about 10 cm/yr (Engebretson and others, 1984, their Table 6). The identity of the subducting plate in the early Tertiary is difficult to establish, but it was likely either the Kula/Pacific Plate or the Philippine Sea Plate. According to Engebretson and others' calculations, in the early Eocene the Kula Plate had a convergence rate with respect to Japan of about 13 cm/year; after the cessation of spreading at 43 Ma (Lonsdale, 1988), the fused Kula/Pacific Plate had a convergence rate with respect to Japan of about 5 cm/year until the beginning of the Oligocene. The Philippine Sea Plate, which may have been the plate subducting beneath southwest Japan after about 43 Ma, also appears to have had a convergence rate of about 10 cm/year prior to mid-Oligocene time (30 Ma) after which convergence slowed considerably and the Shikoku Basin started to open (Louden, 1977; Seno and Maruyama, 1984).

Thus, despite the uncertainties in these figures and in which plate was subducting beneath southwest Japan in the Oligocene, it appears that accretion of strata in the Eocene section occurred at relatively fast convergence rates; but that in the late Eocene or early Oligocene the rate of convergence was approximately halved. Equally important, Taira (1985) and Taira and others (1988) used the age difference between accreted strata and associated mélange material throughout the Shimanto Belt to show that that steadily younging crust had been approaching the trench of southwest Japan since the Cretaceous and that the oceanic crust being subducted during accretion of the Eocene prism was <10 m.y. old (see Fig. 12). The combination of young crust and relatively slow convergence rates seems more than adequate to have produced a geothermal gradient of 30 to 40°C/km in the Eocene section in late Eocene to early or middle Oligocene time.

Interpretation of Stage 3 post-metamorphic deformation

Stage 3 deformation of the Eocene section may be divided into two types: (1) south-side uplift on easterly striking faults, especially the Shiina-Narashi fault; and (2) across-strike shortening and local block rotation on north-striking structures, primarily the Muroto flexure. Uplift of the Eocene section may have occurred in two phases with the first phase resulting in a widespread, generally early Miocene unconformity (Sakai, 1988), and the second phase following the middle Miocene thermal event and being visible in the present pattern of faulting.

Following the offscraping models of Seely and others (1974) and Karig and Sharman (1975), we attribute the first type of uplift, south-side uplift, to sequential underthrusting of younger Oligocene-Miocene rocks beneath the Eocene prism. Mori and Taguchi (1988) found a similar pattern of uplift in their vitrinite reflectance analyses of samples collected from an ~80-km transect through the entire Cretaceous through Oligocene-Miocene Shimanto Belt on the Muroto Peninsula. They identified four sections (including one corresponding to the study area) bounded by high-angle faults with southward (trenchward) increasing coal rank. They interpreted this pattern to reflect episodic uplift due to underthrusting and suggested that this fossilized pattern of uplift is still occurring today, as seen in the present coseismic deformation related to underthrusting of sediment in the Nankai Trough (see Mori and Taguchi, 1988, their Fig. 13).

The fault patterns related to underthrusting have clearly been subsequently modified by continued shortening of the prism. For example, at the time of underthrusting of the Oligocene-Miocene sequence, the Shiina-Narashi fault (or an earlier manifestation of the boundary between the two sequences) was presumably dipping northward at a much lower angle. Landward rotation of this initially low-dipping boundary to subvertical probably occurred together with landward rotation of the strata and fold axial planes in both the Eocene and Oligocene-Miocene accreted rocks.

Latest movement on the subvertical Shiina-Narashi fault, as indicated by the observed offset in coal rank across the fault, is at least as young as late Miocene. The clearly tectonic nature of this contact between the Eocene section and the Hioki tectonic mélange of the Oligocene-Miocene section exposed on the east coast supports this interpretation. The Shiina-Narashi fault is deformed by the Muroto flexure, but it is not clear whether or not its latest movement postdates folding by the flexure.

Across-strike shortening, the youngest type of Stage 3 deformation, has been accommodated by the Muroto flexure. As noted in the Introduction, the Muroto flexure appears to be similar in nature to other major north-trending cross folds spaced fairly evenly along strike in the Shimanto Belt (Sugiyama, 1989a, b). Sugiyama (1989b) has suggested that these features reflect along-strike shortening, which was initiated when the relative motion of the Philippine Sea Plate with respect to southwest Japan changed from north-northwest to west-northwest in the middle Pliocene. The Muroto flexure may be active today as parallel structures are presently forming offshore (Okamura, 1990; Sugiyama, 1989a).

CONCLUSIONS

Illite crystallinity and vitrinite reflectance data from the Eocene section allow us to make the following important conclusions. (1) Peak temperature conditions were maintained after all accretion-related structures, including regional-scale folds and the OST, were formed. (2) Peak heating also appears to have over-

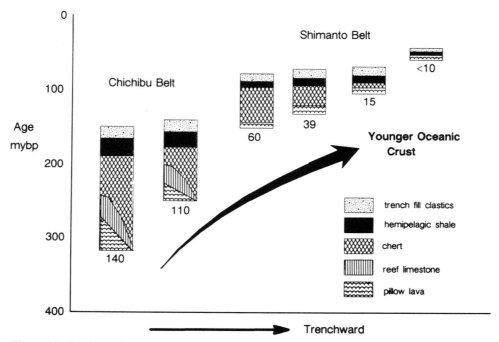

Figure 12. Stratigraphic columns in successively younger parts of the Shimanto Belt showing the decreasing age difference between oceanic crustal rocks and clastics—indicating the steady approach of a spreading ridge to the trench. (After Taira, 1985, Fig. 4; and Taira and others, 1988, Fig. 21.)

lapped in time with both phases of cleavage development (and by implication the OST and the D_2 folding) because the best-developed cleavage is always found in the highest-grade rocks. Structural arguments and radiometric dating support our interpretation that the cleavages were related to progressive deformation during accretion of the Eocene rocks in the Eocene and Oligocene. (3) The grade of metamorphism is diagenetic to anchizone on the basis of illite crystallinity; paleotemperature estimates based on vitrinite reflectance data are ~200 to ~300°C (see also Laughland and Underwood, this volume; and Underwood, Byrne, Hibbard, DiTullio, and Laughland, this volume). (3) Significant post-metamorphic deformation resulted in relative uplift of the southern, western, and most of the interior portions of the study area; this deformation was accommodated by east- and north-striking high-angle faults. (4) Shallow burial ($\leqslant 10$ km) combined with peak paleotemperature estimates leads to an estimated paleogeothermal gradient of 30 to 40°C/km. This gradient may have been caused by a combination of the subduction of young oceanic crust of the Kula/Pacific Plates and significant slowing of the convergence rate in late Eocene or early Oligocene.

These results are important because they demonstrate that illite crystallinity and vitrinite reflectance are useful tools in unraveling thermal and deformation histories in low-grade and structurally complex terranes lacking more traditional measures of metamorphism. Moreover, they illustrate how some constraints can be placed on deformation even if only one method is used and only relative differences in grade are considered. Finally, we suggest that the apparent structural and metamorphic complexity revealed in the small Muroto field area may be typical of the interiors of long-lived accretionary prisms that must continually adjust to changes in the rate and direction of convergence as well as in the nature of the subducting plate itself.

ACKNOWLEDGMENTS

This study constituted part of DiTullio's Ph.D. thesis. She would like to acknowledge the receipt of a Harold T. Stearns grant from GSA and a Sigma Xi Grant in Aid of Research. Much of the field work was funded by NSF grants #EAR85-09461 and EAR87-20743 to Tim Byrne and Dan Karig. The X-ray analysis of illite crystallinity carried out in Japan was greatly assisted by Drs. Hada, Higashi, and Yoshikura at Kochi University. Major funding for this study came from NSF grant #EAR87-06784 to Mike Underwood. Acknowledgment is also made to the Donors of the Petroleum Research Fund, administered by the American Chemical Society, for partial support of this research (Grant No. 19187-ACZ to Underwood). Careful reviews by Mike Underwood, Darrel Cowan, and Mark Cloos greatly improved the manuscript.

REFERENCES CITED

Agar, S., Cliff, R. A., Duddy, I. R., and Rex, D. C., 1989, Short paper: Accretion and uplift in the Shimanto Belt, SW Japan: Journal of the Geological Society, London, v. 146, p. 893–896.

Barker, C. E., 1983, The influence of time on metamorphism of sedimentary organic matter in liquid-dominated geothermal systems, western North America: Geology, v. 11, p. 384–388.

—— , 1988, Geothermics of petroleum systems: Implications of the stabilization of kerogen thermal maturation after a geologically brief heating duration at peak temperature, *in* Magoon, L., ed., Petroleum systems of the United States: U.S. Geological Survey Bulletin 1870, p. 26–29.

—— , 1989, Temperature and time in the thermal maturation of sedimentary organic matter, *in* Naeser, N. D., and McCulloh, T. H., eds., Thermal history of sedimentary basins, Methods and case histories: New York, Springer-Verlag, p. 73–98.

Blenkinsop, T. G., 1988, Definition of low-grade metamorphic zones using illite crystallinity: Journal of Metamorphic Geology, v. 6, p. 623–636.

Bostick, N. H., 1984, Comment *on* 'Influence of time on metamorphism of sedimentary organic matter in liquid-dominated geothermal systems, western North America': Geology, v. 12, p. 689–691.

Cloos, M., 1983, Comparative study of mélange matrix and metashales from the Franciscan subduction complex with the basal Great Valley sequence, California: Journal of Geology, v. 91, p. 291–306.

Davis, D. M., and Langseth, M. G., 1986, Some hydrologic issues in accretionary prisms [abs.]: EOS Transactions of the American Geophysical Union, v. 67, p. 242.

DeLong, S. E., Schwarz, W. M., and Anderson, R. N., 1979, Thermal effects of ridge subduction: Earth and Planetary Science Letters, v. 44, p. 239–246.

DiTullio, L. D., 1989, Evolution of the Eocene accretionary prism in SW Japan: Evidence from structural geology, thermal alteration and plate reconstructions [Ph.D. thesis]: Providence, Brown University, 161 p.

DiTullio, L. D., and Byrne, T. B., 1990, Deformation in the shallow levels of an accretionary prism: The Eocene Shimanto Belt of southwest Japan: Geological Society of America Bulletin, v. 102, p. 1420–1438.

Engebretson, D. C., Cox, A., and Gordon, R. G., 1984, Relative motions between oceanic and continental plates in the Pacific Basin: Geological Society of America Special Paper 206, 59 p.

Ernst, W. G., 1970, Tectonic contact between the Franciscan mélange and the Great Valley Sequence—Crustal expression of a late Mesozoic Benioff zone: Journal of Geophysical Research, v. 75, p. 886–901.

Frey, M., 1987, Very low-grade metamorphism of clastic sedimentary rocks, *in* Frey, M., ed., Low temperature metamorphism: Glasgow, Blackie & Son Ltd., p. 9–58.

Hibbard, J. P., 1988, Evolution of anomalous structural fabrics in an accretionary prism: The Oligocene-Miocene portion of the Shimanto Belt at Cape Muroto, southwest Japan [Ph.D. thesis]: New York, Cornell University, 227 p.

Hibbard, J. P., and Karig, D. E., 1990, Structural and magmatic responses to spreading ridge subduction: An example from southwest Japan: Tectonics, v. 9, p. 207–230.

Ishii, K., 1988, Grain growth and re-orientation of phyllosilicate minerals during the development of slaty cleavage in the South Kitakami Mountains, northeast Japan: Journal of Structural Geology, v. 10, p. 145–154.

James, T. S., Hollister, L. S., and Morgan, W. J., 1989, Thermal modeling of the Chugach Metamorphic Complex: Journal of Geophysical Research, v. 94, p. 4411–4423.

Karig, D. E., and Sharman, G. F., III, 1975, Subduction and accretion in trenches: Geological Society of America Bulletin, v. 86, p. 377–389.

Kemp, A.E.S., Oliver, G.H.J., and Baldwin, J. R., 1985, Low-grade metamorphism and accretion tectonics: Southern Uplands terrain, Scotland: Mineralogical Magazine, v. 49, p. 335–344.

Kisch, H. J., 1980, Illite crystallinity and coal rank associated with lowest grade metamorphism of the Transverse graywacke in the Helvetic zone of the Swiss Alps: Eclogae Geologica Helvetiae, v. 73, p. 753–777.

—— , 1987, Correlation between indicators of low-grade metamorphism, *in* Frey, M., ed., Low temperature metamorphism: Glasgow, Blackie & Son Ltd., p. 227–324.

Knipe, R. J., 1981, The interaction between deformation and metamorphism in slates: Tectonophysics, v. 78, p. 249–272.

Kreutzberger, M. E., and Peacor, D. R., 1988, Behavior of illite and chlorite during pressure solution of shaly limestone of the Kalkberg Formation, Catskill, New York: Journal of Structural Geology, v. 10, p. 803–811.

Kubler, B., 1964, Les argiles, indicateurs de metamorphisme: Revue de l'Institut Francais du Pétrole et Annales des Combustible Liquides, v. 9, p. 1093–1112.

——, 1967, La crystallinite de L'illite et les zones tout á faitsuperieures du metamorphisme, in Etages tectoniques, Colloque de Neuchatel, 1966: Neuchatel, Suisse, á la Baconniere, p. 105–121.

——, 1968, Evaluation quantitative du metamorphisme par la cristallinite de l'illite: Centre de Recherches de Pau (Société Nationale des Pétroles d'Aquitaine), Bulletin, v. 2, p. 385–397.

——, 1970, Crystallinity of illite: Detection of metamorphism in some frontal parts of the Alps: Fortschritte der Mineralogie, v. 47 (Beih. 1), p. 39–40.

Landis, C. A., and Coombs, D. S., 1967, Metamorphic belts and orogenesis in southern New Zealand: Tectonophysics, v. 4, p. 501–518.

Lonsdale, P., 1988, Paleogene history of the Kula Plate: Offshore evidence and onshore implications: Geological Society of America Bulletin, v. 100, p. 733–754.

Louden, K. E., 1977, Paleomagnetism of DSDP sediments, phase shifting of magnetic anomalies, and rotations of the West Philippine Basin: Journal of Geophysical Research, v. 82, p. 2989–3002.

Mackenzie, J. S., Taguchi, S., and Itaya, T., 1990, Cleavage dating by K-Ar isotopic analysis in the Paleogene Shimanto Belt of eastern Kyushu, SW Japan: Journal of Mineralogy, Petrology and Economic Geology, v. 85, p. 161–167.

Merriman, R. J., and Roberts, B., 1985, A survey of white mica crystallinity and polytypes in pelitic rocks of Snowdonia and Llyn, North Wales: Mineralogic Magazine, v. 49, p. 305–319.

Moore, J. C., and Allwardt, A., 1980, Progressive deformation of a Tertiary trench slope, Kodiak Islands, Alaska: Journal of Geophysical Research, v. 85, p. 4741–4756.

Moore, J. C., Mascle, A., and the ODP Leg 110 Scientific Party, 1988, Tectonics and hydrogeology of the northern Barbadoes Ridge: Results from Ocean Drilling Program Leg 110: Geological Society of America Bulletin, v. 100, p. 1578–1593.

Mori, K., and Taguchi, K., 1988, Examination of the low-grade metamorphism in the Shimanto Belt by vitrinite reflectance: Modern Geology, v. 12, p. 325–339.

Okamura, M., 1990, Geologic structure of the upper continental slope off Shikoku and Quaternary tectonic movements of the Outer Zone of southwest Japan: Journal of the Geological Society of Japan, 96, p. 223–237.

Otofuji, Y., and Matsuda, T., 1987, Amount of clockwise rotation of S.W. Japan—Fan shape opening of the southwestern part of the Japan Sea: Earth and Planetary Science Letters, v. 85, p. 289–301.

Oxburgh, E. R., and Turcotte, D. L., 1970, Thermal structure of island arcs: Geological Society of America Bulletin, v. 81, p. 1665–1668.

Price, L. C., 1983, Geologic time as a parameter in organic metamorphism and vitrinite reflectance as an absolute paleogeothermometer: Journal of Petroleum Geology, v. 6, p. 5–38.

Reck, B. H., 1987, Implications of measured thermal gradients for water movement through the northeast Japan accretionary prism: Journal of Geophysical Research, v. 92, p. 3683–3690.

Roberts, B., and Meriman, R. J., 1985, The distinction between Caledonian burial and regional metamorphism in metapelites from North Wales: An analysis of isocryst patterns: Journal of the Geological Society of London, v. 142, no. 4, p. 615–624.

Sakai, H., 1987, Active faults in the Muroto Peninsula: Journal of the Geological Society of Japan, v. 93, p. 513–516.

——, 1988, Origin of the Misaki Olistostrome Belt and re-examination of the Takachiho orogeny: Journal of the Geological Society of Japan, v. 94, p. 945–961.

Seely, D. R., Vail, P. R., and Walton, G. G., 1974, Trench slope model, in Burk, C. A., and Drake, C. L., eds., The geology of continental margins: New York, Springer-Verlag, p. 249–260.

Seno, T., and Maruyama, S., 1984, Paleogeographic reconstruction and origin of the Philippine Sea: Tectonophysics, v. 102, p. 53–84.

Sugiyama, Y., 1989a, Bend of the zonal structure of island arcs and oblique subduction as the cause of bending; Part 1, Bending structures off the Outer Zone of Southwest Japan and "plate boundary" earthquakes: Bulletin of the Geological Survey of Japan, v. 40, p. 533–541.

——, 1989b, Bend of the zonal structure of island arcs and oblique subduction as the cause of bending; Part 2, Bending structures of the outer zone of Southwest Japan and the history of relative motion of the Philippine Sea Plate with respect to southwest Japan: Bulletin of the Geological Survey of Japan, v. 40, p. 543–564.

Sweeny, J. J., and Burnham, A. K., 1990, Evaluation of a simple model of vitrinite reflectance based on chemical kinetics: Bulletin of the American Association of Petroleum Geologists, v. 74, p. 1559–1570.

Taira, A., 1985, Sedimentary evolution of Shikoku subduction zone: Shimanto Belt and Nankai Trough, in Nasu, N., Uyeda, S., Kushiro, I., Kobayashi, K., and Kagami, H., eds., Formation of Active Ocean Margins, OJI International Seminar, Nov. 21–23, 1983: Tokyo, Terra Scientific Publishing Company, p. 835–851.

Taira, A., Tashiro, M., Okamura, M., and Katto, J., 1980, The geology of the Shimanto Belt in Kochi Prefecture, Shikoku, Japan, in Taira, A., and Tashiro, M., eds., Geology and paleontology of the Shimanto Belt, Kochi: Rinya-Kosakai Press, p. 319–389.

Taira, A., Okada, H., Whitaker, J.H.McD., and Smith, A. J., 1982, The Shimanto Belt of Japan: Cretaceous to lower Miocene active margin sedimentation, in Leggett, J. K., ed., Trench-forearc geology: Sedimentation and tectonics on modern and ancient active margins: London, Geological Society of London Special Publication 10, p. 5–26.

Taira, A., Katto, J., Tashiro, M., Okamura, M., and Kodama, K., 1988, The Shimanto Belt in Shikoku, Japan—Evolution of Cretaceous to Miocene accretionary prism: Modern Geology, v. 12, p. 5–46.

Teichmuller, M., 1987, Organic material and very low-grade metamorphism, in Frey, M., ed., Low temperature metamorphism: Glasgow, Blackie & Son Ltd., p. 114–161.

Tissot, B. P., Pelet, R., and Ungerer, P. H., 1987, Thermal history of sedimentary basins, maturation indices, and kinetics of oil and gas generation: Bulletin of the American Association of Petroleum Geologists, v. 71, p. 1445–1466.

Underwood, M. B., and Howell, D. G., 1987, Thermal maturity of the Cambria slab, an inferred trench-slope basin in central California: Geology, v. 15, p. 216–219.

Underwood, M. B., Laughland, M. M., Wiley, T. J., and Howell, D. G., 1989, Thermal maturity and organic geochemistry of the Kandik Basin region, east-central Alaska: U.S. Geological Survey Open-File Report 89-353, 41 p.

Van den Beukel, J., and Wortel, R., 1988, Thermo-mechanical modeling of arc-trench regions: Tectonophysics, v. 154, p. 177–193.

Vrolijk, P., 1987, Rapid tectonically-driven fluid flow in the Kodiak accretionary complex, Alaska: Geology, v. 15, p. 466–469.

Wang, C., and Shi, Y., 1984, On the thermal structure of subduction complexes: A preliminary study: Journal of Geophysical Research, v. 89, p. 7709–7718.

Waples, D. W., 1980, Time and temperature in petroleum formation: Application of Lopatin's method to petroleum exploration: Bulletin of the American Association of Petroleum Geologists, v. 64, p. 916–929.

Yamano, M., Uyeda, S., Aoki, Y., and Shipley, T. H., 1982, Estimates of heat flow derived from gas hydrates: Geology, v. 10, p. 339–342.

MANUSCRIPT ACCEPTED BY THE SOCIETY APRIL 24, 1992

Geological Society of America
Special Paper 273
1993

The thermal imprint of spreading ridge subduction on the upper structural levels of an accretionary prism, southwest Japan

J. P. Hibbard
Department of Marine, Earth, and Atmospheric Sciences, Box 8208, North Carolina State University, Raleigh, North Carolina 27695
M. M. Laughland* and S. M. Kang
Department of Geological Sciences, University of Missouri, Columbia, Missouri 65211
D. Karig
Department of Geological Sciences, Snee Hall, Cornell University, Ithaca, New York 14853

ABSTRACT

Upper Oligocene to lower Miocene accreted rocks and slope-basin strata of the Shimanto accretionary prism at Cape Muroto, southwest Japan, recently have been interpreted as recording the subduction of the active Shikoku back-arc basin spreading ridge. The Shimanto paleotrench was suborthogonal to the trend of the active spreading ridge and subduction apparently started at ~15 Ma. This event resulted in the formation of a deep structural embayment into the prism, accompanied by mid-ocean ridge basalt (MORB) magmatism in the core of this embayment. Our investigation of the thermal history of these rocks by means of vitrinite reflectance and illite crystallinity indicates that they have experienced high geothermal gradients compared to other accretionary prisms.

Mean vitrinite reflectance ($\%R_m$) values determined for the accreted strata range from 0.9 to 3.7%, corresponding to a peak paleotemperature range of ~140 to 280°C; $\%R_m$ values for slope-basin strata cluster at ~1.6%, or paleotemperature of ~200°C. Illite crystallinity studies support the reflectance data. Locally, high $\%R_m$ values associated with clastic dikes along major faults within the accreted strata suggest that the faults served as conduits for high temperature fluids. Based on structural-stratigraphic relationships and porosity measurements, maximum burial of the slope-basin strata is estimated to have been 2 to 4 km; combined with the paleotemperature data, this estimate indicates a *minimum* paleogeothermal gradient in the slope-basin on the order of 70°C/km. Paleogeothermal gradients were likely higher in the accreted strata.

The regional thermal pattern overprints structures related to initial phase of accretion and subsequent intraprism deformation, but thermal maturity has been affected by the embayment and grades smoothly away from the aureoles of the mafic intrusions in the core of the embayment. The embayment and the intrusions are direct results of spreading ridge subduction. Thus, we consider the unusually high paleogeothermal gradients at Cape Muroto to reflect the thermal imprint of active spreading ridge subduction.

*Present address: Mobil Research and Development, Exploration, and Producing Technical Center, 3000 Pegasus Park Dr., Dallas, Texas 75265-0232.

Hibbard, J. P., Laughland, M. M., Kang, S. M., and Karig, D., 1993, The thermal imprint of spreading ridge subduction on the upper structural levels of an accretionary prism, southwest Japan, *in* Underwood, M. B., ed., Thermal Evolution of the Tertiary Shimanto Belt, Southwest Japan: An Example of Ridge-Trench Interaction: Boulder, Colorado, Geological Society of America Special Paper 273.

INTRODUCTION

In recent years, it has become increasingly apparent that subduction of an active spreading ridge significantly influences the evolution of the overriding forearc. A number of studies indicate that the forearc above such a triple junction records the kinematic, magmatic, and thermal influence of the active spreading ridge (Marshak and Karig, 1977; DeLong and Fox, 1977; Herron and others, 1981; Weissel and others, 1982; Barker, 1982; McLaughlin and others, 1982; Moore and others, 1983; Forsythe and Nelson, 1985; Forsythe and others, 1986; Cande and Leslie, 1986; Perfit and others, 1987; Hibbard and Karig, 1990a). The extent of such modifications upon the forearc appear to be related to the geometry of the triple junction, as well as the relative plate velocities. Although TRT (trench-ridge-trench) triple junctions are seemingly "freakish" manifestations of convergent margin geology, they are important, for as noted by DeLong and Fox (1977) and cogently echoed by Nelson and Forsythe (1989), the closing of any ocean basin requires the subduction of at least one spreading ridge. From a review of the references cited above, it is clear that spreading ridge subduction has been a significant event in the evolution of many Tertiary circum-Pacific convergent margins.

The meager available data about the effects of spreading ridge subduction mainly center upon kinematic and magmatic modifications to the forearc (e.g., Marshak and Karig, 1977; Moore and others, 1983; Forsythe and Nelson, 1985; Cande and Leslie, 1986; Hibbard and Karig, 1990a), but there is an obvious dearth of data concerning the thermal structure of the forearc during spreading ridge subduction. Knowledge of the thermal effects of ridge subduction is limited to studies of a fossil accretionary complex in southern Alaska, heat-flow studies at active convergent margins where young oceanic crust is being subducted, and theoretical modelling. Low-pressure/high-temperature metamorphism of the Chugach metamorphic complex of Alaska has been attributed to an elevated geothermal gradient ($>35°C/km$) in the mid- to deep structural levels of an accretionary prism induced, at least in part, by Paleocene subduction of either a spreading ridge or very young oceanic crust (Sisson and others, 1989; James and others, 1989). Anomalously high, near-surface heat-flow measurements along the active Nankai accretionary prism (Yamano and others, 1984; Kinoshita and Yamano, 1985; Nagihara and others, 1989) and the Washington-Oregon convergent margin (Shi and others, 1988) have been attributed to the subduction of young, hot oceanic crust beneath these areas.

More holistic views of the potential thermal effects of ridge subduction on accretionary prisms derive from numerical modelling (DeLong and others, 1978; James and others, 1989); these models are based upon conductive heating from the subducting slab and predict the expected results of elevated geothermal gradients in the accretionary prism. However, these models can only be considered as rough approximations because the thermal structure of accretionary prisms appears to result from the complex interplay of many factors, including radiogenic heating, accretion and erosion in the prism, internal strain heating, frictional heating on the décollement, and advection by fluid flow, in addition to heat conduction from the mantle and the velocity of subduction (Wang and Shi, 1984; Reck, 1987; Dahlen and Barr, 1989). While it is intuitive that the first-order effect of spreading ridge subduction is to increase conductive heat flow in the overriding prism, the effects of such elevated heat flow on other processes, such as frictional heating on the décollement and fluid flow are unknown. Clearly, thermal field data are needed to constrain the theoretical modelling and more accurately portray the thermal effects of ridge subduction on accretionary prisms.

Recently, investigation of the Shimanto accretionary complex at Cape Muroto, Shikoku Island, southwest Japan (Fig. 1), has brought to light a young, extinct, TRT triple junction. The cape is underlain by upper Oligocene to lower Miocene accreted rocks that lie almost directly inboard from the extension of the fossil Shikoku back-arc basin spreading ridge, which was active from ca. 26 to 14 Ma (Chamot-Rooke and others, 1987; Fig. 1). The upper Oligocene to lower Miocene Shimanto rocks are at subgreenschist grade and record, at upper structural levels of the accretionary complex, subduction of the active Shikoku basin spreading ridge. The mechanical and magmatic effects of ridge subduction on this portion of the prism have been reported elsewhere (Hibbard and Karig, 1987, 1990a; Byrne and Hibbard, 1987; Hibbard and others, 1992); in this paper, we examine, in detail, by means of organic and inorganic metamorphism, the thermal affects of ridge subduction on the upper structural level of the prism at Cape Muroto.

GEOLOGIC SETTING

The upper Oligocene to lower Miocene rocks at Cape Muroto constitute the Nabae subbelt, which is the youngest and most outboard division of the Cretaceous to Miocene Shimanto accretionary prism of southwest Japan (Taira and others, 1988). The fossil prism extends from the Boso Peninsula, near Tokyo, southwestward to the Ryuku Islands (Fig. 1). It is separated from predominantly Mesozoic accretionary terranes to the north by a thrust fault, the Butsuzo Tectonic Line. Within the prism, the Nabae subbelt is separated from bordering Eocene to lower Oligocene rocks of the Murotohanto subbelt (Taira and others, 1980) to the north by the steeply northwest-dipping Shiina-Narashi fault (Hibbard and others, 1992; Fig. 2); the fault is interpreted as a steep thrust with the Nabae subbelt forming the footwall. To the south, the Nabae subbelt presumably continues offshore to merge with the active Nankai accretionary complex, which is overriding the inactive Shikoku back-arc basin, now a portion of the Philippine Sea Plate. At present, the Nabae subbelt lies almost directly along the trend of the extinct Shikoku Basin spreading ridge (Fig. 1). In this section we introduce the stratigraphic and structural aspects of the subbelt pertinent to the thermal data and summarize the evidence for subduction of the Shikoku Basin spreading ridge beneath the subbelt at ca. 15 Ma.

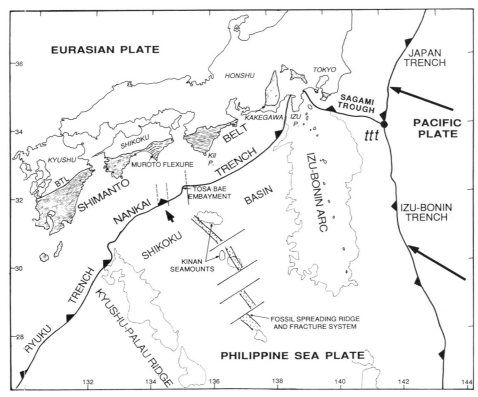

Figure 1. Modern plate setting and major geological elements of southwest Japan. Fossil Shikoku Basin spreading ridge from Chamot-Rooke and others (1987). Heavy arrows indicate approximate relative plate motions; BTL, Butsuzo Tectonic Line; ttt, trench-trench-trench triple junction; and fine lines on the Philippine Sea Plate are 4,000-m bathymetric contours.

Lithostratigraphy

The Nabae subbelt comprises three major tectonostratigraphic units: (1) accreted strata and mélange of the Nabae complex, (2) slope-basin strata of the Shijujiyama Formation, and (3) plutonic rocks of the Maruyama Intrusive Suite.

Nabae complex. Accreted strata of the Nabae complex underlie most of the Nabae subbelt and are divided into two coherent units, the Tsuro and Misaki assemblages, and two mélanges, the Sakamoto and Hioki mélanges (Fig. 2). Biostratigraphic data loosely constrain the age of the complex as late Oligocene–early Miocene (reviewed in Hibbard and others, 1992); the tightest age constraint is provided by Aquitanian foraminifera from the Sakamoto mélange (Saito, 1980).

The Tsuro assemblage comprises a basal varicolored hemipelagic member overlain by medium-bedded turbidites. The two mélanges are composed of mainly clastic sedimentary blocks encased in a ubiquitous dark gray mudstone-shale matrix. The mélange units appear to form a continuous structural substrate to the Tsuro assemblage. The Misaki assemblage resembles the Tsuro rocks, but lacks the basal hemipelagic unit; the structural position of the Misaki rocks with respect to the other units is ambiguous.

Shijujiyama Formation. Slope-basin strata comprise a basal mudstone and mafic volcanic-volcaniclastic unit, the Shiina member, overlain by a sequence of thick-bedded massive sandstone. Most biostratigraphic data indicate a late Oligocene–early Miocene age for the formation; however, early Miocene foraminifera and radiolaria have been reported from the basal Shiina member (Ishikawa, 1982). The slope-basin strata structurally overlie the Hioki mélange and are less deformed than the accreted strata (Fig. 2); the contact between the two units is a structural gradation that has been interpreted as a tectonically modified unconformity (Katto and Arita, 1966; Hibbard and others, 1992).

Maruyama Intrusive Suite. The suite consists of numerous mafic dikes that intrude accreted rocks of the Nabae complex along the eastern side of Cape Muroto. An Rb/Sr biotite–whole-rock age and a K/Ar date determined for the gabbro body in the suite at Cape Muroto cluster at approximately 14 Ma (Hamamoto and Sakai, 1987; Kodama and Takigami, personal communication, 1987). Petrochemical analysis indicates that the Maruyama Suite represents T-type mid-ocean ridge basalts (MORBs), geochemically equivalent to some mafic rocks dredged from the Shikoku Basin (Hibbard and Karig, 1990a).

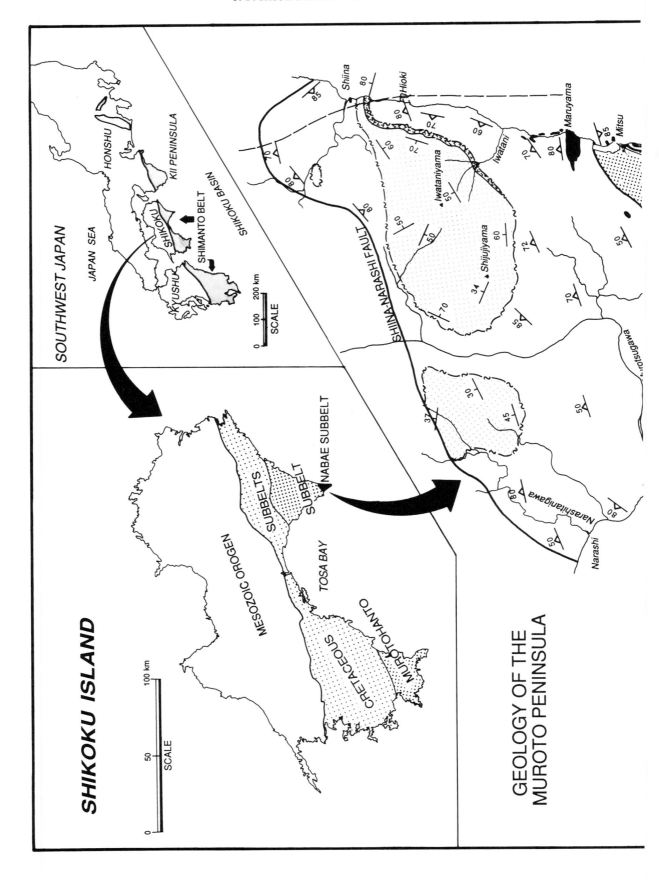

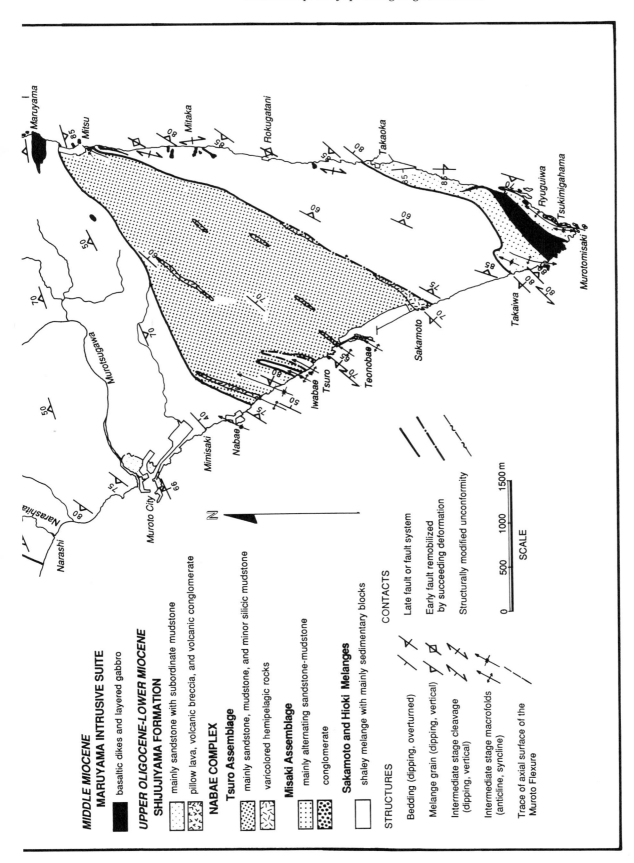

Figure 2. Location and general geology of the Nabae subbelt.

MIDDLE MIOCENE

MARUYAMA INTRUSIVE SUITE

basaltic dikes and layered gabbro

UPPER OLIGOCENE-LOWER MIOCENE

SHIJUJIYAMA FORMATION

mainly sandstone with subordinate mudstone

pillow lava, volcanic breccia, and volcanic conglomerate

NABAE COMPLEX

Tsuro Assemblage

mainly sandstone, mudstone, and minor silicic mudstone

varicolored hemipelagic rocks

Misaki Assemblage

mainly alternating sandstone-mudstone

conglomerate

Sakamoto and Hioki Melanges

shaley melange with mainly sedimentary blocks

CONTACTS

Late fault or fault system

Early fault remobilized
by succeeding deformation

Structurally modified unconformity

STRUCTURES

Bedding (dipping, overturned)

Melange grain (dipping, vertical)

Intermediate stage cleavage
(dipping, vertical)

Intermediate stage macrofolds
(anticline, syncline)

Trace of axial surface of the
Muroto Flexure

SCALE

0   500   1000   1500 m

Structural evolution

Three broad phases of tectonism affected rocks of the Nabae subbelt: (1) an early offscraping event involving landward-vergent thrusting of the Tsuro assemblage and possibly the Misaki assemblage over the mélanges, (2) intermediate stage regional shortening of the thrust sheet and mélanges by continued thrusting and folding, and (3) a late, complex, tectonomagmatic event. The accreted strata have been subjected to all three of these phases, whereas the Shijujiyama Formation has experienced only the latter two of these and the Maruyama Intrusive Suite was deformed during part of the last phase (Hibbard and others, 1992).

Structures related to the early offscraping of the Tsuro assemblage are inhomogeneously developed; the regional mélange fabric, landward-vergent thrusts, and isoclinal and sheath folds represent the most diagnostic structures of this phase. A unique "contact mélange," distinct from the regional mélange units, is developed locally at the tectonic contact between the Tsuro assemblage and the regional mélange (Hibbard and others, 1992); the contact mélange contains blocks of basalt and varicolored hemipelagites that are not found in the regional mélange. Early phase structures in the Misaki assemblage are similar to those in the Tsuro assemblage. The folds in both units indicate that the landward-vergent thrusting was directed suborthogonally to the Shimanto Belt and hence, suborthogonal to the Shimanto paleotrench. The polarity of accretion of the regional mélange terrane is ambiguous.

The subsequent intermediate stage of deformation involved tight asymmetric (towards the land) upright folding around horizontal axes, thrusting, and a variably developed pressure solution cleavage that is now southeast dipping. Apparently, most of the Shijujiyama Formation was deposited atop the Hioki mélange sometime late during this deformation, because the formation appears to be only openly folded and lacks cleavage. The late stage tectonomagmatic event is characterized by the generation of a regional orocline, the Muroto flexure (Hibbard and others, 1992), pervasive faulting, rotation of intermediate phase folds to vertical orientations in the mechanically softer mélanges, and mafic magmatism in the core of the regional flexure.

Early Miocene tectonics

Most reconstructions of the plate configuration for southwest Japan at 15 Ma indicate that there has been minimal lateral migration of features on the Philippine Sea Plate along the southwest Japan margin since that time (Seno and Maruyama, 1984; Hibbard and Karig, 1990a, b; Fig. 3). These reconstructions depict the Muroto area as being juxtaposed against the TRT triple junction formed by the Shikoku Basin spreading ridge and the Shimanto paleotrench in the earliest middle Miocene time. Based on structural-stratigraphic data given above, the Nabae subbelt was accreted to the Shimanto accretionary prism and deformed sometime between 22 and 14 Ma; this indicates that

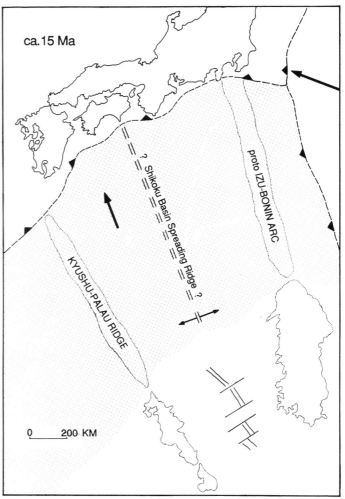

Figure 3. Plate reconstruction for southwest Japan at circa 15 Ma; mainly after Seno and Maruyama (1984). Bold arrows indicate approximate relative plate motions. Note that no allowances made for the opening of the Sea of Japan back-arc basin to the north of southwest Japan.

there was active subduction during the opening of the Shikoku Basin.

This tectonic scenario, considered in conjunction with the geology at Cape Muroto, has led us to interpret the Nabae subbelt as recording the effects of suborthogonal subduction of the Shikoku Basin spreading ridge (Hibbard and Karig, 1990a, b; Hibbard and others, 1992). In particular, the Muroto flexure and attendant deformation is viewed as resulting from the mechanical indentation of the Shimanto accretionary prism in the early Miocene. In addition, the localization of the Maruyama Intrusive Suite MORBs in the core of the flexure is interpreted as representing spreading ridge magmatism during subduction of the active ridge. The landward-vergent accretion of the coherent assemblages may relate to spreading ridge subduction (Byrne and Hibbard, 1987), although the strong overprint of later deformation renders any direct correlation between the two events ambiguous.

Recognition of this triple TRT triple junction and establishment of the mechanical and magmatic responses to subduction of the active spreading ridge (Hibbard and Karig, 1990a) permit further investigation into this tectonic situation. It is timely, now, to establish the thermal effects of subduction of the Shikoku Basin spreading ridge on the Nabae subbelt. In the following sections we will present the metamorphic data for the area, determine the timing and depth of metamorphism and discuss our results in the context of accretionary prism thermal structure.

THERMAL IMPRINT ON THE NABAE SUBBELT

Field and petrographic observations indicate that sedimentary rocks of the Nabae subbelt were subject to diagenesis to very low grade metamorphism. The thermal history of the subbelt is investigated here by both organic and inorganic means. Vitrinite reflectance measurements were conducted on dispersed organic material extracted from clastic rocks of the Nabae complex and Shijujiyama Formation and illite crystallinity studies were undertaken on the finer-grained rocks of these units. In addition, a few samples from the Murotohanto subbelt, immediately to the north of the Shiina-Narashi fault, were analyzed for comparison of the thermal history of the subbelts across the fault. Laboratory methods are described in detail elsewhere in this volume (Underwood, Laughland, and Kang; Laughland and Underwood).

In this study, quantitative interpretation of the thermal maturity of the Nabae subbelt relies mainly upon the vitrinite reflectance data; Laughland and Underwood, Laughland, and Kang (this volume) have discussed the various correlations of mean random reflectance with absolute paleotemperature and calculated paleotemperatures for the Nabae data by a number of different methods. They consider the correlation of Barker (1988; T [°C] = 104[ln%R_m] + 148) to be most accurate for the range of reflectance values found in the Nabae subbelt; this correlation is used for all paleotemperature interpretations given here. Illite crystallinity has yet to be calibrated directly against temperature for a broad range of physical conditions, thus the data here chiefly serve as a semiquantitative check on the reflectance data.

Vitrinite reflectance results and paleotemperature

The vitrinite reflectance data for the area are given in Laughland and Underwood (this volume) and summarized in Figures 4, 5, 6, 7. These data clearly show that there is a major, abrupt decrease in reflectance values as the Shiina-Narashi fault is traversed from north (the Murotohanto subbelt) to south (the Nabae subbelt). Our detailed data support and confirm the original interpretation of a major change in %R_m values in this area of the Shimanto accretionary prism by Mori and Taguchi (1988), based on limited data from both subbelts. Mean reflectance values determined here for the Murotohanto subbelt adjacent to the fault range from 3.3 to 3.9%, with an average of 3.7%. In

contrast, the immediately adjacent rocks of the Hioki mélange and the Shijujiyama Formation show average mean reflectance values of 1.6%. In terms of paleotemperature, these values equate to a temperature differential of approximately 85 to 90°C as recorded by strata across the fault (also see Laughland and Underwood, this volume).

The Nabae complex shows a wide range of vitrinite reflectance values, from 0.9 to 3.7% (~140 to 280°C; Figs. 4, 5). At the scale of the subbelt, this variation is manifest as two orderly trends, one across the strike, and the other along the strike, of the subbelt. Most obviously, %R_m values generally increase southward across the strike of the complex. This trend is evident in the general increase in average %R_m values of the subdivisions of the complex from north to south (Hioki mélanage average = 1.6%, Tsuro assemblage average = 1.8%, Sakamoto mélange average = 1.7%, Misaki assemblage average = 2.9%; Fig. 5). However, it is best seen in the north-south profiles along each side of the Muroto peninsula, in which strike-parallel variation is eliminated from the data (Fig. 6). On the east side of the peninsula, there is an increase from an average %R_m of 1.3% in the Hioki mélange, through an average of 1.6% in the Sakamoto mélange to an average value of 2.9% in the Misaki assemblage; this corresponds to an overall paleotemperature difference of approximately 85°C along the profile. The west side of the peninsula shows an increase in average mean reflectivity from 1.8% in both the Hioki mélange and the Tsuro assemblage, through an average of 2.1% in the Sakamoto mélange, to an average of 2.9% in the Misaki; this equates to an overall paleotemperature difference of approximately 50°C. These paleotemperature differences actually may have been greater than recorded by the organic material, as garnet and biotite are found near the contact of the Misaki assemblage with the gabbro body at Cape Muroto (also see Laughland and Underwood, this volume). In both cross-strike profiles of the subbelt, it appears that the regional southward increase in reflectance values is gradational and is largely independent of the boundaries between components of the complex.

There is also an along-strike increase in %R_m in the subbelt from east to west. This is best displayed by samples from the Hioki and Sakamoto mélanges (Fig. 7). The Hioki mélange shows a change in average %R_m from 1.3 % on the east coast to 1.8% on the west coast. These along-strike changes in %R_m correspond to paleotemperature differences of approximately 37°C in the Hioki mélange and 30°C in the Sakamoto mélange. The nature of the along-strike variation in %R_m (gradational or abrupt), is difficult to assess because of the sparcity of inland data. However, most of the lower reflectance values on the east coast appear to be confined to the immediate vicinity of the core of the Muroto flexure along the east coast of the peninsula.

In contrast to the variations in %R_m in the underlying Hioki mélange, slope-basin strata of the Shijujiyama Formation exhibit a remarkably tight cluster of reflectance values throughout the outcrop area of the formation, with a range of 1.4 to 1.8% (~180 to 210°C; Figs. 4, 5) and an average %R_m of 1.6% (paleotemperature ~200°C). The westward increase in %R_m evident in the

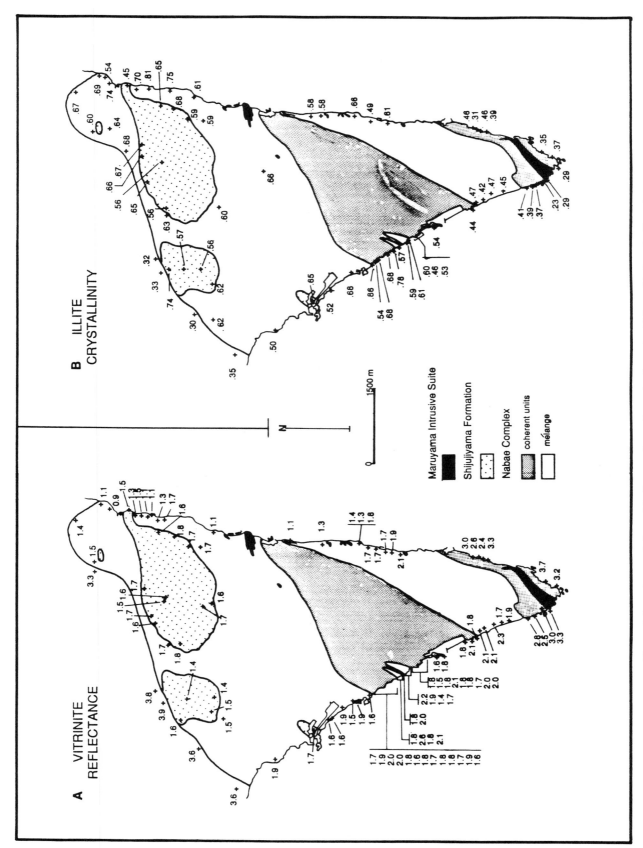

Figure 4. Locations and thermal indicator values of samples from the Nabae subbelt and adjacent portions of the Murotohanto subbelt analyzed in this study; A, R_m values; B, values from illite crystallinity analyses in units of $\Delta°2\Theta$.

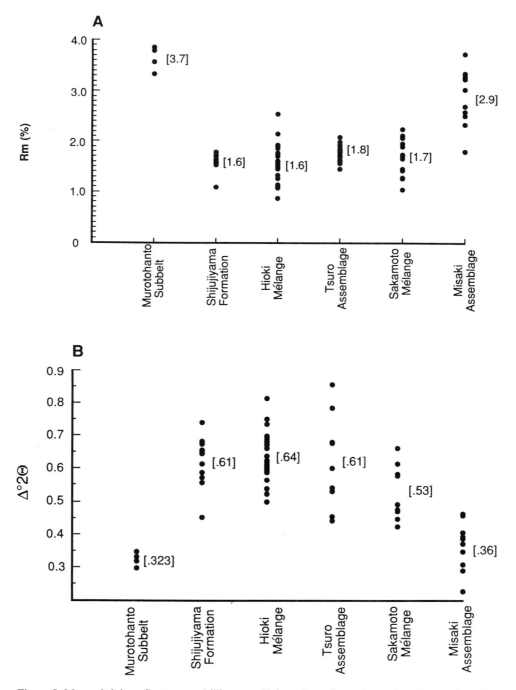

Figure 5. Mean vitrinite reflectance and illite crystallinity values of samples analyzed by stratigraphic unit. Note the left to right format of the stratigraphic units corresponds to a north to south across-strike trend across the area. A, $\%R_m$ values for stratigraphic units showing the large contrast in $\%R_m$ between the Murotohanto subbelt and adjacent rocks of the Nabae subbelt and the overall general southward increase in $\%R_m$ within the Nabae subbelt; average values for each stratigraphic unit given in brackets. B, illite crystallinity values for stratigraphic units showing same general trends as reflectance data; average values for each unit given in brackets.

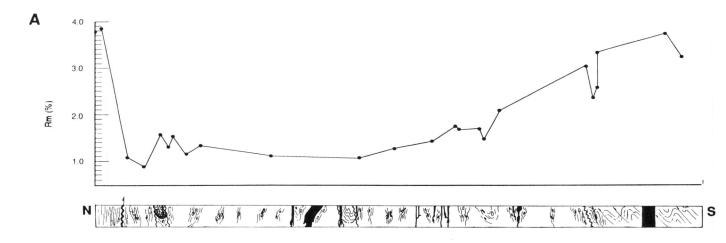

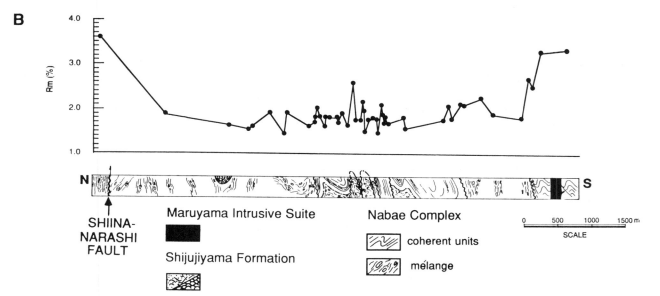

Figure 6. Across-strike, north-south profiles of geology and vitrinite reflectance values for samples from the Nabae subbelt and immediately adjacent Murotohanto subbelt. A, profile along the east coast of the Muroto Peninsula; B, profile along the west coast of the Muroto Peninsula.

underlying Hioki mélange is absent in the Shijujiyama Formation data; moreover, if the Shijujiyama data are divided into eastern and western portions along the Murotsugawa Valley, average R_m for the western Shijujiyama Formation is slightly lower than that of the eastern portion of the formation (1.5% versus 1.7%).

In addition to these first-order observations of the vitrinite reflectance data, there are a number of second-order anomalies evident in the data. Most of these anomalies have a number of nonunique explanations. Two of these anomalies are worthy of note here and will be interpreted in a succeeding section. Detail sampling around the windows of mélange within the Tsuro assemblage along the west coast of the peninsula (Fig. 2) has revealed a major anomalous reflectance value in each window,

very near the contact between the assemblage and the underlying mélange (Figs. 4, 6, 8). Reflectance values of 2.6% and 2.2% in the mélange at the contact contrast sharply with surrounding values in both the mélange and assemblage; these anomalous values indicate a paleotemperature on the order of 20 to 40°C higher than those typical of the surrounding area.

Illite crystallinity data

The results of our illite crystallinity investigation are presented in Underwood, Laughland, and Kang (this volume) and summarized in Figures 4, 6, 7; the crystallinity index is equal to peak width at half height for (001) illite peak (~10Å) after

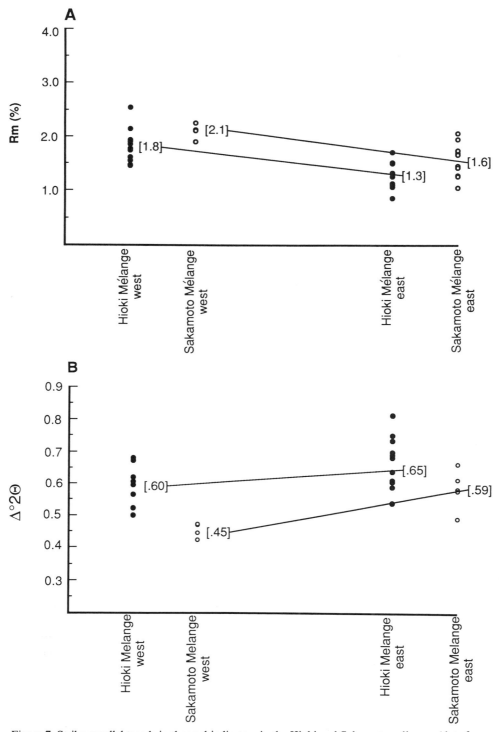

Figure 7. Strike-parallel trends in thermal indicators in the Hioki and Sakamoto mélanges (data from west and east coasts of each unit). Left to right format of units in figure corresponds to a west-to-east trend in thermal values. Bracketed numbers are averages of each unit. A, vitrinite reflectance values decrease from west to east; B, illite crystallinity values show a corresponding increase in $\Delta°2\Theta$ for each unit.

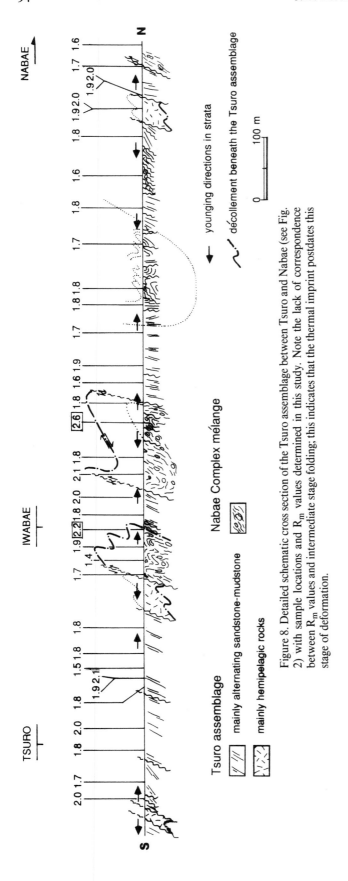

Figure 8. Detailed schematic cross section of the Tsuro assemblage between Tsuro and Nabae (see Fig. 2) with sample locations and R_m values determined in this study. Note the lack of correspondence between R_m values and intermediate stage folding; this indicates that the thermal imprint postdates this stage of deformation.

treatment with ethylene glycol and it is measured in units of $\Delta°2\theta$. Recently, it has been generally agreed upon that the anchizone is defined by crystallinity values ranging from 0.42 to $0.25\Delta°2\theta$ (Frey, 1987; Kisch, 1987; Blenkinsop, 1988), with diagenetic zone values $>0.42\ \Delta°2\theta$. Based upon this zonation, all of the Nabae subbelt is in the zone of diagenesis, with the exception of the Misaki assemblage, which is mainly anchimetamorphic, with one sample reaching epizonal conditions. Samples collected from the Murotohanto subbelt in this study are all in the anchizone.

A single occurrence of pumpellyite in thin section was found in a sample from Sakamoto (Fig. 2), very close to samples that record illite crystallinity values of 0.42 to $0.47\Delta°2\theta$, close to the low-temperature end of the anchizone, and mean reflectance values of 1.8 to 2.2%. Kisch (1987) has noted that the pumpellyite facies correlates approximately with the anchizone.

All of the major trends outlined by vitrinite reflectance are supported by the crystallinity data. The crystallinity data clearly show the Shiina-Narashi fault to be a major boundary in the thermal structure of the area. The overall increase in vitrinite reflectance southward across strike of the subbelt, as well as the westward along-strike increase in reflectance values, is mirrored by a corresponding decrease in crystallinity values (Figs. 6, 7). As well, the lack of any such trends in the Shijujiyama Formation is corroborated by the crystallinity data (Fig. 4).

On the basis of compilation of fluid inclusion and oxygen isotope data, Kisch (1987) has indicated that the onset of anchizone conditions takes place at temperatures of ~200°C. In the Nabae subbelt, samples with illite crystallinity values in the area of the lower limit of the anchizone (0.42 to $0.45\Delta°2\theta$) have reflectance values between 1.6 and 2.1%. Using the paleotemperature scheme of Barker (1988), these reflectance values correspond to paleotemperatures of approximately 200 to 225°C; thus, there is excellent agreement between the organic and inorganic paleotemperature indicators (also see Underwood, Laughland, and Kang, this volume).

Timing of the thermal imprint

Vitrinite reflectance values are a measure of peak heating. Considering that the unusual structural-magmatic history of the Nabae subbelt has been related to subduction of a spreading ridge, it is intuitive that introduction of such a heat source would also yield peak thermal conditions within the accretionary edifice. In this section we review the evidence in support of this hypothesis. Establishing the timing of peak heating with respect to local tectonic events is also important in determining the stratigraphic-structural conditions of metamorphism and hence, paleogeothermal gradients. On the basis of the smoothly varying nature of the regional thermal pattern, it is assumed that peak heating was synchronous throughout the subbelt.

The broad-scale trends of the thermal data outlined above show little regard for stratigraphic and structural boundaries; this strongly suggests that peak metamorphic conditions were attained either late syn- or post-kinematically with respect to all deformation in the subbelt. The thermal structure shows two major trends with relation to the deformation scheme of the subbelt: (1) the reflectance values decrease smoothly away from the Maruyama Intrusive Suite, suggesting that peak heating of the subbelt was synchronous with emplacement of the suite; and (2) the lowest reflectance values in the subbelt are largely confined to the trace of the axial surface of the Muroto flexure (Fig. 4) suggesting that the thermal evolution is genetically related to the flexure. Both the Maruyama Intrusive Suite and the Muroto flexure are manifestations of the late-phase, spreading-ridge collision; thus, it appears that peak metamorphism of the subbelt is related to this event.

There is a broad, qualitative correlation between the development of a strong pressure solution cleavage and elevated reflectance values in the subbelt; this is particularly evident in the Tsuro assemblage at Sakamoto, the Misaki assemblage at Takaoka, and the Sakamoto mélange at Mitaka. The cleavage is axial planar to intermediate phase folds, and thus, one might infer that peak heating in these areas is related to the intermediate phase deformation. However, relationships at Mitaka preclude such an interpretation. Here, the Sakamoto mélange is cut by basaltic dikes (1 to 2 m thick) of the Maruyama Intrusive Suite; in areas removed from the dikes (> ~1 m), the pressure solution cleavage is only subtly distinct from the mélange fabric, even though it crosscuts the fabric at a high angle, whereas adjacent to the dikes (< ~1 m), the cleavage is very intensely developed. This relationship is also evident in thin section; samples located away from the dikes show subtle opaque solution seams whereas in samples taken close to the dikes, opaque solution residues are much more intensely developed. These observations strongly suggest that locally, the intermediate stage pressure solution cleavage has been enhanced during later stage thermal activity.

A detailed vitrinite reflectance study of the Tsuro assemblage also was undertaken to help constrain the timing of peak heating (Fig. 8). The section of the assemblage along Tosa Bay represents a thrust sheet of varicolored hemipelagic rocks and overlying turbidites that has been emplaced upon the regional mélange terrane during early phase obduction; this thrusting resulted in the formation of the unique contact mélange zone between the thrust sheet and the regional mélange (Hibbard and others, 1992). Subsequently, both structural units were folded and cleaved during the intermediate phase of deformation and pervasively faulted during the late-stage event. Two windows of the mélange substrate are exposed in the middle of the section, mainly as a result of intermediate-phase folding.

Figure 8 shows that peak heating clearly overprints stratigraphy, early phase structural stacking, and intermediate-phase folding. In particular, the windows of mélange mainly show reflectance values either equivalent to or lower than those of the overlying Tsuro assemblage, indicating a post-early thrusting establishment of the regional thermal pattern. The major syncline in the north-central portion of the section exhibits increasing reflectance values both up-section and up-structure, demonstrating a post-intermediate stage peak metamorphism. The relationship be-

tween peak heating and the late pervasive fault system is difficult to establish, although some of the small variations in reflectance values may have resulted from shuffling of the section along the many late faults. These data lead to the interpretation that peak temperatures were attained late- to post-kinematically with respect to the major deformation phases, as suggested by regional relationships outlined above.

Two anomalous reflectance values associated with the structural windows in the Tsuro section appear to run counter to this broad interpretation. The high values are found in mélange, near the north contacts of the windows with the Tsuro assemblage (Fig. 8). In both cases, the samples analyzed show evidence of fluid flow of the sediments. In the southerly window, a sample from a shale diatreme that injects a folded hemipelagite block in the contact mélange zone (Fig. 9A) yields a $\%R_m$ of 2.2%. From the northerly window a shale sample near the late fault contact with the overlying Tsuro assemblage is injected by small random dikes (millimeter to centimeter scale) and irregular bodies of fine-grained sandstone (Fig. 9B) and yields a $\%R_m$ of 2.6%. In both cases, the fluidization of the sediment clearly predates the intermediate phase cleavage and can be related to early phase thrusting (Hibbard and others, 1992). The coincidence of fluidized sediments near a major early phase thrust contact and anomalously high reflectance values strongly suggests that these local samples record a relict thermal event, related to early phase movement on the basal thrust of the Tsuro assemblage, that has survived through the later lower-grade regional thermal overprint.

The sum of the data indicates that peak heating in most of the Nabae subbelt postdates the intermediate stage deformation, and regional trends suggest that it was synchronous with the late-stage magmatism and flexure formation. These late-stage events are considered to be manifestations of subduction of the Shikoku Basin spreading ridge, thus, the regional thermal event recorded in the subbelt is interpreted to also reflect this event. Locally, as at Iwabae, relict thermal peaks related to earlier events are preserved.

Depth of burial

Geologic relationships and porosity measurements of clastic sediments in the subbelt provide rough constraints on the depth of rocks in the subbelt at the time of metamorphism. Peak temperature was imprinted upon the subbelt following the emplacement and deformation of thrust sheets and the deposition of the Shijujiyama slope-basin. The preservation of landward-vergent thrust sheets attests to only shallow burial and minimal erosion of the Nabae complex. The deposition of the lower Miocene Shijujiyama slope basin atop the upper Oligocene to lower Miocene complex also suggests a shallow depth of burial for accreted rocks of the subbelt. The thickness of the slope basin should be an approximate indication of the maximum depth of burial of the subbelt.

A semiquantitative assessment of the thickness of the Shiju-

Figure 9. Examples of fluid features in sedimentary rocks with anomalously high $\%R_m$ values. A, pebble-rich black mudstone dike piercing varicolored hemipelagite strata in large block in contact mélange in structural window at Iwabae. $\%R_m$ value of 2.2% obtained from the mudstone matrix of the dike. B, black mudstone engulfed by irregular bodies of intrusive sandstone in the structural window north of Iwabae. $\%R_m$ value of 2.6% obtained from the mudstone, which appears to be stratigraphically in situ.

jiyama slope basin can be made by comparison with both the depth of known slope basins and the thicknesses of nearby early-middle Miocene basins preserved elsewhere atop the Shimanto Belt, along strike of the Nabae subbelt. Some of the thickest modern trench slope basins reported are from the Sunda arc, where Moore and others (1980) have noted upper-trench slope basins with 2 to 3 km of sediment fill on seismic reflection profiles. Preserved Neogene slope-basin deposits on Nias Island, Indonesia, may be as thick as 3 to 5 km, although the extent of structural repetition in this section is unknown (Moore and Karig, 1980). Thicker slope basins have been encountered in the Aleutian accretionary prism (up to 7 km; J. Sample, personal communication, 1990); however, data from other portions of the Shimanto Belt suggest that the Shijujiyama basin probably never attained such a thickness. Early to middle Miocene slope basins also overlie Tertiary accreted rocks of the Shimanto Belt in western Shikoku Island and on the Kii Peninsula, to the east of Muroto (Fig. 2). The Misaki Group of western Shikoku has been estimated to be ~3,000 m thick (Katto and Taira, 1978; Kimura, 1985) and the Tanabe and Kumano Groups of Kii range from ~1,100 to 4,000 m (Chijiwa, 1988; Hisatomi, 1988). As with the Nias basin, the effects of structural repetition in these areas are unknown. From all data presented, it appears that a reasonable estimate of the maximum burial depth for the Shijujiyama Formation is on the order of 2 to 4 km.

On the basis of paleomagnetic and structural arguments, Hibbard and Karig (1990a) have estimated a 25 to 35° bulk back rotation of the Nabae subbelt. If we consider (1) an original 10° topographic slope to the subbelt, (2) 25° of total back rotation, and (3) that a line connecting the base of the slope basin and rocks at Cape Muroto is horizontal, then simple geometrical calculations indicate that the Misaki assemblage was approximately 1 km deeper than the base of the Shijujiyama Formation. This depth differential could be even less if bulk rotation of the subbelt was distributed along a number of faults in the subbelt, each of which had a seaward side down component of dip-slip motion; this scenario could effectively back-rotate the subbelt yet maintain the same structural level at the prism surface. Such faults have not been identified, but they would be difficult to recognize considering the structural complexity of the subbelt.

Porosity measurements on the sandstones of the subbelt (Fig. 10) can also be used to estimate the depth of burial by comparison with a porosity-depth graph for accretionary prisms compiled by Bray and Karig (1985). Porosity measurements on the Nabae complex yield an average porosity of 1.5 (16 samples) whereas the Shijujiyama Formation has an average porosity of 7.2 (18 samples; Hibbard, 1988; Karig, unpublished data). Comparison with the porosity depth chart gives estimates of a 7- to 15-km burial depth for the Nabae complex and a 2- to 4-km depth for the Shijujiyama Formation. The Shijujiyama data are in good accord with the depth inferred from the slope-basin data, but the Nabae complex burial estimates seem unreasonably deep for an obducted unit.

It is possible that either the porosity data from the Nabae

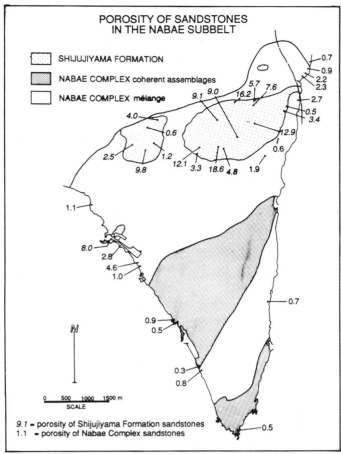

Figure 10. Sample location and porosity values of sandstones analyzed from the Nabae subbelt. Note, key in lower left corner does not indicate averages for units.

complex are inaccurate or that some factor other than burial and mechanical deformation is responsible for such low porosities. Houseknecht (1984) has shown that quartz sandstone porosity decreases with increased thermal maturity mainly as a result of enhanced intergranular pressure solution. Evidence of pressure solution is rampant throughout quartz-rich sandstones of the Nabae complex, not only in the form of grain suturing and nesting, but also as a pressure solution overprint on brittle faults and fractures, including late-phase structures. We suggest that the low porosities of the Nabae complex are a direct result of intense pressure solution, driven by the abnormally high heat flow in the subbelt. Similarly, samples of Shijujiyama sandstones with anomalously low porosities appear to display more affects of pressure solution in thin section than those with higher porosities.

Consideration of the regional geologic relationships, the thickness of known slope-basins, and porosity data all lead us to the conclusion that the maximum depth of burial of the base of the slope basin during metamorphism was approximately 3 km. The southerly portions of the Nabae complex may have been up

to 1 km deeper. These estimates are probably greater than depths actually achieved by rocks now exposed in the subbelt, because estimates of the preserved thickness of the poorly exposed and deformed Shijujiyama Formation are less than 1 km (Hibbard and others, 1992), and there is no evidence to indicate that the basin overlapped the Tsuro and Misaki assemblages and the Sakamoto mélange; both of these obducted assemblages appear to be <1 km thick.

Paleogeothermal gradients

Knowing the maximum temperature and depth of metamorphism of the Nabae subbelt, we can determine a paleogeothermal gradient. One problem with a straightforward determination of the geothermal gradient based on limited data (i.e., temperature/depth and no control on thermal conductivity) is the necessary assumption that the gradient is constant with depth. Also, recent modeling suggests that geothermal gradients are linear for middle to lower depths of accretionary prisms, but this appears untenable for upper structural levels subject to fluid advection (Wang and Shi, 1984; Reck, 1987). With these precaution noted, we present paleogeothermal gradient calculations for the Nabae subbelt only as first approximations of the actual thermal system.

We have determined the paleogeothermal gradients for each of the major units in the subbelts by using average paleotemperatures for each unit and what we estimate to be the two limiting conditions of burial: (1) equal burial of the entire subbelt at 2 km, and (2) 4-km thickness of the Shijujiyama Formation and 25° bulk back-rotation of the subbelt with an original topographic slope of 10° (Table 1; Fig. 11). We also list the calculated gradients for the 3-km burial depth that we favor. In all cases, bottom-water temperature is considered to be 0°C. The range of geothermal gradients for the subbelt within each of these models varies from ~15°C/km (4 km, 25° rotation) to ~45°C/km (2 km, equal burial depth). These variations in geothermal gradient within each model may seem inordinately large, but are within the range of geothermal gradient variations measured over short distances (<10 km) at the surface of the toes of the modern Washington-Oregon and Nankai accretionary prisms (Shi and others, 1988; Yamano and others, 1984; Kinoshita and Yamano, 1985; Nagihara and others, 1989).

Admittedly, these are "broad side of the barn" calculations, but they do illustrate two significant features of the thermal structure of the subbelt: (1) there was a high geothermal gradient at a few kilometers depth during the early-middle Miocene in this area of the accretionary prism, and (2) lower geothermal gradients are associated with the east side of the peninsula, along the core zone of the Muroto flexure in the Nabae subbelt.

DISCUSSION

Thermal trends within the Nabae subbelt

Traditionally, accretionary prisms have been viewed as regions of depressed geothermal gradients, although this view has

TABLE 1. POSSIBLE MIDDLE MIOCENE GEOTHERMAL GRADIENTS AT CAPE MUROTO

Stratigraphic Unit	Geothermal Gradient (°C/km) for Model Depth*		
	All Units at 2 km	All Units at 3 km	Base of Shijujiyama Formation at 4 km with 25° back rotation of complex
Shijujiyama Formation	102	68	51
Hioki mélange (W)	108	72	54
Hioki mélange (E)	90	60	45
Tsuro assemblage	108	72	48
Sakamoto mélange (W)	116	77	46
Sakamoto mélange (E)	102	68	41
Misaki assemblage	134	88	50

*See text for discussions of depth models and geothermal gradients.

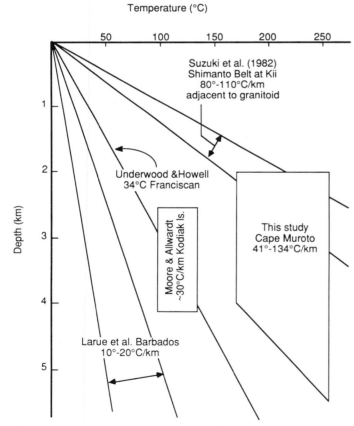

Figure 11. Comparison of geothermal gradients determined for upper structural levels of assorted accretionary prisms (data from Moore and Allwardt,1980; Suzuki and others, 1982; Larue and others, 1985; Underwood and Howell, 1987).

not been entirely upheld by recent studies (Moore and Allwardt, 1980; Suzuki and others, 1982; Larue and others, 1985; Underwood and Howell, 1987; Underwood and others, 1988; Chijiwa, 1988). The interpreted paleogeothermal gradients in the Nabae subbelt are some of the highest recorded within an accretionary prism (Fig. 11). To date, the only other area with equivalent paleogeothermal gradients is in the lower-middle Miocene rocks of the Kumano Group overlying the Shimanto Belt on the Kii Peninsula (Fig. 1). Here, the Miocene paleogeothermal gradient has been interpreted to be on the order of 80 to 110°C/km (Suzuki and others, 1982; Chijiwa, 1988); however, it should be noted that this portion of the prism is also unusual, as it may have been involved in subduction of mafic off-axis magmatic centers (Hibbard and Karig, 1990a, b) and it is also an area of widespread near-surface felsic magmatic activity (Chijiwa, 1988).

In the Nabae subbelt, it appears that the largest contribution of this high heat flow comes from conduction of heat through the prism; at this time it is difficult to assess the relative contributions of other sources that may have participated in the thermal system, such as advection by fluids (Reck, 1987) and friction from the décollement (Dahlen and Barr, 1989).

Interpreted paleogeothermal gradients for the Nabae subbelt rival values measured in accretionary prisms where young oceanic crust is being subducted. In both the active Nankai and Washington-Oregon accretionary prisms, geothermal gradients in excess of 50°C/km are common (Shi and others, 1988; Yamano and others, 1984; Kinoshita and Yamano, 1985; Nagihara and others, 1989). It might be expected that gradients above a subducting spreading ridge should be significantly higher than those for the subduction of young, oceanic crust; this may well be, as the modern measured geothermal gradients are of an ephemeral, near-surface field, subject to change with time and depth.

The origin of the major spatial trends in the thermal maturity of the Nabae subbelt is less clear than that of the overall high heat flow. The southward, across-strike paleotemperature increase is almost certainly a function of the approach toward the major gabbro body at the cape, but may also involve an increase in structural depth due to bulk back-rotation in the subbelt. An explanation for the westward, strike-parallel increase in paleotemperature is less forthcoming; this trend is also apparent in the Murotohanto subbelt and there, has been attributed to differential, along-strike uplift of the prism (Mori and Taguchi, 1988; DiTullio and others, this volume). The lack of any east-west change in paleotemperature in the Shijujiyama Formation appears to preclude this interpretation, at least for the Nabae subbelt, and suggests that the trend is an original feature of the thermal field in the accreted rocks. In addition, the limited data surprisingly suggest that the low paleotemperatures appear to be associated with the east side of the peninsula, near the core of the Muroto flexure.

One possible explanation for depressed heat flow in the core of the flexure relates to the paleomorphology of the flexure. Since the flexure is interpreted to be the result of the subduction of a bathymetric high (the Shikoku Basin spreading ridge), it may well have formed a ridgelike structure within the overriding Shimanto accretionary prism. Based on mathematical modeling of conductive heat flow at the Washington-Oregon convergent margin, Shi and others (1988) have predicted that geothermal gradients are substantially lower at bathymetric ridges than at bathymetric lows; their modeling very closely matches the observed heat flow at the toe of the Washington-Oregon accretionary prism. Thus, if the flexure represented a bathymetric high in the prism, as might be expected, heat flow associated with this high may have been dampened relative to adjacent areas. If this interpretation is valid, then a complex interplay between magma intrusion and bathymetry in the flexure core likely is responsible for the present observed thermal pattern.

The Shijujiyama slope-basin shows reflectance values that are either equivalent to or less than most of those of the neighboring Hioki mélange, except near the core of the Muroto flexure, where slope-basin values are greater. The latter change may, in part, be related to topographic effects, as noted above. It is noteworthy that lowermost slope basins along the Washington-Oregon accretionary prism show elevated heat flow relative to the adjacent, topographically higher, accreted strata (Shi and others, 1988). The overlap in thermal values between the slope basin and subjacent mélange is also characteristic of slope-basin–accreted strata thermal relationships on Sitkinak Island, Alaska (Moore and Allwardt, 1980).

The high reflectance values associated with the fluidized sediments at the décollement to the Tsuro assemblage appear to indicate that high fluid temperatures were present, locally, during the early phases of tectonism. In theory, the elevated values could be due to either friction along the fault or advection by warm fluids. Friction has been interpreted to significantly increase heat flow adjacent to major fault zones (e.g., Graham and England, 1976; Scholz, 1980; Bustin, 1983). Simple calculation[1] indicates that under reasonable to optimal conditions, friction could elevate local temperatures by ~10°C and thus, probably can not account for the 20 to 40°C-temperature spike associated with the fluidized sediments. Instead, we are impressed with the coincidence of high reflectance values with the fluid features in these sediments. Temperature anomalies have been noted in modern accretionary prisms where fluids are being vented (e.g., Kulm and others, 1986; Henry and others, 1989). These modern examples measured near the sea floor, appear to be diluted by cold oceanic waters and the thermal anomalies are on the order of tenths of a degree Centigrade. Much hotter fluids have been interpreted as the source of high thermal anomalies in the body of the Barbados accretionary prism (Moore and others, 1988); also, it is expected that hot fluids would be available in the toe region of accretionary prisms during active spreading ridge subduction.

---

[1] $\Delta T = dV[\mu\rho gd(1-\lambda)k^{-1}$; T = temperature, d = depth (3 km), V = velocity of fault slip (6 cm/yr), μ = coefficient of friction (0.50), ρ = sediment density (2,500 kg/m^3), g = acceleration due to gravity, λ = pore fluid pressure ratio (0.9), k = coefficient of thermal conductivity (2.0 W/m°C). $\Delta T = 10.5$°C.

PRÉCIS AND CONCLUSIONS

The thermal imprint upon the Nabae subbelt, as recorded by organic and inorganic materials, is characterized by high temperatures at shallow structural levels in an accretionary prism. Peak temperatures in the accreted rocks of the Nabae complex range from ~140 to 280°C whereas those of slope-basin strata of the Shijujiyama Formation show a tighter range, from ~180 to 210°C. On the basis of circumstantial evidence, the maximum depth of exposed rocks in the Nabae subbelt is estimated to have been in the range of 2 to 4 km.

Two major trends in thermal patterns are evident in the accreted rocks, including an overall increase in peak temperatures both from north to south and from east to west. These trends are absent in the Shijujiyama Formation data. The north to south increase in %R_m values likely reflects the influence of the large gabbro body at Cape Muroto and may also be related to an increase in structural depth in this direction. An explanation for the east to west trend is less forthcoming; this trend may reflect a complex interplay between paleobathymetry and magmatism.

The regional thermal overprint was coeval with the late-stage tectonothermal event that affected the subbelt. This event has been attributed to the early-middle Miocene subduction of the Shikoku Basin spreading ridge beneath the subbelt; thus, peak heating of the Nabae subbelt is interpreted to reflect the thermal consequences of subduction of an active spreading ridge.

Spatially, the thermal effects of this event may impinge on older portions of the prism, such as the Murotohanto subbelt, situated further inland from the Miocene paleotrench, as well as on coeval portions of the belt along strike from Muroto (see Underwood, Hibbard, DiTullio, and Laughland, this volume). The Oligocene-Miocene portions of the accretionary prism in Kyushu and on the Kii Peninsula also are apparent sites of high paleogeothermal gradients (Suzuki and others, 1982; Chijiwa, 1988; Aihara, 1989). The basin spreading ridge was subducted suborthogonal to the paleotrench (Fig. 3), thus the apparent lateral extent of high heat flow in the Miocene accretionary prism is rather surprising. This thermal pattern may be a reflection of the youthfulness of the entire Shikoku Basin at the time of collision with the southwest Japan convergent margin (Hibbard and Karig, 1990b); in addition, the Shikoku Basin is known to be host to extensive off-axis magmatism (Klein and Kobayashi, 1979), which may have served to maintain high heat flow in older portions of the basin, and thus elevating heat flow in the entire prism during subduction of the basin. It is also noteworthy that anomalously high heat flow persists today in this region (Yamano and others, 1984; Kinoshita and Yamano, 1985; Nagihara and others, 1989).

These conclusions concerning the thermal evolution of the Nabae subbelt are somewhat tenuous, considering the ephemeral nature of heat flow, some of the assumptions made with regard to timing of the imprint, and the uncertainties inherent in the interpretation of organic and inorganic metamorphic indicators. However, they are consistent with the tectonic evolution of the southwest Japan plate margin, as well as our fledgling knowledge of heat flow in accretionary prisms, both modern and ancient. Hopefully, these data will serve to add constraints to our conceptualization of the thermal evolution of convergent margins.

ACKNOWLEDGMENTS

Funding for this study was provided by National Science Foundation grants to D. Karig and T. Byrne and to M. Underwood, a grant from the Petroleum Research Fund of the American Chemical Society to M. Underwood (Grant No. 19187-AC2), and Harold Stearns awards from the Geological Society of America to J. Hibbard. Gussie Ownby is thanked for lab assistance. We are also grateful to Skip Stoddard for a review of an early form of the manuscript and to later reviews by M. Underwood, J. Sample, and D. Fischer, all of which helped to substantially improve the final manuscript. Domo arigato gozaimasu to the family of Teratoshi Tomioka of Mitsu, Kochi prefecture, for their hospitality and comraderie.

REFERENCES CITED

Aihara, A., 1989, Paleogeothermal influence on organic metamorphism in the neotectonics of the Japanese Islands: Tectonophysics, v. 159, p. 291–305.

Barker, C. E., 1988, Geothermics of petroleum systems: Implications of the stabilization of kerogen thermal maturation after a geologically brief heating duration at peak temperature, *in* Magoon, L., ed., Petroleum systems of the United States: U.S. Geological Survey Bulletin 1870, p. 26–29.

Barker, P., 1982, The Cenozoic subduction history of the Pacific margin of the Antarctic Peninsula: Ridge crest–trench interactions: Journal of the Geological Society of London, v. 139, p. 787–801.

Blenkinsop, T., 1988, Definition of low-grade metamorphic zones using illite crystallinity: Journal of Metamorphic Geology, v. 6, p. 623–636.

Bray, C., and Karig, D., 1985, Porosity of sediments in accretionary prisms and some implications for dewatering processes: Journal of Geophysical Research, v. 90, p. 768–778.

Bustin, R., 1983, Heating during thrust faulting in the Rocky Mountains: Friction or fiction?: Tectonophysics, v. 95, p. 309–328.

Byrne, T., and Hibbard, J., 1987, Landward vergence in accretionary prisms: The role of the backstop and thermal history: Geology, v. 15, p. 1163–1167.

Cande, S., and Leslie, R., 1986, Late Cenozoic tectonics of the southern Chile trench: Journal of Geophysical Research, v. 91, p. 471–496.

Chamot-Rooke, N., Renard, V., and LePichon, X., 1987, Magnetic anomalies in the Shikoku Basin: A new interpretation: Earth and Planetary Science Letters, v. 83, p. 214–228.

Chijiwa, K., 1988, Post-Shimanto sedimentation and organic metamorphism: An example of the Miocene Kumano Group, Kii Peninsula: Modern Geology, v. 12, p. 363–387.

Dahlen, F., and Barr, T., 1989, Brittle frictional mountain building 1. Deformation and mechanical energy budget: Journal of Geophysical Research, v. 94, p. 3906–3922.

DeLong, S., and Fox, P., 1977, Geological consequences of ridge subduction, *in* Pittman, W., and Talwani, M., eds., Island arcs, deep sea trenches, and back arc basins: American Geophysical Union, Maurice Ewing Series 1, 221–228.

DeLong, S., Fox, P., and McDowell, F., 1978, Subduction of the Kula Ridge at the Aleutian trench: Geological Society of America Bulletin, v. 89, p. 83–95.

Forsythe, R., and Nelson, E., 1985, Geological manifestations of ridge collision: Evidence from the Golfo de Penas–Taitao Basin, southern Chile: Tectonics, v. 4, p. 477–495.

Forsythe, R., and 7 others, 1986, Pliocene near-trench magmatism in southern

Chile: A possible manifestation of ridge collision: Geology, v. 14, p. 23–27.

Frey, M., 1987, Very low-grade metamorphism of clastic sedimentary rocks, *in* Frey, M., ed., Low temperature metamorphism: London, Blackie, p. 9–58.

Graham, C., and England, P., 1976, Thermal regimes and regional metamorphism in the vicinity of overthrust faults: An example of shear heating and inverted metamorphic zonation from southern California: Earth and Planetary Science Letters, v. 31, p. 142–152.

Hamamoto, R., and Sakai, H., 1987, Rb-Sr age of granophyre associated with the Cape Muroto gabbroic complex: Science Reports of the Department of Geology, Kyushu University, v. 15, p. 1–5.

Henry, P., Lallemant, S., LePichon, X., and Lallemand, S., 1989, Fluid venting along Japanese trenches: Tectonic context and thermal modeling: Tectonophysics, v. 160, p. 277–291.

Herron, E., Cande, S., and Hall, B., 1981, An active spreading center collides with a subduction zone: A geophysical survey of the Chile Margin triple junction, *in* Kulm, L., Dymond, J., Dasch, E., and Hussong, D., eds., Nazca Plate: Crustal formation and Andean convergence: Geological Society of America Memoir 154, p. 683–701.

Hibbard, J., 1988, Evolution of anomalous structural fabrics in an accretionary prism; The Oligocene-Miocene portion of the Shimanto Belt at Cape Muroto, southwest Japan [Ph.D. thesis]: Ithaca, New York, Cornell University, 227 p.

Hibbard, J., and Karig, D., 1987, Sheath-like folds and progressive fold deformation in Tertiary sedimentary rocks of the Shimanto accretionary complex, Japan: Journal of Structural Geology, v. 9, p. 845–857.

—— , 1990a, Structural and magmatic responses to spreading ridge subduction; An example from southwest Japan: Tectonics, v. 9, p. 207–230.

—— , 1990b, An alternative plate model for the early Miocene evolution of the southwest Japan margin: Geology, v. 18, p. 170–174.

Hibbard, J., Karig, D., and Taira, A., 1992, Anomalous structural evolution of the Shimanto accretionary prism at Murotomisaki, Shikoku Island, Japan: Island Arc, 1, p. 130–144.

Hisatomi, K., 1988, The Miocene forearc basin of southwest Japan and the Kumano Group of the Kii Peninsula: Modern Geology, v. 12, p. 389–408.

Houseknecht, D., 1984, Influence of grain size and temperature on intergranular pressure solution, quartz cementation, and porosity in a quartzose sandstone: Journal of Sedimentary Petrology, v. 54, p. 348–361.

Ishikawa, T., 1982, Radiolarians from the southern Shimanto Belt (Tertiary) in Kochi Prefecture, Japan: News of Osaka Micropaleontologists Special Volume 5, p. 399–407.

James, T., Hollister, L., and Morgan, W., 1989, Thermal modeling of the Chugach Metamorphic Complex: Journal of Geophysical Research, v. 94, p. 4411–4423.

Katto, J., and Arita, M., 1966, Geology of the Muroto Peninsula, Shikoku, Japan: Research Report of Kochi University, Natural Science, v. 15, no. 8, p. 59–63.

Katto, J., and Taira, A., 1978, The Misaki Group (Miocene), southwestern Shikoku: Research Report of Kochi University, Natural Science, v. 27, p. 165–180.

Kimura, K., 1985, Sedimentation and sedimentary facies of the Tertiary Shimizu and Misaki Formations in the southwestern part of Shikoku: Journal of the Geological Society of Japan, v. 91, p. 815–831.

Kinoshita, H., and Yamano, M., 1985, The heat flow anomaly in the Nankai Trough area, *in* Kagami, H., Karig, D., Coulbourn, W., and others, eds., Initial Reports of the Deep Sea Drilling Project, vol. 87: Washington, D.C., U.S. Government Printing Office, p. 737–743.

Kisch, H., 1987, Correlation between indicators of very low-grade metamorphism, *in* Frey, M., ed., Low temperature metamorphism: London, Blackie, p. 227–300.

Klein, G., and Kobayashi, K., 1979, Geological summary of the North Philippine Sea, based on Deep Sea Drilling Project Leg 58 results, *in* Klein, G., and Kobayashi, K., eds., Initial Reports of the Deep Sea Drilling Project, v. 58: Washington, D.C., U.S. Government Printing Office, p. 951–962.

Kulm, L., and 13 others, 1986, Oregon subduction zone: Venting, fauna, and carbonates: Science, v. 231, p. 561–566.

Larue, D., Schoonmaker, J., Torrini, R., Lucas-Clark, J., Clark, M., and Schneider, R., 1985, Barbados: Maturation, source rock potential and burial history within a Cenozoic accretionary complex: Marine and Petroleum Geology, v. 2, p. 96–110.

Marshak, S., and Karig, D., 1977, Triple junctions as a cause for anomalously near-trench igneous activity between the trench and volcanic arc: Geology, v. 5, p. 233–236.

McLaughlin, R., Kling, S., Poore, R., McDougall, K., and Beutner, E., 1982, Post-middle Miocene accretion of Franciscan rocks, northwestern California: Geological Society of America Bulletin, v. 93, p. 595–605.

Moore, G., and Karig, D., 1980, Structural geology of Nias Island, Indonesia: Implications for subduction zone tectonics: American Journal of Science, v. 280, p. 193–223.

Moore, G., Curray, J., Moore, D., and Karig, D., 1980, Variations in geologic structure along the Sunda fore arc, northeastern Indian Ocean, *in* Hayes, D., ed., Studies of East Asian tectonics and resources: EOS Transactions of the American Geophysical Union, p. 145–160.

Moore, J. C., and Allwardt, A., 1980, Progressive deformation of a Tertiary trench slope, Kodiak Island, Alaska: Journal of Geophysical Research, v. 85B, p. 4741–4756.

Moore, J. C., Mascle, A., and the staff of ODP Leg 110, 1988, Tectonics and hydrogeology of the northern Barbados Ridge: Results from Ocean Drilling Program Leg 110: Geological Society of America Bulletin 100, 1578–1593.

Moore, J. C., Byrne, T., Plumley, P., Reid, M., Gibbons, H., and Coe, R., 1983, Paleogene evolution of the Kodiak Islands, Alaska: Consequences of ridge-trench interaction in a more southerly latitude: Tectonics, v. 2, p. 265–293.

Mori, K., and Taguchi, K., 1988, Examination of the low-grade metamorphism in the Shimanto Belt by vitrinite reflectance: Modern Geology, v. 12, p. 325–339.

Nagihara, S., Kinoshita, H., and Yamano, M., 1989, On the high heat flow in the Nankai Trough area—A simulation study on a heat rebound process: Tectonophysics, v. 161, p. 33–41.

Nelson, E., and Forsythe, R., 1989, Ridge collision at convergent margins: Implications for Archean and post-Archean crustal growth: Tectonophysics, v. 161, p. 307–315.

Perfit, M., and 6 others, 1987, Geochemistry and petrology of volcanic rocks from the Woodlark Basin: Addressing questions of ridge subduction, *in* Taylor, B., and Exon, N., eds., Marine geology, geophysics, and geochemistry of the Woodlark Basin-Solomon Islands: Houston, Circum-Pacific Council for Energy and Mineral Resources Earth Science Series, v. 7, p. 113–154.

Reck, B., 1987, Implications of measured thermal gradients for water movement through the northeast Japan accretionary prism: Journal of Geophysical Research, v. 92, p. 3683–3690.

Saito, T., 1980, The lower Miocene planktonic foraminifera from Sakamoto, Muroto City, Kochi, *in* Taira, A., and Tashiro, M., eds., Geology and paleontology of the Shimanto Belt: Kochi, Rinyakosaikai Press, p. 227–234.

Scholz, C., 1980, Shear heating and the state of stress on faults: Journal of Geophysical Research, v. 85, p. 6174–6184.

Seno, T., and Maruyama, S., 1984, Paleogeographic reconstruction and origin of the Philippine Sea: Tectonophysics, v. 102, p. 53–84.

Shi, Y., Wang, C., Langseth, M., Hobart, M., and von Huene, R., 1988, Heat flow and thermal structure of the Washington-Oregon accretionary prism—A study of the lower slope: Geophysical Research Letters, v. 15, p. 1113–1116.

Sisson, V., Hollister, L., and Onstott, T., 1989, Petrologic and age constraints on the origin of a low pressure/high temperature metamorphic complex, southern Alaska: Journal of Geophysical Research, v. 94, p. 4392–4410.

Suzuki, S., Oda, Y., and Nambu, M., 1982, Thermal alteration of vitrinite in the Miocene sediments of the Kishu Mine area, Kii Peninsula: Kozau Chishitsu, v. 32, p. 55–65.

Taira, A., Tashiro, M., Okamura, M., and Katto, J., 1980, The geology of the Shimanto Belt in Kochi Prefecture, Shikoku, Japan, *in* Taira, A., and Tashiro, M., eds., Geology and paleontology of the Shimanto Belt: Kochi, Rinyakosaikai Press, p. 319–389.

Taira, A., Katto, J., Tashiro, M., Okamura, M., and Kodama, K., 1988, The Shimanto Belt in Shikoku, Japan—Evolution of Cretaceous to Miocene accretionary prism: Modern Geology, v. 12, p. 1–42.

Underwood, M., and Howell, D., 1987, Thermal maturity of the Cambria slab, an inferred trench-slope-basin in central California: Geology, v. 15, p. 216–219.

Underwood, M., O'Leary, J., and Strong, R., 1988, Contrasts in thermal maturity within terranes and across terrane boundaries of the Franciscan Complex, northern California: Journal of Geology, v. 96, p. 399–415.

Wang, C., and Shi, Y., 1984, On the thermal structure of subduction complexes: A preliminary study: Journal of Geophysical Research, v. 89, p. 7709–7718.

Weissel, J., Taylor, B., and Karner, G., 1982, The opening of the Woodlark Basin, subduction of the Woodlark spreading system, and the evolution of northern Melanesia since mid-Pliocene time: Tectonophysics, v. 87, p. 253–277.

Yamano, M., Honda, S., and Uyeda, S., 1984, Nankai Trough: A hot trench?: Marine Geophysical Researches, v. 6, p. 187–203.

MANUSCRIPT ACCEPTED BY THE SOCIETY APRIL 24, 1992

Geological Society of America
Special Paper 273
1993

Regional and local variations in the thermal history of the Shimanto Belt, southwest Japan

Lee DiTullio* and Shigeki Hada
Department of Geology, Kochi University, Kochi 780, Japan

ABSTRACT

Paleothermal data from the Cretaceous through Miocene Shimanto Belt reveal important variations in thermal history, both along and across strike within the ancient accretionary prism. Two types of variations are considered: those due to different tectonic settings (e.g., tectonic mélange versus coherent accreted rocks versus slope-basin deposits); and those due to different times of accretion. Within the Cretaceous subbelt on Shikoku, shale samples from mélange and coherent rocks show the same level of illite crystallinity, but samples from a probable slope-basin deposit are significantly less crystalline suggesting that the slope-basin strata experienced less heating than underlying accreted rocks. In contrast, within the Tertiary subbelt, mélange shales, coherent accreted strata, and slope-basin deposits all overlap with respect to illite crystallinity and vitrinite reflectance values; moreover, the grade tends to be higher than in the Cretaceous subbelt. This difference in grade may merely reflect different levels of exposure, but is more likely related to the fact that both the Eocene-Oligocene and Oligocene-Miocene subbelts experienced relatively high paleogeothermal gradients, apparently resulting from separate ridge subduction events in the late Eocene/early Oligocene and middle Miocene (DiTullio and others, this volume; Hibbard and others, this volume).

In addition to overall differences in grade between the Cretaceous and Tertiary subbelts, marked variations in metamorphism and deformation style in the Shimanto Belt are apparent along strike between the Shikoku and the neighboring Kii Peninsula and Kyushu. In Kyushu, all the subbelts of the Shimanto Belt are characterized by low dips. In the relatively highly metamorphosed Cretaceous subbelt and in the Kitagawa Group of the Eocene-Oligocene subbelt, this may reflect deformation in an underplating regime. However, in the Eocene to lower Oligocene Hyuga Group, as well as probably originally in the Paleogene Otanashigawa Group and the Oligocene-Miocene Muro Group of Kii Peninsula, moderate- to low-dipping strata are associated with only diagenetic grades of thermal alteration. Although the Hyuga and Muro Groups may represent slope-basin deposits, clearly accreted Eocene rocks such as the Kitagawa and Otanashigawa Groups have experienced markedly less landward rotation than equivalently thermally altered accreted rocks in Shikoku. This difference in landward rotation may be related to landward-vergent structures observed in the Oligocene-Miocene subbelt. Landward-vergent structures may have resulted from a seaward-dipping backstop caused by the high paleogeothermal gradient resulting from subduction of the Shikoku basin spreading ridge beneath the Muroto Peninsula in the middle Miocene (Byrne and Hibbard, 1987). However formed, the landward-vergent structures probably contributed to the enhanced steepening of strata in the Shimanto Belt in Shikoku with respect to other areas along strike.

---

*Present address: 326 12th St., Apt. 2L, Brooklyn, New York 11215.

DiTullio, L., and Hada, S., 1993, Regional and local variations in the thermal history of the Shimanto Belt, southwest Japan, *in* Underwood, M. B., ed.; Thermal Evolution of the Tertiary Shimanto Belt, Southwest Japan: An Example of Ridge-Trench Interaction: Boulder, Colorado, Geological Society of America Special Paper 273.

INTRODUCTION

The other papers in this volume deal with only a small portion of the Shimanto Belt of southwest Japan, namely the well-exposed Eocene-Oligocene and Oligocene-Miocene sections on the Muroto Peninsula of Shikoku Island. However, approximately half of the Shimanto Belt consists of accreted rocks of Cretaceous age (the so-called Lower Shimanto Group; see Taira and others, 1982) and the Shimanto Belt as a whole is exposed for 1,800 km from the Nansei Islands in the west to near Tokyo in the east. Thus, in order to get a more general picture of metamorphism in an accretionary setting, it is appropriate to extend our analysis as far as possible both across and along strike. In this paper, we begin with the Muroto Peninsula and then synthesize a variety of published paleothermal data and new illite crystallinity data in an attempt to examine the types and possible causes of spatial and temporal variations in thermal alteration within the Shimanto accretionary prism. Despite a complicated history of burial, intrusion, and uplift, a coherent pattern of heating can be discerned which reveals much about the tectonostratigraphic settings and deformation history of the area.

In the sections that follow, we first outline pertinent aspects of the geology of the Shimanto Belt. Then, after briefly discussing the illite crystallinity and vitrinite reflectance techniques, we present both published and new data from selected areas. Finally, we discuss the implications of the spatial and temporal variations in the data for identifying deformation environments within the prism and regional tectonic factors that may influence patterns of heating.

GEOLOGIC OVERVIEW OF THE SHIMANTO BELT

The Shimanto Belt occupies the Outer Zone of southwest Japan to the southeast of the Cretaceous arc granites and the Ryoke and Sambagawa paired metamorphic belts (Geological Survey of Japan, 1987). Bordering on the Pacific Ocean on the islands of Kyushu, Shikoku, and Honshu, the Shimanto Belt is in fault contact along the Butsuzo Tectonic Line with a Jurassic accretionary complex to the northwest (the Sanbosan Belt; Fig. 1). As a whole, the Shimanto Belt is only weakly metamorphosed and consists of steeply dipping imbricated and folded sandstone/shale turbidites with intercalated mélange. Extensive micropaleontologic investigations have revealed that individual fault-bounded packets young to the north but that overall the belt

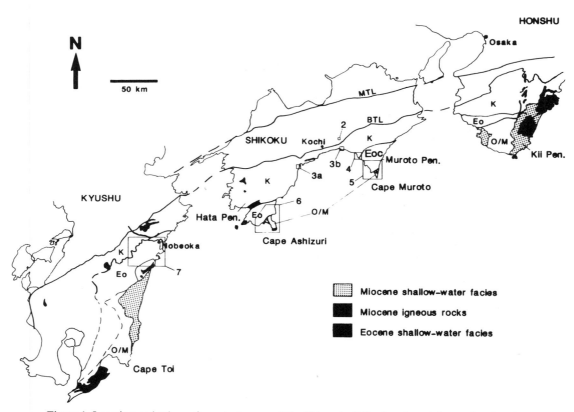

Figure 1. Location and schematic geologic map of the Shimanto Belt of southwest Japan showing the distribution of the Cretaceous, Eocene-Oligocene, and Oligocene-Miocene subbelts and highlighting slope-basin deposits and middle Miocene intrusions. Subbelts of the Shimanto Belt: K, Cretaceous; Eoc, Eocene-Oligocene; O/M, Oligocene/Miocene. Numbers are locations of Figures 2 to 7. See text for discussion. MTL, Median Tectonic Line; BTL, Butsuzo Tectonic Line; Pen, peninsula.

youngs to the south towards the present-day Nankai Trough (Taira and others, 1980, 1982). This pattern, together with lithofacies, structural style, and tectonic setting, have led these workers to interpret the Shimanto Belt as an accretionary prism.

For the purposes of this paper we consider three fault-bounded subdivisions of the Shimanto Belt: the Cretaceous, the Eocene-Oligocene, and the Oligocene-Miocene. These subbelts may be considered to represent three distinct accretionary prisms and are discussed as such later. All of these subbelts contain three major lithotectonic units: tectonic mélange; deformed but coherent turbidite sequences, interpreted as offscraped trench fill; and shallow-marine facies, interpreted as slope-basin deposits. We summarize the geology of each below.

Cretaceous subbelt

In the Cretaceous subbelt of Shikoku, the accreted rocks are represented by Albian to Campanian folded, imbricated, and cleaved turbidites and several narrow belts of mélange. The mélange contains blocks of Valanginian to Santonian chert, limestone, shale, and MORB (mid-ocean ridge basalt)-type pillow basalt, which float in a younger sheared shale matrix of Campanian age (Taira and others, 1980, 1982, 1988). On the Hata Peninsula of western Shikoku, this subbelt is 40 km wider than in the east due to the presence of the Late Cretaceous Nonogawa Group. Contemporaneous with the Nonogawa Group, but of shallow-water facies, is the Uwajima Group in western Shikoku (Teraoka, 1979). Other Late Cretaceous slope-basin deposits are the Nakasuji Group, also in western Shikoku, and the Doganaro and Uwagumi Formations to the east of Kochi City (Aoki and Tashiro, 1982) (see the location of Fig. 2 in Fig. 1).

To the east of Shikoku, on the Kii Peninsula of Honshu, this subbelt is represented by the Late Cretaceous Hidakagawa Group, which consists predominantly of fine-grained turbidites but also includes conglomeratic horizons with clasts of limestone, granite, and felsic volcanics (Kumon and others, 1988). On Kyushu, to the west of Shikoku, the Cretaceous Shimanto subbelt is represented by strata of the Morotsuka Group. All the Cretaceous rocks are east striking, imbricated, folded, and cleaved (Sakai, 1978); however, in contrast to the diagenetic grade found in Shikoku and Kii, the Morotsuka Group in Kyushu is characterized by greenschist-facies metamorphism (Imai and others, 1979).

Strata at Cape Oyama in eastern Shikoku (see the location of Fig. 4 in Fig. 1) have previously been considered Eocene (Geological Survey of Japan, 1987) but micropaleontologic work (Taira and others, 1980; Okamura, personal communication, 1989) indicates that at least the relatively undeformed and homoclinally dipping southern third of the section is latest Cretaceous in age. An Eocene age was inferred because the central part of the section contains a distinctive metaclast conglomerate that is correlated with a similar conglomerate in the Tanokuchi Formation on the Hata Peninsula thought to be Eocene in age (Taira and others, 1980). In addition, a relatively well developed pressure solution cleavage in the northern part of the section is virtu-

ally identical to the cleavage in the Eocene-Oligocene turbidites of the Gyoto syncline, located 10 km to the south (see DiTullio and others, this volume; DiTullio and Byrne, 1990). Although the relationship between the various parts of the Cape Oyama section is not clear, structurally they appear to be part of a continuous sequence. Uncertainties in age for all but the southernmost part lead us to tentatively include this entire section in the Cretaceous subbelt.

Eocene-Oligocene subbelt

Eocene-Oligocene rocks in the Shimanto Belt are similar in lithology and structure to those of the Cretaceous subbelt and consist of faulted, folded, and cleaved turbidites with intercalated mélange; in Shikoku, the age difference between blocks and matrix of the Eocene mélange is less than 10 m.y. (Taira and others, 1988). Rocks of this subbelt on the Hata Peninsula are typically finer grained with more mélange and slump deposits and are not so well exposed as those on the Muroto Peninsula. In both areas, strikes vary between east and northeast and spaced pressure solution cleavage is common. In the southwestern part of the Hata Peninsula, an Eocene/Oligocene? shallow-water facies (Hirata Formation; Taira and others, 1980) is deposited in angular unconformity on rocks of the Cretaceous Shimanto subbelt (Fig. 1).

In Kyushu, the Eocene-Oligocene subbelt is represented by the lower Mid-Paleogene Kitagawa Group (MacKenzie and others, 1990), which is locally highly sheared, and by the generally less deformed and thermally altered Eocene-to-lower-Oligocene Hyuga Group (Sakai and Kanmera, 1981; Nishi, 1988). The former unit is in the hanging wall of the low-angle Nobeoka Thrust, whereas the latter unit is in the footwall of the thrust; to the south, the Hyuga Group is in fault contact with Miocene acid igneous rocks. All of the strata are imbricated and contain southerly vergent folds; intercalated mélange contains block of MORB-type pillow basalt (Sakai and Kanmera, 1981; Nishi, 1988). In Kii, this subbelt is represented by the Paleogene Otanashigawa Group. These rocks are folded and faulted, moderately to steeply north-dipping, and moderately well cleaved; unfortunately, useful thermal data are lacking due to a severe middle Miocene thermal overprint (Chijiwa, 1988).

Oligocene-Miocene subbelt

The Oligocene-Miocene subbelt is different in lithotectonic facies from the older sections of the Shimanto Belt. In particular, much of the Oligocene-Miocene subbelt is considered to be a huge olistostromal deposit that was deposited in and near the former trench during strike-slip movement coincident with the opening of the Shikoku back-arc basin in early Miocene (Sakai, 1988b). In Shikoku, most of the Oligocene-Miocene rocks have been subsequently tectonically deformed. On the Muroto Peninsula, the subbelt consists of roughly equal proportions of mélange and folded coherent rocks. These rocks strike to the north-northeast and dip steeply southeast reflecting the presence of landward-vergent structures (Byrne and Hibbard, 1987; Hibbard and others, this volume). In addition, a small synclinal slope-basin containing shallow-water facies of the Shijujiyama Formation is present. In western Shikoku, on Cape Ashizuri, the easterly striking Shimizu Formation (late Oligocene to late early Miocene; Teraoka, personal communication, 1989) is highly deformed and cleaved; it is unconformably overlain by the nearly undeformed, gently dipping, in-part fluvial facies of the Miocene Misaki Group (Taira and others, 1980; DiTullio, unpublished field notes).

In Kyushu, to the southeast of the Hyuga Group, the Oligocene-Miocene subbelt is represented by the late Eocene to early Miocene Nichian Group, which consists of huge olistostromes with little or no later deformation (Sakai, 1988a). On the Kii Peninsula, this subbelt is represented by the thick, gently folded, and locally slumped Oligocene to early Miocene Muro Group, which may represent a large slope basin. These rocks are in turn unconformably overlain by slope-basin deposits belonging to the late early Miocene to middle Miocene Tanabe and Kumano Groups (Kumon and others, 1988).

Middle Miocene and younger intrusions have affected the thermal metamorphism of Oligocene-Miocene and older strata in all of these areas (see Fig. 1). Everywhere except the Muroto Peninsula these intrusions are granitic. The most extensive thermal overprint from the middle Miocene intrusions appears to be in Kii where Chijiwa reports vitrinite reflectances as high as $6.0\%R_{max}$ (see below). At Cape Muroto, there is one 14-m.y.-old, 400-m-wide gabbroic body; several smaller mafic dikes are exposed along the eastern coast in the Oligocene/Miocene subbelt (Hibbard and Karig, 1990).

PALEOTHERMAL DATA

Paleothermal data discussed in this paper consist entirely of measurements of vitrinite reflectance and/or illite crystallinity in shaly horizons. Briefly, the vitrinite reflectance technique uses the increase in reflectivity of polished organic matter as a measure of thermal maturity. Reflectance is measured in oil from 50 different randomly oriented particles per sample and is typically reported as the average or mean reflectance ($\%R_m$), although sometimes the maximum reflectivity ($\%R_m$) is reported. Both the illite and vitrinite methods are most accurate in low-grade (zeolite facies) rocks; the vitrinite reflectance has the advantages of being influenced by fewer environmental controls and is irreversible. As in the other papers in this volume, we use Barker's (1988) correlation scheme for vitrinite reflectance and temperature as the best with which to make our paleotemperature estimates; Laughland and Underwood (this volume) consider these estimates to have an error of ±30°C. Values in the range of $2\%R_m$ correspond to peak temperatures of ~200°C and values of $4\%R_m$ reflect temperatures of ~300°C. See Laughland and Underwood (this volume) for a full discussion of this method.

The illite crystallinity index uses the width at half height of the illite 001 basal X-ray reflection as an indicator of the degree of metamorphism (Kubler, 1964). Illite crystallinity is measured in $\Delta°2\theta$ (Kubler index) and decreases with increasing crystallinity or grade such that values of between $0.25\Delta°2\theta$ and $0.42\Delta°2\theta$ (the anchizone) roughly correspond to reflectances of between $>4.0\%R_m$ and $2.5\%R_m$ to $3.1\%R_m$, respectively (Kisch, 1987). At low grade, the material being measured is typically a mixed layer clay containing illite and expandable smectite. Overlap from the basal reflection of smectite gives the illite 10Å peak an asymmetric low-angle tail resulting in a spuriously high values of $\Delta°2\theta$ (low crystallinities). It is standard procedure, therefore, to glycolate samples before they are run in order to remove this interference effect (Kisch, 1987). Moreover, typically only the <2 microns fraction is measured to minimize contamination by detrital mica. These procedures were followed for all the samples in this study. Most of the samples from the Muroto study area were run at the University of Missouri using digital XRD data (see Laughland and Underwood, this volume, for details); illite crystallinity values from other areas and the interior of the Muroto Peninsula were measured at Kochi University using analog XRD data (see DiTullio and others, this volume, for details).

In the following section, primarily due to a lack of data in the Oligocene-Miocene rocks, the thermal data from the Eocene-Oligocene and Oligocene-Miocene subbelts are presented together under the heading Tertiary subbelt.

Cretaceous subbelt

The degree of thermal alteration in mélange, coherent accreted rocks, and slope-basin strata of the Cretaceous subbelt in Shikoku was studied with illite crystallinity. Five samples from both folded and homoclinally dipping shales of slope-basin strata of the Upper Cretaceous Uwagumi Formation, east of Kochi City, yielded diagenetic grade crystallinities of, on average, $0.57\Delta°2\theta$ (see Fig. 2). In contrast, mudstones from both coherent accreted rocks (a total of five samples) and mélange (a total of nine samples) in two different areas on the coast, Awa (Fig. 3A) and Tei (Fig. 3B), all yielded slightly higher values, with an average of $0.49\Delta°2\theta$. Similar findings were reported by Hada (1988) with respect to the level of illite crystallinity in the Tei and Awa areas. He found, however, that the coherent and mélange mudstones differ in physical properties, with the latter having higher yield strength and values of Young's modulus, which he interpreted as due to early overpressuring. Thus, even though the mélange was more strongly deformed than the coherent strata, it did not experience a significantly different degree of heating. Both types of accreted rocks were, however, apparently more deeply buried (heated) than presently exposed coeval slope-basin strata.

Vitrinite reflectance data reported by Mori and Taguchi (1988, their Fig. 8 and 9) in eastern Shikoku indicate similar relatively low degrees of thermal maturity for the Cretaceous subbelt as a whole. Their systematic across-strike sampling for some 40 km in two transects reveals an average reflectance of $~2.0\%R_{max}$. The consistency of coal rank across strike suggests that peak heating postdates much of the accretion-related folding

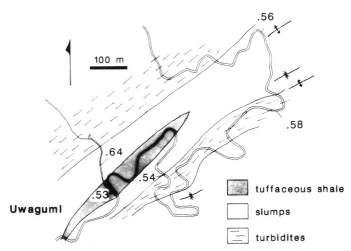

Figure 2. Geologic map of part of the Uwagumi Group, a Cretaceous slope-basin deposit east of Kochi City in Shikoku, showing illite crystallinity sample locations and values. Note that these values are significantly lower than for Cretaceous accreted rocks in Figure 3. See text for discussion.

The section at Cape Oyama is of similar grade to other accreted rocks of the Cretaceous subbelt in eastern Shikoku: both the northern cleaved shale-rich section and the southernmost homoclinal section have an average crystallinity of $0.49\Delta°2\theta$ (Fig. 4). Mori and Taguchi (1988) report one reflectance value of $\sim3.0\%R_m$ for the northern section and one reflectance value of $\sim2.5\%R_m$ for the section containing the metaclast conglomerate near the Namura River. At least two anticline/syncline pairs of a few hundred meters wavelength are present that do not appear to influence the pattern of thermal alteration. The lowest grade part of this section appears to be in the very northernmost 800 m, with an average crystallinity of $0.57\Delta°2\theta$. This area is separated from more cleaved and somewhat higher grade rocks (average crystallinity of $0.44\Delta°2\theta$) to the south near Kono town, by a late high-angle fault visible in outcrop (Fig. 4).

In Kyushu, Imai and others (1979) and Toriumi and Teruya (1988) report greenschist-facies mineral assemblages in the Cretaceous Morotsuka Group. Aihara (1989) indicates vitrinite reflectances in the range ~3.0-$0.5\%R_m$ for these rocks (see Fig. 7, below). No illite samples were taken from this part of the Cretaceous subbelt.

Tertiary subbelt

Thermal alteration in accreted coherent rocks of the Tertiary subbelt was examined in detail in the Muroto Peninsula and in reconnaissance in the Hata Peninsula and Kyushu. The best-studied sections of both the Eocene-Oligocene and Oligocene-Miocene rocks are on the Muroto Peninsula where recent structural studies make possible a detailed consideration of the relationship of thermal alteration to deformation (see Hibbard

and imbrication. However, the data also show interesting variations caused by late- or post-metamorphic faults. For example, two sequences of southward increasing thermal maturity are evident, which probably reflect upturned imbricate packages (Mori and Taguchi, 1988). In addition, the western transect exposes rocks of slightly higher rank than the coastal transect about 20 km to the east; this suggests some differential uplift perhaps accommodated by a north-trending, west-side-up fault such as that in the axial zone of the Muroto flexure to the south (see below).

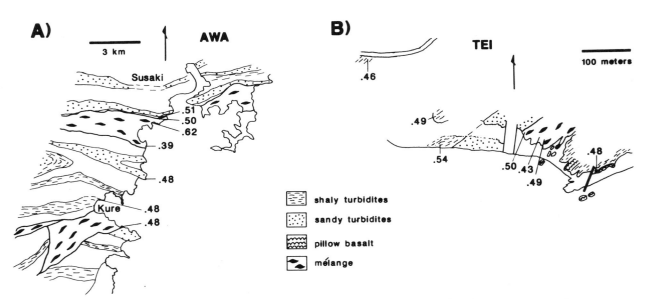

Figure 3. Geologic maps showing illite crystallinity sample locations and values for two exposures of Cretaceous mélange and coherent rocks in Shikoku. A) Awa mélange to the southwest of Kochi City. B) Tei mélange to the southeast of Kochi City. Note that the crystallinity of the mélange and coherent strata is indistinguishable. See text for discussion.

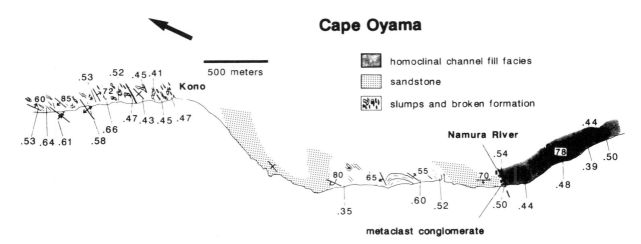

Figure 4. Geologic strip map of the Late Cretaceous strata exposed on the coast at Cape Oyama showing illite crystallinity sample locations and values. See text for discussion.

and others, this volume; and DiTullio and others, this volume). The major results of those studies are summarized here to facilitate comparisons with other parts of the Shimanto Belt (Fig. 5).

Within the Eocene-Oligocene subbelt of the Muroto Peninsula, the level of thermal alteration ranges from diagenetic ($\sim 0.6\Delta°2\theta$ and $\sim 2.0\%R_m$) in the north and locally in the east, to anchizone and locally epizone ($\sim 0.30\Delta°2\theta$ and $\sim 4.0\%R_m$) in the rest of the area. In this subbelt, two phases of folding and cleavage development are recognized: an east-striking phase, which was confined to the Shiina structural domain; and a later, northeast-striking phase, which deformed the Gyoto structural domain (DiTullio and Byrne, 1990; Fig. 5). The pattern of thermal alteration cuts across both sets of regional-scale folds, indicating that at least the final phase of peak heating postdated these structures. However, a consistent association was also observed between the best-developed cleavages of both the early and late phases of deformation and the most thermally mature rocks. DiTullio and others (this volume) argue that this apparently paradoxical relationship is possible because a deformation regime dominated by shortening and subject to peak temperatures within the prism was occupied by the Shiina and Gyoto domains at different times. They conclude that peak heating persisted from at least the first phase of cleavage development through the second phase of folding and cleavage development.

Other important aspects of the pattern of thermal alteration in the Muroto Peninsula include its relationship to different lithotectonic units and late faulting. High-angle post-metamorphic faults have uplifted the southern part of this area and downdropped the northeastern section. The most important faults are the east-trending Shiina-Narashi fault, which separates the Eocene-Oligocene from the Oligocene-Miocene rocks to the south, and a north-trending fault associated with the axis of the Muroto flexure near the east coast (Fig. 5; DiTullio and others, this volume). Mélange and coherent rocks have essentially the

same values of thermal alteration as measured by both illite crystallinity and vitrinite reflectance (see Fig. 8, below). Isoclinally folded and steeply south dipping strata at Cape Hane were suggested by DiTullio and Byrne (1990) on structural and lithologic grounds to represent a possible slope-basin deposit; however, unlike the situation in the Cretaceous subbelt, the level of thermal alteration is equivalent to and not lower than that of at least the accreted rocks exposed on the east coast near Sakihama.

Similar relationships occur in the adjacent Oligocene-Miocene subbelt, although the level of thermal alteration is generally lower (except for contact metamorphism near the gabbro at the tip of the peninsula). Here, extensive mélange and coherent accreted strata are deformed into landward-vergent folds, which again apparently bear no relation to the dominantly diagenetic grades of thermal alteration (~ 0.7 to $0.45\Delta°2\theta$ and ~ 1 to $2.0\%R_m$). The grade of the mélange is indistinguishable from that of the coherent strata and the shallow-water sediments of the Shijujiyama slope basin are of grades equivalent to, and locally higher than, those of the accreted rocks (see Hibbard and others, this volume; Laughland and Underwood, this volume).

In neighboring Hata Peninsula to the west, the deformation history is not as well known and only sparse illite data are available, but patterns of thermal alteration in Eocene strata exposed on the coast near the Shimanto River appear similar to those on Muroto. The illite samples indicate mostly diagenetic grade, with one sample from mélange in the Hiromi Complex just north of the Shimanto River reaching into the anchizone (0.42 to $0.25\Delta°2\theta$CI) (Fig. 6). To the south and west, large middle Miocene granitic masses are exposed that, at least on Cape Ashizuri, have clearly affected thermal patterns. For example, the level of thermal alteration in the essentially undeformed Miocene Misaki slope-basin strata ($0.40\Delta°2\theta$) is very high considering its minimal burial and its only ~ 3 km thickness. This suggests that deposition of the Misaki Group strata preceded or overlapped with intrusion

MUROTO

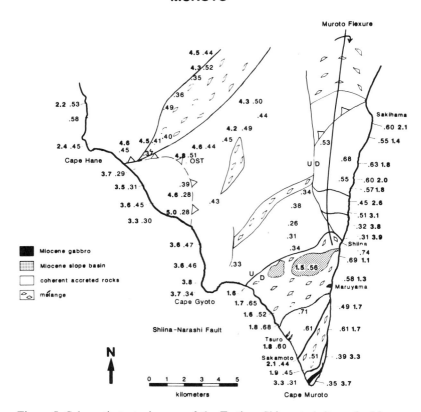

Figure 5. Schematic tectonic map of the Tertiary Shimanto belt on the Muroto Peninsula summarizing illite crystallinity and vitrinite reflectance data (bold numbers) presented in DiTullio and others (this volume) and Hibbard and others (this volume). Note the anchizone grade of most of the Eocene-Oligocene rocks; the large drop in grade from Eocene-Oligocene to Oligocene-Miocene rocks across the Shiina-Narashi fault; the similar or even higher level of metamorphism in the Shiju-jiyama slope-basin strata versus underlying mélange; and the high grade at the southern tip of the peninsula near the gabbroic intrusion. See text for discussion.

of the 13-Ma granite. The Misaki Group is separated from slightly higher grade ($\sim 0.30\Delta°2\theta$) adjacent Eocene accreted rocks by the late high-angle Misaki fault. Agar and others's (1989) fission track data from the Eocene rocks near the Shimanto River indicate that they cooled through $\sim 110°C$ at about 10 Ma; this probably reflects unroofing shortly after reheating by the granite.

Finally, in Kyushu, vitrinite reflectance data from Aihara (1989; personal communication, 1989) and new illite data from near Nobeoka City are in agreement with structural interpretations of the Nobeoka fault as an important thrust and indicate a relatively low grade of metamorphism for the Paleogene Hyuga Group. Cretaceous Morotsuka Group rocks in the hanging wall of the thrust have vitrinite reflectance values of 3.0 to 5.0%R_m (locally 6.0%R_m); whereas in the footwall, Hyuga Group rocks have values of only 1.5 to 3.0%R_m (Fig. 7). Consistent with this, our illite data indicate diagenetic crystallinity values of ~ 0.5%R_m

for Hyuga Group rocks and epizone values of 0.2%R_m for two illite samples from the Kitagawa Group just north of the Nobeoka thrust. Mackenzie and others (1990) indicate only lower anchizone crystallinity values in the Kitagawa Group about 2 km north of the thrust, but these values are still higher than the diagenetic grade of the Hyuga Group. They also report a K-Ar date of 48 Ma on cleavage from this part of the Kitagawa Group; this age agrees well with the inferred age of the earlier east-striking cleavage in the Eocene-Oligocene subbelt on the Muroto Peninsula.

For the Oligocene-Miocene rocks, Aihara reports vitrinite reflectances of between 1 and 2%R_m for olistoliths in the Nichinan Group and we got illite crystallinity values of $0.55\Delta°2\theta$ for the Honjo olistolith in the Cape Toi area versus $0.63\Delta°2\theta$ for nearby shallower water facies (Sakai, 1988a). These values are consistent with the inferred shallow burial of the olistostromal Nichinan Group; they are also equivalent to the grade of

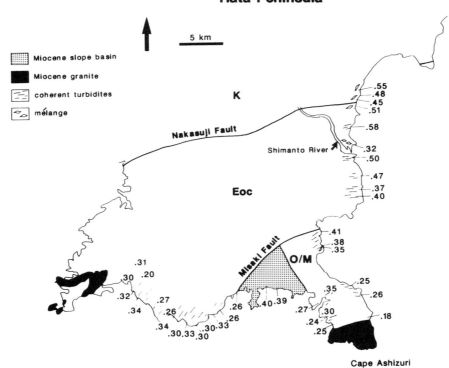

Figure 6. Generalized geologic map of the Hata Peninsula, western Shikoku, showing illite crystallinity sample locations and values. Note the anchizone value of mélange in the Eocene-Oligocene subbelt north of the Shimanto River; the lower crystallinity of the Misaki Group slope-basin strata versus accreted Eocene rocks across the Misaki fault; and the contact metamorphism adjacent to the granite at Cape Ashizuri. See text for discussion.

Oligocene-Miocene accreted rocks in Shikoku (see Fig. 8). In total, the paleothermal data, although sparse, appear to indicate equivalent grades amongst the Tertiary Shimanto strata in Kyushu. The fairly uniform distribution of grade (except at the Nobeoka Thrust) suggests that peak heating postdated accretion-related deformation. However, in contrast to the relatively elevated thermal maturity of slope-basin deposits in Shikoku, Aihara (1989) reports low reflectances ($\leq 0.5\% R_m$) for the Miocene Miyazaki Group, an extensive slope basin in eastern Kyushu, and we have one illite value of $0.57 \Delta°2\theta$ consistent with a diagenetic grade. These values indicate a grade significantly lower than in the Shijujiyama (with respect to reflectance) and Misaki (with respect to crystallinity) slope basins in Shikoku and may indicate either that the middle Miocene thermal overprint was less severe in Kyushu than farther to the east, or that the Miyazaki Group is younger than the igneous activity and thereby escaped the thermal overprint.

DISCUSSION

Relationship of deformation to thermal alteration/metamorphism and implications for deformation environments within the prism

Our survey of the Shimanto Belt suggests some important correlations between style of deformation and degree of regional metamorphism that may be used to infer possible environments of deformation within an accretionary prism. The paleothermal data are summarized by area, subbelt, and lithotectonic unit in Figure 8. For example, the highest grade and presumably most deeply buried rocks discussed are found in Kyushu in the Cretaceous hanging wall of the Nobeoka Thrust. Structures in these rocks, including the thrust itself, display low to moderate dips and strong asymmetry. This style of deformation suggests a kinematic environment dominated by simple shear without much vertical

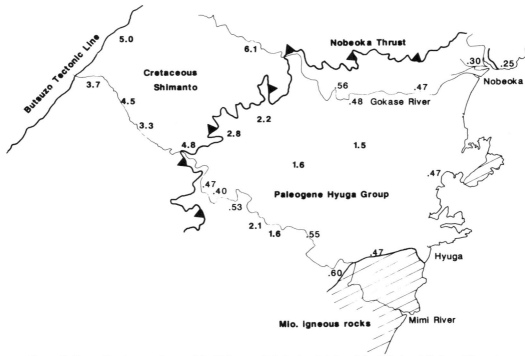

Figure 7. Generalized map of part of the Shimanto Belt in the vicinity of the Mimi and Gokase Rivers in Kyushu showing illite crystallinity and vitrinite reflectance (bold lettering) sample locations and values. The vitrinite reflectance data is from Aihara (personal communication, 1989). Note the dramatic change in grade across the Nobeoka Thrust and the relatively low grade of the Hyuga Group compared to the Eocene-Oligocene Shimanto in Muroto.. See text for discussion.

thickening—perhaps reflecting a regime of underplating (Needham and McKenzie, 1988).

In contrast, most of the coherent accreted rocks in the Shimanto Belt are only slightly to moderately metamorphosed (diagenetic to anchizone) generally with tight folds, moderately well developed pressure solution cleavages, and steep dips. These rocks have clearly experienced significant landward rotation and vertical thickening—as is also indicated by extension directions associated with cleavage development (DiTullio and Byrne, 1990). These relationships suggest that intermediate levels of the prism (~anchizone) are characterized by a higher proportion of pure shear reflected in vertical thickening (e.g., Karig, 1986).

The lowest grade rocks in the Shimanto Belt are, for the most part, in slope-basin strata. In the Cretaceous, the slope-basin strata are lower grade than accreted strata; but in younger rocks, the various lithotectonic units all overlap in grade. However, the least-deformed rocks do appear to come from the shallower levels of the prism as with the probable slope basin Muro Group on Kii or the olistostromal Nichinan Group in Kyushu. An exception are mélanges formed in the décollement zone at the base of the prism

where sediments are subjected to strong shear and layer-parallel extension (Byrne, 1984; Fisher and Byrne, 1987); the low-grade Hioki mélange south of the Shiina-Narashi fault on the Muroto Peninsula is a possible example. In general though, the shallowest levels of the prism are apparently characterized by very little deformation (especially shortening deformation) and no pressure solution cleavage development.

These observations from the Shimanto Belt suggest a generalized model of deformation environments within an accretionary prism (Fig. 9). In such a model, the base of the prism is dominated by simple shear and ductile behavior of strata. At shallow levels with little or no metamorphism, this ductility is provided by unlithified, probably overpressured sediments (Hada, 1988; Fisher and Byrne, 1987). At deeper levels, higher temperatures and pressures result in deformation accommodated by crystal plastic deformation, which produces tectonic fabrics. Intermediate levels of the prism are characterized by approximately anchizone metamorphic conditions and a higher degree of pure shear reflected in folding, cleavage formation, and landward rotation of strata. In the highest levels of the prism, slope basins may

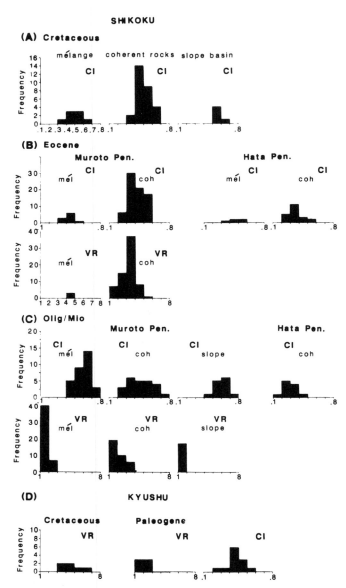

Figure 8. Histograms of paleothermal data from selected areas. (A) Crystallinity values from slope basin (slope), coherent accreted (coh), and mélange (mél) units of the Cretaceous Shimanto Belt in Shikoku Island. (B) Crystallinity and vitrinite reflectance values from coherent accreted and mélange units of the Eocene-Oligocene subbelt on the Muroto Peninsula and crystallinity values from the Hata Peninsula of Shikoku Island. (C) Crystallinity and vitrinite reflectance values from slope basin, coherent accreted, and mélange units of the Oligocene/Miocene subbelts of the Muroto and Hata Peninsulas of Shikoku Island. (D) Vitrinite reflectance values from the Cretaceous subbelt and vitrinite reflectance and crystallinity values from the Paleogene subbelt of Kyushu Island.

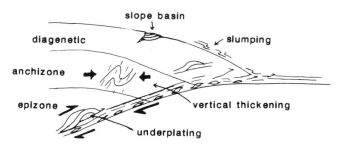

Figure 9. Schematic model showing possible zones of thermal alteration/ metamorphism and typical accompanying structural style within an accretionary prism. Zones are defined on the basis of crystallinity index as follows: diagenetic = >0.42$\Delta°2\theta$; anchizone = 0.42 to 0.25$\Delta°2\theta$; and epizone = <0.25$\Delta°2\theta$ (Kisch, 1987).

develop and subsequently experience shortening; unstable slopes can also produce olistostromal deposits that may eventually reach the trench and become reaccreted (Sakai, 1988b).

Temporal and spatial variations in thermal alteration/metamorphism

One of the most fundamental variations in thermal alteration/metamorphism within the Shimanto Belt is between the Cretaceous subbelt (except in Kyushu) and the Tertiary subbelts. Primarily, it appears that the mélange and coherent units of the Cretaceous Shimanto Belt were subjected to indistinguishable degrees of peak heating whereas inferred slope-basin deposits were distinctly cooler. However, in the Tertiary these relationships disappear: mélange shales, coherently accreted strata, and slope-basin deposits are all approximately the same somewhat higher grade. In addition, the Cretaceous subbelt is generally of lower grade than the Tertiary rocks. Although this may only reflect a difference in level of exposure of the different parts of the Shimanto Belt, the two patterns are more likely related and may reflect the fact that paleogeothermal gradients were high in both the late Eocene and middle Miocene, apparently due to separate ridge subduction events (see DiTullio and others, this volume; Hibbard and others, this volume).

Another important observation is the spatial variation in grade and deformation that is found within the Tertiary Shimanto subbelts. For example, in Kyushu, the Hyuga Group rocks as a whole show among the lowest degrees of thermal alteration reported in the Eocene-Oligocene subbelt and, perhaps not coincidently, relatively low dips; younger Nichinan and Miyazaki Group rocks are also relatively undeformed and only weakly thermally altered. Tertiary strata in Kii are also low dipping and have only a weak cleavage suggesting that they may also experience relatively low temperatures during deformation; but at present corroborating data are lacking and may be impossible to find because of a severe middle Miocene thermal overprint (see Chijiwa, 1988). By comparison, most of the coeval rocks in Shikoku dip more steeply and many are of anchizone grade (probably approximately ~100 °C hotter). This may simply reflect a deeper level of exposure in Shikoku, but more likely the difference is related to the unusually high paleogeothermal gradients experienced in this area during accretion-related deformation.

In particular, the steep dips in the Tertiary Shimanto subbelt on Shikoku may be related to ridge subduction in the middle Miocene. Not only do the Shikoku Tertiary rocks appear to have experienced the most landward rotation, the Oligocene-Miocene strata in Muroto have landward-vergent structures. These aspects of the Tertiary prism may be related in a manner similar to that discussed by Byrne and Hibbard (1987) in their model of a modified lithification front resulting in a forward-dipping backstop. In their model, the lithification front dips seaward following temperature contours resulting from ridge subduction. A seaward-dipping backstop results in landward-vergent structures that presumably would contribute to increased landward-rotation of previously accreted rocks. The subduction of the Shikoku Basin spreading ridge was nearly orthogonal and centered beneath Muroto (see Hibbard and others, this volume), which may explain why similar features were not developed farther to the east and west.

CONCLUSIONS

This survey of thermal alteration data from the Shimanto Belt reveals several significant relationships between thermal grade and deformation in the accretionary prism of southwest Japan. First, based on paleotemperature estimates from vitrinite reflectance data, accreted trench turbidites, like most of those that make up the bulk of all three subbelts, were not subjected to temperatures in excess of ~300°C and typically experienced peak temperatures of ~100 to 250°C. Second, different lithotectonic units sometimes differ in thermal grade. This has been most successfully applied in the case of Cretaceous slope-basin deposits; apparently mélange and coherent accreted rocks experience similar thermal conditions and, at least in the Shimanto Belt, cannot be distinguished on this basis. The validity of this relation depends on the relative burial depths of the different units, possibly on the paleogeothermal gradient, and most importantly on the absence of later thermal overprinting such as has affected much of the Tertiary Shimanto subbelt. Third, spatial and/or temporal variations in metamorphism record important regional tectonic events (ridge subduction) that may have influenced deformation style as well as paleotemperatures. Finally, a general comparison of degree of metamorphism and structural style supports earlier ideas that intermediate levels of the prism experience more pure shear deformation than deeper, warmer levels of the prism that are dominated by simple shear.

ACKNOWLEDGMENTS

We would like to acknowledge the financial support of Monbusho research scholarship to DiTullio. Reviews by Michael Underwood, Rebecca Dorsey, and Jane Tribble substantially improved the manuscript.

REFERENCES CITED

Agar, S. M., Cliff, R. A., Duddy, I. R., and Rex, D. C., 1989, Short paper: Accretion and uplift of the upper structural levels of an accretionary prism: Southwest Japan: Journal of the Geological Society of London, v. 146, p. 893–896.

Aihara, A., 1989, Paleogeothermal influence on organic metamorphism in the neotectonics of the Japanese Islands: Tectonophysics, v. 159, p. 291–305.

Aoki, T., and Tashiro, M., 1982, A stratigraphical study of the Cretaceous Shimanto Belt (The "Dogonaro" and Uwagumi Formations) at Uwagumi, Kagami-machi, Kami-gun, Kochi Prefecture, Shikoku: Research Reports of Kochi University, v. 31, p. 1–24.

Barker, C. E., 1988, Geothermics of petroleum systems: Implications of the stabilization of kerogen thermal maturation after a geologically brief heating duration at peak temperature, *in* Magoon, L. B., ed., Petroleum systems of the United States: U.S. Geological Survey Bulletin 1870, p. 26–29.

Byrne, T., 1984, Early deformation in mélange terranes in the Ghost Rocks Formation, Kodiak Islands, Alaska, *in* Raymond, L. A., ed., Mélanges: Their nature, origin and significance: Geological Society of America Special Paper 198, p. 21–52.

Byrne, T., and Hibbard, J., 1987, Landward vergence in accretionary prisms: The role of the backstop and thermal history: Geology, v. 15, p. 1163–1167.

Chijiwa, K., 1988, Post-Shimanto sedimentation and organic metamorphism: An example of the Miocene Kumano Group, Kii Peninsula: Modern Geology, v. 12, p. 363–387.

DiTullio, L. D., and Byrne, T. B., 1990, Deformation in the shallow levels of an accretionary prism: The Eocene Shimanto Belt of southwest Japan: Geological Society of America Bulletin, v. 102, p. 1420–1438.

Fisher, D., and Byrne, T., 1987, Structural evolution of underthrusted sediments, Kodiak Islands, Alaska: Tectonics, v. 6, p. 775–793.

Geological Survey of Japan, 1987, Geologic Map of Japan: Geological Survey of Japan, scale 1:1,000,000.

Hada, S., 1988, Physical and mechanical properties of sedimentary rocks in the Cretaceous Shimanto Belt: Modern Geology, v. 12, p. 341–359.

Hibbard, J. P., and Karig, D. E., 1990, Structural and magmatic responses to spreading ridge subduction: An example from southwest Japan: Tectonics, v. 9, p. 207–230.

Imai, I., Teraoka, Y., Okumura, K., Ono, K., 1979, Geology of the Mikado District: Kagoshima, Geological Survey of Japan Quadrangle Series, no. 52, scale 1:50,000.

Karig, D. E., 1986, Kinematics and mechanics of deformation across some accreting forearcs, *in* Advances in earth and planetary sciences; Formation of active ocean margins (Proceedings of the Oji Symposium on the Formation of Ocean Margins, Tokyo, Nov. 21–23, 1983): p. 155–177.

Kisch, H. J., 1987, Correlation between indicators of very low-grade metamorphism, *in* Frey, M., Low temperature metamorphism: New York, Chapman and Hall, p. 227–246.

Kubler, B., 1964, Les argiles, indicateurs de metamorphisme: Revue de l'Institut Francais du Pétrole et Annales des Combustible Liquides, v. 9, p. 1093–1112.

Kumon, F., and 8 others, 1988, Shimanto Belt in the Kii Peninsula, southwest Japan: Modern Geology, v. 12, p. 71–96.

Mori, K., and Taguchi, K., 1988, Examination of low-grade metamorphism in the Shimanto Belt by vitrinite reflectance: Modern Geology, v. 12, p. 325–339.

Needham, D. T., and Mackenzie, J. S., 1988, Structural evolution of the Shimanto Belt accretionary complex in the area of the Gokase River, Kyushu, SW Japan: Journal of the Geological Society of London, v. 145, p. 85–94.

Nishi, H., 1988, Structural analysis of part of the Shimanto accretionary complex, Kyushu, Japan, based on planktonic foraminiferal zonation: Tectonics, v. 7, p. 641–652.

MacKenzie, J. S., Taguchi, S., and Itaya, I., 1990, Cleavage dating by K-Ar isotopic analysis in the Paleogene Shimanto Belt of eastern Kyushu, S.W. Japan: Journal of Mineralogy, Petrology, and Economic Geology, v. 85, p. 161–167.

Sakai, T., 1978, Geologic structure and stratigraphy of the Shimantogawa Group in the middle reaches of the Gokase River, Miyazaki Prefecture: Kyushu University Science Reports, Department of Geology, v. 13, p. 23–28 (in Japanese with English abstract).

Sakai, T., and Kanmera, K., 1981, Stratigraphy of the Shimanto Terrain and tectonostratigraphic setting of greenstones in the northern part of Miyazaki Prefecture, Kyushu: Kyushu University, Science Reports, Department of Geology, v. 19, p. 31–48 (in Japanese with English abstract).

Sakai, H., 1988a, Toi-misaki olistostrome of the Southern Belt of the Shimanto terrane, South Kyushu; Part II, Deformation structure of huge submarine slides and their processes of formation: Journal of the Geological Society of Japan, v. 94, p. 837–853 (in Japanese with English abstract).

—— , 1988b, Origin of the Misaki Olistostrome Belt and re-examination of the Takachiho orogeny: Journal of the Geological Society of Japan, v. 94, p. 945–961 (in Japanese with English abstract).

Taira, A., Katto, J., Tashiro, M., and Okamura, M., 1980, The geology of the Shimanto Belt in Kochi Prefecture, Shikoku, *in* Taira, A., and Tashiro, M., eds., Geology and paleontology of the Shimanto Belt: Kochi, Rinyakosaikai Press, p. 319–389 (in Japanese with English abstract).

Taira, A., Okada, H., Whitaker, J.H.McD., and Smith, A. J., 1982, The Shimanto Belt of Japan: Cretaceous–lower Miocene active margin sedimentation, *in* Leggett, J. K., ed., Trench-forearc geology: Geological Society of London Special Publication, no. 10, p. 5–26.

Taira, A., Katto, J., Tashiro, M., Okamura, M., and Kodama, K., 1988, The Shimanto Belt in Shikoku, Japan—Evolution of Cretaceous to Miocene accretionary prism: Modern Geology, v. 12, p. 5–46.

Teraoka, Y., 1979, Provenance of the Shimanto geosynclinal sediments inferred from sandstone compositions: Journal of the Geological Society of Japan, v. 85, p. 753–769.

Toriumi, M., and Teruya, J., 1988, Tectono-metamorphism of the Shimanto Belt: Modern Geology, v. 12, p. 303–324.

Manuscript Accepted by the Society April 24, 1992

Geological Society of America
Special Paper 273
1993

Pyrolysis of organic matter in the Nabae subbelt, Shimanto accretionary complex, southwest Japan

J. P. Hibbard
Department of Marine, Earth, and Atmospheric Sciences, Box 8208, North Carolina State University, Raleigh, North Carolina 27695
D. K. Larue*
Department of Geology, University of Puerto Rico, Mayaguez, Puerto Rico 00708

ABSTRACT

A Rock-Eval pyrolysis study of the Nabae subbelt, the Oligocene-Miocene portion of the Cretaceous to Miocene Shimanto accretionary complex on Shikoku Island, southwest Japan, has yielded ambiguous data concerning the character of the organic component of rocks in the subbelt. The poor data are attributed to the high level of organic maturity in the subbelt. This high organic maturity has been interpreted to be the result of an episode of spreading ridge subduction in the local area. Total organic carbon (TOC) ranges from less than 1 to 30%; average TOC in all stratigraphic units is >0.5%, suggesting that the Nabae subbelt may have been a good hydrocarbon source prior to the thermal disturbance. Several organic-rich samples, some with high hydrogen index values, in spite of high thermal maturation, may indicate local preservation of type II organic matter.

INTRODUCTION

Rock-Eval pyrolysis is recognized as a standard method to characterize both the type and degree of maturation of organic matter (e.g., Espitalie and others, 1977; Clementz, 1978). However, this method has been employed sparingly in the investigation of the vast sedimentary repositories represented by accretionary complexes (Larue, 1986; Larue and others, 1985; Underwood, 1987). This Rock-Eval study of a portion of the Shimanto accretionary complex evolved from two distinct goals of the authors; (1) to add to the sparse knowledge of the organic component of accretionary prisms (DKL), and (2) to help elucidate the structural-stratigraphic relationships of a portion of an accretionary complex with the use of a relatively novel tool, Rock-Eval pyrolysis (JPH). The study shows that rocks of the Nabae subbelt have a high thermal maturity, which (1) allows for only gross characterization of included organic matter, and (2) renders Rock-Eval pyrolysis ineffective as an aid in clarifying structural-stratigraphic relationships.

REGIONAL GEOLOGY

The Oligocene-Miocene rocks at Cape Muroto, Shikoku Island, constitute the Nabae subbelt, which is the youngest and most outboard division of the Cretaceous to Miocene Shimanto accretionary complex of southwest Japan (Taira and others, 1988). The subbelt is faulted against the Eocene-Oligocene Murotohanto subbelt to the north; the details of the regional geology are presented elsewhere in this volume (see Underwood, Hibbard, and DiTullio; Hibbard and others). Briefly, the subbelt consists of three major lithotectonic elements, including: (1) Oligocene–early Miocene accreted strata of the Nabae complex, including the coherent Tsuro and Misaki assemblages and the Hioki and Sakamoto mélanges (Fig. 1); (2) lower to middle Miocene, probable slope-basin, strata of the Shijujiyama Formation; and (3) middle Miocene mafic dikes of the Maruyama Intrusive Suite (Fig. 1). These elements have been heterogeneously affected by three broad phases of deformation (Hibbard and Karig, 1990a; Underwood, Hibbard, and DiTullio, this vol-

*Present address: Exxon Production Research Company, P.O. Box 2189, Houston, Texas 77252-2189.

Hibbard, J. P., and Larue, D. K., 1993, Pyrolysis of organic matter in the Nabae subbelt, Shimanto accretionary complex, southwest Japan, *in* Underwood, M. B., ed., Thermal Evolution of the Tertiary Shimanto Belt, Southwest Japan: An Example of Ridge-Trench Interaction: Boulder, Colorado, Geological Society of America Special Paper 273.

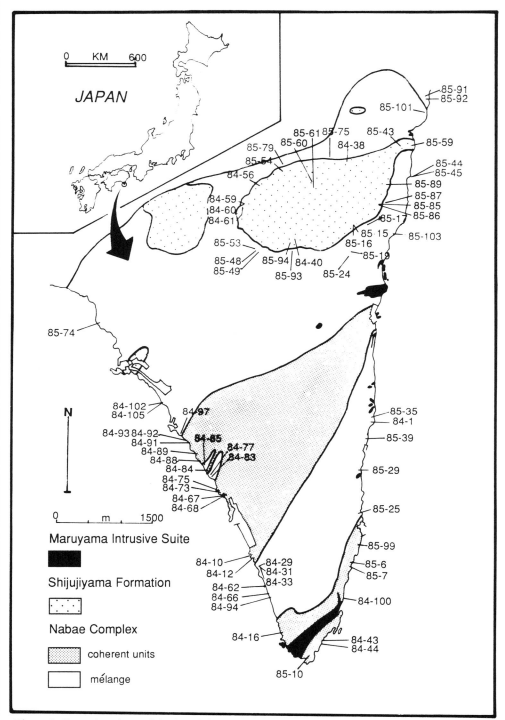

Figure 1. General geology of the late Oligocene–Miocene Nabae subbelt and location of samples for Rock-Eval pyrolysis.

ume; Hibbard and others, this volume). One of the primary goals of the present study with respect to the local geology was to determine if accreted strata of the complex could be distinguished from overlying slope-basin strata by Rock-Eval analysis of the organic component of these lithotectonic packages.

DATA AND INTERPRETATION

We analyzed 67 Nabae subbelt samples by Rock-Eval pyrolysis, including 16 samples from probable slope-basin strata of the Shijujiyama Formation and 51 samples from accreted strata of the Nabae complex (Fig. 1, Table 1). In the Nabae complex, the breakdown of samples by stratigraphic unit includes 12 samples from the Tsuro assemblage, 8 samples from the Misaki assemblage, 21 samples from the Hioki mélange, and 10 samples from the Sakamoto mélange (Fig. 1, Table 1). In addition, two samples from the Murotohanto subbelt, collected just north of the Nabae subbelt on the east side of the peninsula, were analyzed.

Rock-Eval pyrolysis provides data concerning total organic carbon (TOC) in the sample, as well as volatiles formed during programmed heating of the sample. The data are presented on a pyrogram, on which three peaks can generally be identified. The areas under these peaks, denoted by S, are interpreted to represent the following: S_1 is a measure of the bitumen present in the sample; S_2 reflects hydrocarbons formed during cracking of the kerogens during heating; S_3 peak represents a measure of the amount of CO_2 in the sample released during heating. Additionally, the analysis yields a maximum temperature, termed T_{max}, at which kerogens crack most rapidly.

Pyrograms for the samples from the Nabae subbelt have subdued peaks, indicating that the samples were devolatilized. TOC contents range from 0.1 to 30.34%. Coal-like material was found in all of the units. It is worthy of note that all units in the Nabae subbelt show an average TOC > 0.50%, which according to Tissot and Welte (1978) qualify them as good *potential* source rocks. Average TOC of the Tsuro assemblage and the Sakamoto mélange appears to be significantly greater than that of other units. However such differences in average TOC between units are thought to reflect sampling bias; the most time and detail of mapping were expended in these units, consequently more organic-rich samples were identified and procured from them relative to other units.

S_1 values are low and S_2 values are variable (generally 0.1 to 1.47 mg hydrocarbon/g of rock, with the exception of samples 85-24 and 85-103, Table 1). T_{max} values are unreliable because of the jagged nature of the S_2 peaks on the pyrograms. A comparison of T_{max} with R_m values determined independently for the subbelt samples (Laughland and Underwood, this volume) shows a poor correlation between these parameters (Fig. 2).

It is well known that clay minerals can affect the clarity and amplitude of the S_2 peak (Espitalie and others, 1980; Katz, 1983). Thus, conceivably, such poor correlation could be improved if kerogen concentrates were analyzed instead of whole-rock samples. We tested this possibility by running ten samples of kerogen concentrates; however, no improvement was noted in these data and thus they are not included here.

It is also recognized that Rock-Eval pyrolysis is not dependable at the high organic maturities present in the Nabae subbelt samples (R_m values ~1.7 to 2.0%), because of previous devolatilization of the samples and consequently poor and variable definition of the S_2 peak. Problems with the pyrolysis data are also indicated by the hydrogen index (HI = S_2/TOC) and oxygen index (OI = S_3/TOC) data (Fig. 3). Many HI and OI data points fall in a field greater than that expected for a given minimum maturation of R_m of 1.0% (Tissot and Welte, 1978).

Several organic-rich samples retain high HI values in spite of the overall high maturity of the rocks (Shijujiyama Formation: 85-54, 85-86; Tsuro assemblage: 84-12, 84-84, 84-92, 84-93; Hioki mélange: 84-83, 84-102, 85-24, 85-49). Such high HI values may indicate that the organic material was originally type II. Type II organic matter is defined as that which is enriched in hydrogen relative to oxygen. Type II kerogen is usually related to marine sediments where an autochthonous organic matter, derived from a mixture of phytoplankton, zooplankton, and microorganisms (bacteria), has been deposited in a reducing environment, or deposited very rapidly in an oxidizing environment (Tissot and Welte, 1978).

Other samples with lower HI values may reflect type III organic matter, which refers to kerogen with a relatively low initial H/C ratio. Type III kerogen is derived essentially from continental plants and contains significant identifiable vegetal debris. Alternatively, those samples with low HI values could represent bacterially degraded type II organic matter. Plant stems and debris were evident in two outcrops of the Shijujiyama Formation; thus, samples from this unit with lower HI values likely contain some organic matter from a continental source.

CONCLUSIONS

Achievement of the goals of this study—to characterize the organic component of sedimentary rocks of an accretionary complex by Rock-Eval pyrolysis and to use these data as an aid to structural-stratigraphic studies—is largely precluded by the high thermal maturity of Nabae subbelt samples. The Rock-Eval pyrolysis data presented here corroborate the conclusion of vitrinite reflectance and illite crystallinity studies in the area (Hibbard and others, this volume)—that the sedimentary rocks of the area exhibit a high level of thermal maturity. The high level of thermal maturity can be ascribed to pervasive diagenesis and anchimetamorphism at a high structural level in the Shimanto accretion-

J. P. Hibbard and D. K. Larue

TABLE 1. ROCK-EVAL PYROLYSIS DATA FOR WHOLE-ROCK SAMPLES FROM THE NABAE SUBBELT AND ADJOINING AREAS WITH COMPARATIVE R_m VALUES

Sample	TOC (wt.%)	S_2 (mg/g rk)	HI	OI	T_{max} (°C)	R_m(%)*	Sample	TOC (wt.%)	S_2 (mg/g rk)	HI	OI	T_{max} (°C)	R_m(%)*
Shijujiyama Formation							85-10	0.2		4			
84-40	0.7	0.3	48		546	1.7	85-99	0.4	0.1	17	27		3.0
84-56	0.6	0.2	38	28	541	1.7							
	0.6		1	31		1.7	**Hioki mélange**						
84-59	0.9	0.4	46	21	537	1.8	84-77	0.5	0.4	68	25	532	1.7
84-60	0.6	0.1	17	143	515	1.7		0.7	0.5	73	10	470	1.7
84-61	0.4	0.1	18	2	520		84-83	0.8	0.8	105	23	513	
85-15	0.3	0.1	29					0.7	0.3	41	24	515	
85-16	0.6		5	138		1.7	84-85	0.5	0.1	18	58	527	
85-43	0.5	0.1	19	102			84-97	0.4	0.2	58	48	504	
85.54	2.0	5.0	252			1.6	84-102	0.4	0.5	147			
85-59	0.3	0.3	96			1.5		2.7	0.5	19	17	556	
85-60	0.3		7			1.6		2.6	0.5	20	17	554	
85-61	0.4	0.3	68			1.5	84-105	0.6	0.2	38	18	526	
85-85	0.4		9				85-17	0.2		5	138		
85-86	0.4	0.4	108	78		1.8	85-19	0.4		2	69		1.7
85-89	0.5	0.3	60			1.6	85-24	7.2	23.9	334	9	449	
85-94	0.5		2	86			85-44	0.4	0.3	71	126		
							85-45	0.4	0.3	74	103		1.3
Tsuro assemblage							85-48	0.4	0.3	64			
84-10	0.1	0.1			576		85-49	0.7	0.8	108			
84-12	8.5	1.5	17	8	572		85-53	0.5	0.1	23			
	0.1	0.2	184	30	580		85-74	0.4		3			1.9
84-67	0.7	0.4	49	6	529	1.7	85-75	0.3		3			1.7
84-68	0.1	0.4			451		85-79	0.4	0.1	63			
84-73	0.5	0.2	45	143	524	2.0	85-87	0.3		12	71		
84-75	6.4	1.1	16	10	582	2.1	85-93	0.4	0.1	20			
	9.1	0.9	9	7	575	2.1	85-101	0.3	0.3	81	138		1.1
84-84	0.5	0.6	128	46		1.8	85-103	3.9	3.6	94	14	468	1.1
	2.1	0.9	28	8		1.8							
	0.6	0.5	75	46		1.8	**Sakamoto mélange**						
84-88	0.6	0.1	22	0	587		84-29	0.5	0.2	34	34	515	
	10.0	1.5	14	8	575			0.5	0.3	50	36	541	
	9.7	1.4	14	7	579		84-31	0.5	0.2	28	45	448	
84-89	0.7	0.4	46	24	460	1.7	84-33	2.5	0.5	20	17	555	2.1
84-91	6.7	0.7	9	6	579	2.0		0.8	0.1	18	5	518	2.1
	7.2	1.1	15	5	509	2.0	84-62	0.4	0.3	64	23	525	
	7.3	0.7	9	6	573	2.0	84-66	7.0	0.9	12	69		
84-92	0.5	0.8	169	40		1.8	84-94	0.5	0.1	19	36	543	1.7
84-93	0.6	0.9	143	32	589	1.8		0.5	0.1	29	31	548	1.7
	0.6	0.8	139	22	590	1.8	85-25	0.6	0.1	22	109	442	
							85-29	30.3	0.5	2	5	531	2.1
Misaki assemblage							85-35	0.5	0.3	71	147		
84-16	0.2	0.1	28	61	497		85-39	2.2	0.2	11	29	500	1.7
84-43	0.4	0.3	59	0	482								
84-44	0.4	0.3	69	19	569	3.7	**Murotohanto subbelt**						
84-11	0.2	0.1	41	133			85-91	0.2		4	30		
85-6	2.1		0	59		2.4	85-92	0.2		20	5		
85-7	2.4		2	43		2.6							

*From Laughland and Underwood, this volume.

TOC, total organic carbon; S_2, hydrocarbons formed during cracking of the kerogens during heating; HI, hydrogen index (HI = S_2 / TOC); OI, oxygen index (OI = S_3 / TOC); S_3, peak amount of CO_2 in sample released during heating; T_{max}, maximum temperature at which kerogens crack most rapidly; R_m, vitrinite reflectance.

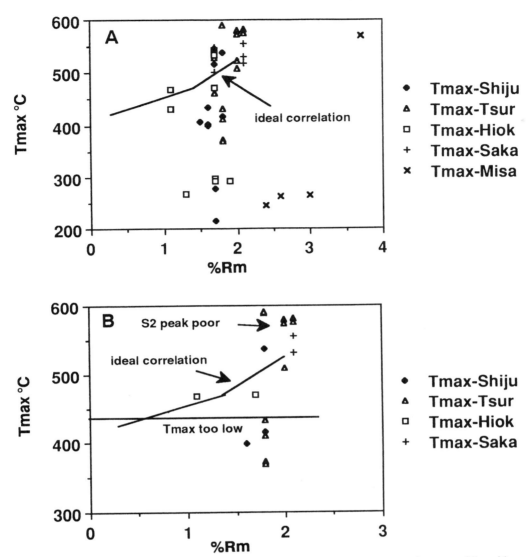

Figure 2. A comparison of T_{max}, the maximum temperature at which kerogens crack most rapidly, with R_m, vitrinite reflectance values, determined independently for subbelt samples. A, all data plotted for T_{max} and R_m. Ideal correlation line from data in Dow (1978); note poor fit. B, T_{max} data for S_2 values greater than 0.5 plotted against R_m, in recognition that T_{max} is defined on the basis of S_2. T_{max} values that are too low for given maturity ($R_m > 1.0\%$) are shown below the line. T_{max} values that are too high compared to R_m data generally have a poorly defined S_2 peak. S_2, hydrocarbons formed during cracking of the kerogens during heating.

ary complex during an episode of spreading ridge subduction in the middle Miocene (Hibbard and Karig, 1990a, b; Hibbard and others, this volume); thus the thermal maturity evident in the Nabae subbelt is not considered typical of accretionary complexes, in general.

We have recognized the possible presence of type II organic matter in several stratigraphic units and type III organic material may occur in all units. However, the quality of the data is not sufficient to distinguish stratigraphic units on the basis of organic matter type. Average TOC values >0.50% in all units indicate that prior to the present degree of maturation, rocks of the Nabae subbelt were likely good hydrocarbon sources.

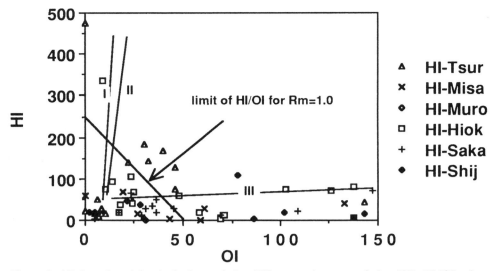

Figure 3. All data plotted for the hydrogen index (HI) versus the oxygen index (OI). HI/OI values greater than 250/50 are probably unreliable, because high maturation should preclude significant amounts of volatiles.

ACKNOWLEDGMENTS

This study was supported, in part, by funds from a National Science Foundation grant to D. Karig and T. Byrne, and Harold T. Stearns awards from the Geological Society of America to J. Hibbard. Whole-rock analyses were undertaken at Stanford University; the kerogen concentrate analyses were carried out by Jerry Clayton, U.S. Geological Survey. The manuscript has benefitted from constructive reviews by J. Clayton, R. Kettler, and M. Underwood. The family of Teratoshi Tomioka is gratefully acknowledged for their hospitality and comraderie during the field component of this study.

REFERENCES CITED

Clementz, D., 1978, Effect of oil and bitumen saturation on source-rock pyrolysis: American Association of Petroleum Geologists, v. 63, p. 2227–2232.

Dow, W., 1978, Petroleum source beds on continental slopes and rises: American Association of Petroleum Geologists Bulletin, v. 62, p. 1584–1606.

Espitalie, J., Madec, M., and Tissot, B., 1977, Source rock characterization method for petroleum research: Offshore Technology Conference, 9th, Houston, Texas, p. 439–444.

——— , 1980, Role of mineral matrix in kerogen pyrolysis: Influence on petroleum generation and migration: American Association of Petroleum Geologists Bulletin, v. 64, p. 59–66.

Hibbard, J., and Karig, D., 1990a, Structural and magmatic responses to spreading ridge subduction; An example from southwest Japan: Tectonics, v. 9, p. 207–230.

——— , 1990b, An alternative plate model for the early Miocene evolution of the southwest Japan margin: Geology, v. 18, p. 170–174.

Katz, B. J., 1983, Limitations of 'Rock-Eval' pyrolysis for typing organic matter: Organic Geochemistry, v. 4, p. 195–199.

Larue, D., 1986, Organic matter in limestone and mélange matrix from the Franciscan and Cedros subduction complexes, in Abbot, P., ed., Cretaceous stratigraphy of western North America: Society of Economic Paleontologists and Mineralogists, Pacific Section, v. 46, p. 211–221.

Larue, D., Schoonmaker, J., Torrini, R., Lucas-Clark, J., Clark, M., and Schneider, R., 1985, Barbados: Maturation, source rock potential and burial history within a Cenozoic accretionary complex: Marine and Petroleum Geology, v. 2, p. 96–110.

Taira, A., Katto, J., Tashiro, M., Okamura, M., and Kodama, K., 1988, The Shimanto Belt in Shikoku, Japan—Evolution of Cretaceous to Miocene accretionary prism: Modern Geology, v. 12, p. 1–42.

Tissot, B., and Welte, D., 1978, Petroleum formation and occurrence: Springer Verlag, New York, 538 p.

Underwood, M., 1987, Thermal maturity and hydrocarbon potential of Franciscan terranes in coastal northern California: Accreted basement to the Eel River basin, in Tectonics, sedimentation and evolution of the Eel River and associated basins of northern California: San Joaquin Geological Society Miscellaneous Publication 37, p. 89–98.

MANUSCRIPT RECEIVED BY THE SOCIETY APRIL 24, 1992

Geological Society of America
Special Paper 273
1993

Evolution of the Shimanto accretionary complex: A fission-track thermochronologic study

Noriko Hasebe, Takahiro Tagami, and Susumu Nishimura
Department of Geology and Mineralogy, Faculty of Science, Kyoto University, Kyoto 606, Japan

ABSTRACT

To place thermotectonic constraints on the evolution of the Cretaceous to Neogene Shimanto Belt, we carried out fission-track (FT) analyses of detrital apatite and zircon collected from both sandstone turbidites and blocks in mélanges on the Muroto Peninsula, Shikoku. Eight FT apatite ages show good agreement around 10 Ma, with all data except for one passing the χ^2 test. These results, in conjunction with depositional ages of Cretaceous to early Oligocene, demonstrate that the Shimanto Belt was heated hotter than ~125°C (apatite total annealing temperature) and subsequently cooled below ~100°C at ~10 Ma. The cooling pattern is attributed to higher thermal gradients and/or uplift caused by rapid subduction of the newly formed Shikoku Basin at ~15 Ma. On the other hand, 23 zircon samples show a large range of sample ages from 150 to 17 Ma, with all failing the χ^2 test except for one from a mélange. Hence, most parts of the Shimanto Belt have not been heated above the zircon partial annealing zone (ZPAZ; ~190 to 260°C), whereas some parts of the mélanges have experienced temperatures above 260°C. FT age spectra of zircon single-grain data show that in the Northern Shimanto Belt (NSB) the youngest peak in each sample is younger than its depositional age, suggesting the maximum temperature was in the ZPAZ. In contrast, all age peaks from the Southern Shimanto Belt (SSB), are consistently older than depositional ages, providing no evidence of heating up to ZPAZ. These contrasting patterns probably reflect systematic differences in the maximum temperature reached during the evolution of the Shimanto Belt. The age-temperature paths estimated for individual tectonic units suggest successive accretion, growth, and uplift of offscraped imbricate packages as well as underplating of mélanges beneath them in an accretionary wedge.

INTRODUCTION

The evolution of accretionary prisms is an important geologic process in the context of the formation and development of continental crust. In recent years, details have been better understood by scientific drilling and multichannel seismic reflection profiling, throwing light on how sediments accrete at the frontal toe (e.g., Kagami and others, 1983; Karig, 1986; Moore and Watkins, 1981; Moore, 1989; Moore and others, 1991; Westbrook and Smith, 1983). However, some aspects of prism evolution remain ambiguous: How does an individual imbricate slice behave during subduction? How does underplating proceed? and, What tectonic settings are responsible for the formation of mé-

langes? On-land studies on accretionary prisms with a well-established geological framework should provide valuable information to help answer these questions.

The Shimanto Belt, lying along the Pacific coastal range of Southwest Japan subparallel to the modern Nankai Trough (Fig. 1), is a particularly well studied on-land accretionary prism (Sakai and Kanmera, 1981; Taira and others, 1980a). It preserves an accretionary history from Cretaceous to Miocene time. Many regional field surveys have been carried out, including sedimentological, biostratigraphical, and structural studies (Agar and others, 1989; Cowan, 1990; DiTullio and Byrne, 1990; Hibbard and Karig, 1987; Kumon, 1983; Sakai, 1988; Taira and others, 1980a). Petrologically, the Shimanto Belt consists of unmeta-

Hasebe, N., Tagami, T., and Nishimura, S., 1993, Evolution of the Shimanto accretionary complex: A fission-track thermochronologic study, *in* Underwood, M. B., ed., Thermal Evolution of the Tertiary Shimanto Belt, Southwest Japan: An Example of Ridge-Trench Interaction: Boulder, Colorado, Geological Society of America Special Paper 273.

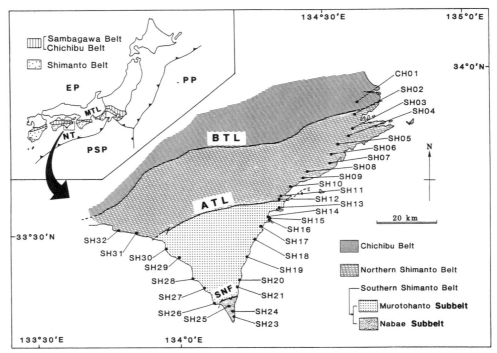

Figure 1. Map showing the location of the Shimanto Belt and sampling sites on the Muroto Peninsula. The map was modified from Taira and Tashiro (1987), Taira and others (1989), and the geological map published by the Regional Forestry Office of Kochi Prefecture (Katto and others, 1977). MTL, Median Tectonic Line; BTL, Butsuzo Tectonic Line; ATL, Aki Tectonic Line; SNF, Shiina-Narashi fault; NT, Nankai Trough; EP, Eurasia Plate; PSP, Philippine Sea Plate; PP, Pacific Plate.

morphosed to low-grade metamorphosed coherent turbidites and mélanges, suggesting rather shallow maximum burial depth (Toriumi and Teruya, 1988). Therefore, thermal indicators in the low-temperature range (i.e., 100 to 300°C) should provide valuable information on the thermal history of the Shimanto Belt and may further constrain the tectonic evolution. In this context, fission-track (FT) thermochronology should be effective (Gleadow and others, 1983; Green and others, 1989; Harrison and McDougall, 1980; Hurford, 1986; Kamp and others, 1989; Naeser, 1979; Wagner and others, 1977).

In this study, 31 sandstone samples were collected from the Muroto Peninsula, Shikoku Island (Fig. 1), and apatite and zircon grains were analyzed by the FT method. The Cretaceous to Tertiary Shimanto Belt is widely exposed on the Muroto Peninsula, and unlikely to have been affected by later thermal events. Although many Tertiary felsic plutonic rocks intruded the Shimanto Belt (Shibata, 1978), no such plutons or Quaternary volcanism are known in the Muroto Peninsula. The only event that could affect the Shimanto Belt thermally is the gabbroic intrusion at the toe of Cape Muroto (Yoshizawa, 1953). Other regions of the Muroto Peninsula are considered to preserve the initial thermal record of accretionary history and are thus suitable for the purpose of our study.

GEOLOGICAL SETTING AND SAMPLING

Southwest Japan is divided by the Median Tectonic Line (M.T.L.) into the Inner Zone to the north and the Outer Zone to the south (Fig. 1). The Outer Zone consists of the Sambagawa Metamorphic Belt, the Chichibu Belt, and the Shimanto Belt from north to south. The Sambagawa Belt and Chichibu Belt are regarded as the Jurassic subduction complexes (Banno and Sakai, 1989; Ozawa and others, 1985). The Shimanto Belt has a tectonic contact with the Chichibu Belt along the Butsuzo Tectonic Line (B.T.L.), and is mainly composed of coherent turbidites, tectonic mélanges, and slope-basin deposits. It is divided by the Aki Tectonic Line (A.T.L.) into the Northern Shimanto Belt of Cretaceous age and the Southern Shimanto Belt of Tertiary age (Fig. 1). Toriumi and Teruya (1988) studied the low-grade metamorphism of the Northern Shimanto Belt petrologically, and suggested the maximum temperature-pressure condition of 200 to 300°C and 3 to 5 kb. The temperature condition estimated from the vitrinite reflectance data on the Muroto Peninsula (DiTullio and others, this volume; Hibbard and others, this volume; Laughland and Underwood, this volume; Mori and Taguchi, 1988) was 140 to 320°C according to the correlation of Barker (1988).

The Shimanto Belt was intruded by the great mass of the

Miocene granitic rocks (Shibata, 1978). This has been attributed to the rapid subduction of the newly formed hot Shikoku Basin (Takahashi, 1986). On the Muroto Peninsula, however, only a gabbroic intrusion having a Rb-Sr whole-rock biotite isochron age of 14.4 ± 0.4 Ma (Hamamoto and Sakai, 1987) has been found at the toe producing hornfels aureola of ~40 m width (Yoshizawa, 1953), correlated chemically with tholeiitic off-ridge volcanism in the Shikoku Basin (Miyake, 1983; Takahasni, 1986). Alternatively, the intrusion was considered to have resulted from the opening-ridge subduction of the Shikoku Basin (Hibbard and Karig, 1990a). Although these possible events could have overprinted the initial thermal records of pre-Miocene accretion, the vitrinite reflectance data from the Eocene Shimanto Belt are consistent to the structural features related to accretionary processes (DiTullio and others, this volume; see also Laughland and Underwood, this volume). Hence the Miocene thermal event would have not substantially affected the pre-Miocene thermal structure of the Cretaceous-Eocene Shimanto Belt currently outcropped, and was rather localized to the Miocene Shimanto Belt.

The Northern Shimanto Belt has been divided into five tectonolithologic units along the eastern side of the Muroto Peninsula (Kumon, 1983); these include, from north to south, the Hinotani unit, the Akamatsu unit, the Taniyama unit, the Hiwasa unit, and the Mugi unit (Table 1). The Hinotani and Hiwasa units are turbidites composed mainly of sandstone and sandy flysch. The Akamatsu, Taniyama, and Mugi units are mélanges, which consist of shale and muddy flysch together with intersheared blocks of chert, greenstone, and acidic tuff. The five units are

TABLE 1. UNITS IN THE SHIMANTO BELT
WITH THEIR DEPOSITIONAL AGES AND SAMPLING SITES

Belt	Subbelt	Unit	Depositional Age	Ma**	Sampling Site‡
CCB			~Jurassic	208.0 to 145.6	CH01
NSB		Hinotani[†]	Albian to Turonian	112.0 to 88.5	SH02
		Akamatsu[§]	Tithonian to Coniacian	152.1 to 86.6	SH03
		Taniyama[§]	Albian to Santonian	112.0 to 83.0	SH04
		Tei Mélange[§]	Turonian to Campanian	90.4 to 74.0	SH32
		Hiwasa[†]	Late Cretaceous	97.0 to 65.0	SH05 to SH09
		Mugi[§]	Late Cretaceous	97.0 to 65.0	SH10 to SH12, SH31
SSB	Murotohanto	Oyamamisaki[†]	Eocene	56.5 to 35.4	SH30
		Naharigawa[†]	Middle to Late Eocene	50.0 to 35.4	
		(A)			SH13 to SH16,
		(B)			SH17, SH18, SH29,
		(C)			SH19, 28
		Muroto[†]	Late Eocene to Early Oligocene	38.6 to 29.3	
		(Shiina)			SH20, SH21, SH26
		(Gyoto)			SH27
		Sakihama Mélange[§]	Paleocene to Early Eocene	65.0 to 50.0	
SSB	Nabae	Hioki Mélange[§]	Oligocene to Early Miocene	35.4 to 16.3	SH25
		Shijujiyama Formation[†]	Early Miocene	23.3 to 16.3	
		Tsuro[†]	Late Oligocene to Early Miocene	29.3 to 16.3	SH24
		Sakamoto Mélange[†]	Late Oligocene to Early Miocene	29.3 to 16.3	
		Misaki[†]	Late Oligocene to Early Miocene	29.3 to 16.3	SH23

*CCB = Chichibu Belt; NSB = Northern Shimanto Belt; SSB = Southern Shimanto Belt. Units of NSB and SSB after DiTullio and Byne (1990), Hibbard and Karig (1990b), Kumon (1983), and Taira and others (1980a).
[†]Coherent.
[§]Mélange.
**Numerical ages after Harland and others (1989).
‡SH03, SH04, SH05, SH10 through 12, SH25, and SH31 and 32 are collected from sandstone blocks in shale, SH30 from conglomerate, and others from massive sandstone. SH23 was collected about 30 m north of the contact with the Muroto-Misaki gabbroic intrusion.

divided by north-dipping high-angle thrusts (Kimura and Mukai, 1991). The depositional age of each unit is known from radiolarian fossils yielded in shale or chert (Kumon, 1983; Matsugi and others, 1987; Suyari, 1986) and is summarized in Table 1. The Tei mélange that crops out at the western side of the Muroto Peninsula has contained units of different ages. Chert has yielded the Valanginian to Cenomanian radiolaria, and the shale matrix, the Turonian to Campanian radiolaria (Taira and others, 1980b).

The Southern Shimanto Belt has been divided into two subbelts; the Murotohanto subbelt and the Nabae subbelt, separated by the Shiina-Narashi fault (Fig. 1; Taira and others, 1980a). The Murotohanto subbelt consists of the Eocene to lower Oligocene sediments, and is further subdivided into four units from north to south: the Oyamamisaki unit, the Naharigawa unit, the Sakihama mélange, and the Muroto unit. The Oyamamisaki unit is composed of turbidites. Its biostratigraphic age is not known precisely, because it lacks age diagnostic fossils. Taira and others (1980a) have assigned it Cretaceous to Eocene age. The Naharigawa unit is also turbidites and is of middle to late Eocene age as indicated by molluscan fauna (Katto and Tashiro, 1979), nannofossils (Okada and Okamura, 1980), and radiolarian fossils (Taira and others, 1980b). It includes three members labeled A, B, and C. The Sakihama mélange, unlike the Cretaceous mélanges, does not include ribbon chert. The age of the shaly matrix is middle to late Eocene. The Muroto unit is turbidites, and more shaly than the Naharigawa unit. Its depositional age is late Eocene to early Oligocene (Taira and others, 1980a). DiTullio and Byrne (1990) divided the Muroto unit into two structural domains, the hanging-wall Gyoto and footwall Shiina domains, juxtaposed along an out-of-sequence thrust.

The Nabae subbelt has been divided into five units from north to south: the Hioki mélange, the Shijujiyama Formation, the Tsuro assemblage, the Sakamoto mélange, and the Misaki assemblage (Hibbard and Karig, 1990b). The Hioki mélange is adjacent to the Murotohanto subbelt and contains variety of lithologies (greenstones, sandy breccias, shales, lenses of sand and mud, etc.) of different ages, chaotically mixed together. The age of the shaly matrix is the late Oligocene to early Miocene (Taira and others, 1980a). Sakamoto mélange have yielded Oligocene–early Miocene radiolaria (Taira and others, 1980a) and early Miocene foraminifera (Saito, 1980). The Tsuro and Misaki assemblages contain mainly turbidite facies. The radiolaria (Okamura, 1980) and molluscs (Katto and Tashiro, 1979) indicate a late Oligocene to early Miocene age. The Shijujiyama Formation is surrounded by the Hioki mélange and contains an unusual coarse sandstone facies with molluscan fossils (Taira and others, 1980a) and a basal mudstone and mafic volcanic-volcaniclastic unit (Hibbard and Karig, 1990b). Taira and others (1988) interpret this unit to be a lower Miocene shelf-slope basin deposit. An alternative interpretation is that this represents an olistolith that has moved downslope from its original neritic environment (Sakai, 1988).

Sampling sites are shown in Figure 1 and Table 1. Although our analyses were carried out on sandstone samples, the depositional age of each unit was estimated from fossils, mainly radiolar-

ia, included in adjacent shale. In coherent turbiditic units, sandstone and shale compose successive upward-fining sequences, indicating no time interval between their depositions. In the case of mélange units, the relationship between the depositional age of sandstone block and that of shaly matrix encounters inherent uncertainties. However, lenticular shape of sandstone blocks is likely to suggest that they were formed by deformation of interbedded sandstone layers in shale due to layer-parallel extension (Cowan, 1985). Hence it can be reasonably assumed that sandstone blocks have the same depositional ages with shaly matrix.

EXPERIMENTAL METHOD

Mineral separation and sample preparation procedures described by Tagami and others (1988) were carried out using about 5-kg rock samples for each sampling locality. Zircons and apatites were dated by the external detector method (e.g., Naeser, 1976).

After mineral separation using conventional heavy liquid and magnetic techniques, zircons and apatites were handpicked under the binocular microscope and mounted in TOYOFRON® PFA teflon sheet and epoxy resin, respectively. They were polished with diamond pastes and etched chemically in KOH:NaOH eutectic etchant at $225.0 \pm 0.5°C$ for zircons, and in 0.6% HNO_3 at $32.5 \pm 0.5°C$ for apatites, until fission tracks appeared well etched under the microscope. As for zircons, the appropriate etching time of each grain varies because of the accumulation of α-damage (Gleadow, 1981). Because zircons of widely varying detrital ages need to be analyzed, each zircon mount was etched stepwise at ~5-hr intervals prior to irradiation. After each step-etch, grains in the optimal etching condition (Sumii and others, 1987) were searched for and counted to determine spontaneous track densities. This procedure was repeated until most grains suitable for measurement were etched and counted optimally. As a result, optimal etching time for each zircon grain ranges from 10 to 30 hr in the samples analyzed successfully. This "multi-etch" technique is rather laborious but provides high resolution in determining multimodal ages of detrital zircons. Note that the induced track densities in some slightly overetched grains, which had reached the optimal condition in the early stage of etching, can be measured correctly, as confirmed by age standard analyses (e.g., Tagami, 1987).

Samples were irradiated at the Irradiation Pit (IP) facility of TRIGA II Reactor at Musashi Institute of Technology (MITR) and Thermal Column Pneumastic Tube (TC-Pn) facility and Heavy Water (D_2O) facility of Kyoto University Research Reactor (KUR-1). These facilities are well characterized and their "Cd-ratios" for Au are 13 to 15 for IP at MITR; 200 for Tc-Pn at KUR; and 5000~ for D_2O at KUR; which satisfies the criteria recommended by Hurford (1990). During irradiation, induced fission tracks were recorded on the external detector, a low-uranium mica sheet that was firmly attached to each mount. The induced tracks were revealed by etching the mica in 46% HF at $32.5 \pm 0.5°C$ for 4 to 5 minutes.

Etched tracks were counted in transmitted light using dry 100× objective at a total true magnification of 925× throughout. FT ages were calculated by the zeta age calibration approach (Hurford and Green, 1983). Statistical error on a FT age was given by the "conventional analysis" (Green, 1981). For zircons and apatites irradiated at MITR, we adopted the ζ values of 348.4 ± 8.3 (2σ) and 319.6 ± 17.4, respectively (Tagami, 1987); and for those at KUR, 358.9 ± 14.0 and 309.6 ± 23.4 (Tagami and Hasebe, in preparation). Age standards were irradiated at each irradiation run and analyzed along with unknown samples. These standards yielded results concordant with their reference ages (Table 2).

RESULTS AND INTERPRETATION

Apatite ages

Eight sites yield enough apatite grains for age determination: one from the Chichibu Belt (CH01), one from the northernmost site (SH02) of the Northern Shimanto Belt, and the others were from the Eocene Shimanto Belt, including two sites (SH19, 28) from the C member of the Naharigawa unit and four (SH20, 21, 26, 27) from the Muroto unit (Fig. 1). All apatites have rounded shapes indicating their detrital origin. Analytical results are shown in Table 3. The FT apatite ages show a good agreement at around 10 Ma for all samples. In addition, all FT data except CH01 pass the χ^2 test (Galbraith, 1981; Green, 1981) at the 5% criterion, and CH01 shows probability of 4%, which nearly passes the χ^2 test. These results indicate that the variability in the track count data is limited to the inherent variability in the radiometric decay process. Each sample appears to represent a unique grain-age population. In general, sandstones consist of detrital minerals supplied from various source rocks, and thus contain crystals of variable ages older than the timing of deposition. In the present case, the FT ages are much younger than their Cretaceous to Eocene depositional ages, with a unique FT age population for each sample (see Tables 1 and 3). This suggests that the samples have been heated over the apatite partial annealing zone (APAZ > ~70 to 125°C, after Gleadow and others, 1983; see also Naeser, 1981) to totally reset their ages. So, these are really cooling ages after their total annealing.

Zircon ages

Although zircons were separated from all the rock samples collected, sandstone blocks in mudstone matrix usually contain only minor zircon. In several cases there were not enough for age determination. Obtained zircons show much variation in shape, from euhedral to rounded, suggesting multiple source of grains and thus initial multimodal grain-age distributions. Twenty-three sites were anlayzed successfully (Table 4). Most samples have extremely low χ^2 probabilities [P(χ^2)], less than 0.1%, reflecting a mixture of different grain-age populations in these samples. SH04, SH32, SH19, and SH26 yield relatively high P(χ^2) values of 1 to 5%, suggesting that these four samples can be regarded, to some extent, as composed of single grain-age populations. Considering the multimodal grain-age population generally found in unannealed sandstones, these four samples would be best ex-

TABLE 2. FISSION TRACK AGE CALCULATIONS OF SOME AGE STANDARDS IRRADIATED WITH UNKNOWN SAMPLES IN THIS STUDY

Sample*	$\zeta \pm 2\sigma$ ($\times 10^6$/cm²)	ρs	Ns	ρi ($\times 10^6$/cm²)	Ni	ρd ($\times 10^6$/cm²)	Nd	$T \pm 2\sigma$ (Ma)	n	P(χ^2) (%)	RA ± 2 (Ma)
Apatite											
DURN1-AP1	319.6 ± 17.4	0.14	178	0.35	444	50.17	2,332	32.1 ± 6.1	11	30	31.4 ± 0.5
DURN1-AP3	309.6 ± 23.4	0.17	618	0.65	2,347	78.66	2,925	32.1 ± 4.0	6	60	31.4 ± 0.5
DURN1-AP4	309.6 ± 23.4	0.15	633	0.41	1,777	56.72	2,109	31.3 ± 4.0	6	50	31.4 ± 0.5
FCT87-AP1	319.6 ± 17.4	0.19	256	0.56	761	50.17	2,332	27.0 ± 4.3	22	99	27.77 ± 0.04
FCT87-AP2	309.6 ± 23.4	0.24	145	0.35	215	28.41	3,301	29.7 ± 6.8	10	85	27.77 ± 0.04
Zircon											
BM4ll-ZR2	358.9 ± 14.0	1.06	883	1.00	832	9.24	1,074	17.6 ± 2.1	12	8	16.3 ± 0.2
FCT87-ZR1	348.2 ± 8.3	4.74	1,355	4.06	1,161	12.87	2,093	26.2 ± 2.5	8	45	27.77 ± 0.04
FCT87-ZR3	358.9 ± 14.0	4.95	1,380	3.68	1,026	10.56	1,227	25.5 ± 2.7	12	15	27.77 ± 0.04
MDCIII-ZR1	358.9 ± 14.0	13.25	1,016	2.93	225	12.22	1,420	99.0 ± 15.9	10	97	98.7 ± 0.6

*DURN1 = Durango; FCT87 = Fish Canyon Tuff; BM4ll = Buluk Member Tuff; MDCIII = Mount Doromedary Complex.

Note: ζ = Zeta value for NBS-SRM612 dosimeter glass used for age calculation; ρs = Density of spontaneous tracks; Ns = Number of spontaneous tracks counted to determine ρs; ρi = Density of induced tracks in a sample; Ni = Number of induced tracks counted in a muscovite external detector to determine ρi; ρd = Density of induced tracks in NBS-SRM612 dosimeter glass; Nd = Number of induced tracks counted in a muscovite external detector to determine ρd; T = FT age calculated from pooled Ns and Ni for all grains counted; n = Number of counted grains (or fields in case of DURN1 samples); P(χ^2) = Probability of χ^2 for N degrees of freedom (N = n-1) quoted to the nearest 5 or 10 percent except for those less than 5 percent and more than 95 percent (Galbraith, 1981); RA = Reference age after Green (1985), Miller and others (1985), Hurford and Watkins (1987).

TABLE 3. APATITE FISSION TRACK ANALYTICAL RESULTS

Sample	ρs (x10^6/cm^2)	Ns	ρi (x10^6/cm^2)	Ni	ρd (x10^6/cm^2)	Nd	T ± 2σ (Ma)	n	P(χ^2) (%)	Sampling Site
CH01	0.29	119	2.45	1,018	0.5437	2,527	10.2 ± 2.1	7	4	Chichibu Belt
SHO2	0.21	189	1.48	1,331	0.5017	2,332	11.4 ± 1.9	16	85	Hinotani U.
SH19	0.34	77	1.12	251	0.2841	3,301	13.5 ± 3.7	10	99	Naharigawa U., C
SH28	0.16	23	0.76	110	0.2841	3,301	9.2 ± 4.3	4	40	Naharigawa U., C
SH20	0.41	49	2.47	299	0.2841	3,301	7.2 ± 2.3	6	95	Muroto U. (Shiina)
SH21	0.22	106	1.08	525	0.2841	3,301	8.9 ± 2.0	12	95	Muroto U. (Shiina)
SH26	0.10	23	0.53	122	0.2841	3,301	8.3 ± 3.8	6	90	Muroto U. (Shiina)
SH27	0.29	38	1.31	170	0.2841	3,301	9.8 ± 3.6	7	60	Muroto U. (Gyoto)

Note: ρs = Density of spontaneous tracks; Ns = Number of spontaneous tracks counted to determine ρs; ρi = Density of induced tracks in a sample; Ni = Number of induced tracks counted in a muscovite external detector to determine ρi; ρd = Density of induced tracks in NBS-SRM612 dosimeter glass; Nd = Number of induced tracks counted in a muscovite external detector to determine ρd; T = Fission track age calculated from pooled Ns and Ni for all grains counted; n = Number of counted grains; P(χ^2) = Probability of χ^2 for N degrees of freedom (N = n-1) quoted to the nearest 5 or 10% except for those less than 5% and more than 95% (Galbraith, 1981); U. = Unit; C = C member of the unit.

plained by the existence of moderate thermal overprints after their deposition to more or less anneal fossil tracks. Overall, however, P(χ^2) values of zircon analysis are consistently low, which demonstrates that most samples were not heated above the zircon partial annealing zone (ZPAZ) by post-depositional events.

The thermal annealing characteristics and closure temperature of fission-tracks in zircon have been better understood by laboratory annealing experiments (Tagami and others, 1990; see also Kasuya and Naeser, 1988), natural borehole analysis (Zaun and Wagner, 1985), and orogenic cooling study (Hurford, 1986). Although further fundamental researches are needed, we here adopt the range of the ZPAZ studied by Zaun and Wagner (1985), who looked at the stability of spontaneous fission tracks in zircon using the drill core from the Urach III borehole. They showed that tracks were effectively annealed between 158°C and 224°C for an annealing duration of 10^8 years. The results would indicate that the ZPAZ has temperature limits of ~190 to 260°C with closure temperature of ~225 ± 30°C for a cooling rate of 10°C/m.y., on the basis of both the range of ZPAZ for cooling rate of 0.1°C/m.y. and the activation energy (Zaun and Wagner, 1985). Note that the rate of an order of 10°C/m.y. can approximately be applied to the thermal history of the Shimanto Belt, as will be shown in Figure 4. The ZPAZ adopted would be accompanied by the error of ±50°C because of the uncertainties in the zone of ZPAZ observed, the cooling (or heating) rate of the Shimanto Belt, and the activation energy.

To investigate the degree of track annealing in our zircon data, we also examined the FT age spectrum using single-grain ages (Hurford and others, 1984). Each age spectrum is formed by summing up probability distributions of grain age data, commonly expressed as a mean and its error, measured for individual crystals. This approach possesses significant advantage over other conventional treatments, such as an age histogram, in that peaks on the spectrum can be used to objectively identify different grain-age populations in sample. Note that the observed age spectrum does not necessarily represent the parent frequency distribution because, in practice, it can be difficult to extract and combine all grain populations in a given sample, especially when a population of grains has track densities too high to measure reliably (i.e., ⩾3 × 10^7/cm^2). In interpreting FT age spectra in the context of thermal history analysis, the most important is the timing of deposition relative to the FT peak ages. Fossils usually provide an age range for the age of the depositional event (Table 1). The youngest limit of this range is used here to represent the minimum depositional age of the sample. Another point to be noted is that in drawing FT age spectra, the peak height of younger grains tend to be overestimated (see the equation of Hurford and others, 1984). Hence, the youngest peak formed solely by a grain (i.e., SH17, 20, 24, and 30 in Fig. 3b) shall not be used for discussion.

At deposition, individual detrital grains retain a variety of FT ages older than the timing of deposition, and related to their respective provenances. When the rock has been heated over the FT total stability zone after deposition, all grain ages are progressively reduced, resulting in a shift of the initial age spectra toward the timing of the thermal event. In this regard, we can classify the observed patterns of spectrum into three types using the degree of track annealing. (A) If a rock was not heated up to the PAZ, the age spectrum is characterized by multiple peaks, all of which are older than the depositional age (corresponding to the schematic pattern A of Fig. 2). (B) If a rock has been heated into the PAZ, the spectrum may retain several age peaks, and indistinguishable from case A by its shape only. In case B, however, some peaks could be younger than the depositional age (Fig. 2B). (C) If a rock was heated above the PAZ, the tracks in each grain are totally annealed. Hence, all grains should have a statistically concordant age and expected to pass the χ^2 test. The calculation age is younger than the depositional age and indicates the time that

TABLE 4. ZIRCON FISSION TRACK ANALYTICAL RESULTS

Sample	ρs (x10^6/cm^2)	Ns	ρi (x10^6/cm^2)	Ni	ρd (x10^6/cm^2)	Nd	T ± 2σ (Ma)	n	P(χ^2) (%)	Sampling Site
CH01	9.63	3,469	1.72	620	0.1514	2,462	147.7 ± 14.6	11	<0.1	Chichibu Belt
SH02	6.41	3,106	1.31	635	0.1514	2,462	129.1 ± 12.8	16	<0.1	Hinotani U.
SH03	11.03	1,884	2.01	342	0.1514	2,462	145.4 ± 18.4	6	<0.1	Akamatsu U.
SH04	6.91	948	2.24	307	0.1514	2,462	81.5 ± 11.4	7	5	Taniyama U.
SH32	11.98	1,281	2.99	320	0.0969	2,026	68.6 ± 9.9	12	1	Tei Mélange
SH06	6.08	1,271	1.98	414	0.1514	2,462	81.0 ± 9.9	7	<0.1	Hiwasa U.
SH07	4.98	585	1.51	1,922	0.1514	2,462	86.7 ± 9.2	10	<0.1	Hiwasa U.
SH09	7.89	2,027	1.64	422	0.1287	2,093	107.8 ± 12.7	9	<0.1	Hiwasa U.
SH31	11.33	3,186	1.99	560	0.097	2,028	97.6 ± 11.4	22	<0.1	Mugi U.
SH30	8.94	3,596	2.62	1,054	0.0978	2,045	58.8 ±5.9	29	<0.1	Oyamamisaki U.
SH13	5.26	1,663	1.08	340	0.1287	2,093	109.7 ± 14.2	11	<0.1	Naharigawa U., A
SH17	7.18	6,043	1.20	1,017	0.1006	2,104	112.4 ± 10.9	27	<0.1	Naharigawa U., B
SH18	8.35	6,918	1.53	1,266	0.1026	2,145	100.8 ± 9.4	29	<0.1	Naharigawa U., B
SH29	9.35	3,944	1.55	654	0.0939	1,964	113.9 ± 12.4	20	<0.1	Naharigawa U., B
SH19	9.87	929	0.95	89	0.1086	2,271	215.0 ± 49.9	6	1	Naharigawa U., C
SH28	8.12	2,592	1.54	491	0.0948	1,983	100.9 ± 12.1	21	<0.1	Naharigawa U., C
SH20	8.67	4,073	1.30	611	0.1024	2,142	127.8 ± 14.2	20	<0.1	Muroto U. (Shiina)
SH21	7.52	4,337	1.24	713	0.0941	1,969	105.0 ± 11.3	28	<0.1	Muroto U. (Shiina)
SH26	8.07	2,120	1.73	454	0.1081	2,260	94.0 ± 11.6	14	1	Muroto U. (Shiina)
SH27	8.54	3,433	1.63	653	0.1100	2,300	107.8 ± 11.7	19	<0.1	Muroto U. (Gyoto)
SH25	2.84	2,943	1.82	1,891	0.1051	2,199	30.4 ± 2.8	47	<0.1	Hioki Mélange
SH24	7.38	4,431	1.73	1,037	0.0774	1,618	59.9 ± 6.2	34	<0.1	Tsuro Assemblage
SH23	2.77	1,616	2.50	1,463	0.0871	1,822	17.1 ± 1.8	28	<0.1	Misaki Assemblage

Note: ρs = Density of spontaneous tracks; Ns = Number of spontaneous tracks counted to determine ρs; ρi = Density of induced tracks in a sample; Ni = Number of induced tracks counted in a muscovite external detector to determine ρi; ρd = Density of induced tracks in NBS-SRM612 dosimeter glass; Nd = Number of induced tracks counted in a muscovite external detector to determine ρd; T = Fission Track age calculated from pooled Ns and Ni for all grains counted; n = Number of counted grains; P(χ^2) = Probability of χ^2 for N degrees of freedom (N = n-1) quoted to the nearest 5 or 10% except for those less than 5% and more than 95% (Galbraith, 1981); U. = Unit; A, B, C = A, B, or C members of the unit.

the samples cooled through the effective closure temperature (which depends on the mineral and cooling rate) after the thermal event (Fig. 2C).

In this study, two patterns of zircon age spectrum are recognized (Fig. 3): (I) the youngest age of the peaks is younger than the timing of deposition (all in Fig. 3a except for SH02 and SH06; and SH23 in Fig. 3b); (II) the youngest peak is older than the timing of deposition (SH02 and SH06 in Fig. 3a; and all in Fig. 3b except for SH23). The former pattern should correspond to the case B, and the youngest peak age place a maximum limit on the timing of a post-depositional thermal event. For pattern II, there are two alternative interpretations; heating below the PAZ (case A) or that into the PAZ (case B), which are indistinguishable by an age spectrum analysis.

With respect to the shape of the age spectra, two samples from mélanges (SH04 and SH32) appear to have sharply peaked patterns than others (Fig. 3a), probably as a consequence of their relatively high P(χ^2) values (Table 4). These characteristic fea-

tures, in addition to their spectra showing pattern I, can be attributed to higher maximum temperature, probably close to the bottom of ZPAZ, recorded in the two samples. However, as for the other two samples having relatively high P(χ^2) values (SH19 and SH26 in Fig. 3b), neither the position of age peaks nor the form of age spectra provides an additional evidence of heating up to ZPAZ. The age spectra of SH23 and SH25 (Fig. 3b) have remarkably unimodal patterns with relatively young peak ages, although they do not show the evidence of complete annealing after deposition, as can be seen in their low P(χ^2) values of <0.1% (Table 4).

THERMAL HISTORY OF THE SHIMANTO BELT

The Cretaceous (Northern) Shimanto Belt

Turbidites. The Hinotani and Hiwasa units are coherent turbidites in the Northern Shimanto Belt. The Hiwasa unit yielded three zircon (SH06, SH07, SH09) and no apatite data,

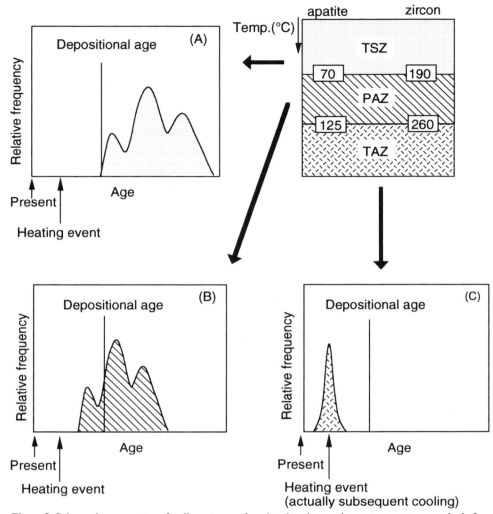

Figure 2. Schematic age spectra of sedimentary rocks related to the maximum temperature reached after deposition. TSZ, total stability zone for fission tracks; PAZ, partial annealing zone; TAZ, total annealing zone. A, spectrum for zircon population with maximum temperature within the total stability zone; B, spectrum for zircon population with maximum temperature within the partial annealing zone; C, spectrum for zircon population with maximum temperature within the total annealing zone.

two of which show pattern I, suggesting partial annealing of tracks. The maximum temperature estimated from the Hiwasa unit should have been within the ZPAZ. A set of zircon and apatite ages was obtained from the Hinotani unit: the apatite age indicates cooling around 11 Ma (Table 3) subsequent to the heating above the APAZ, whereas the zircon age spectrum shows the pattern II. Zircon spectra of pattern II do not exclude the possibility of the absence of partial annealing, these data therefore suggest that the major part of turbidite units of the Northern Shimanto Belt may have been heated up to the ZPAZ after their deposition, and then cooled through the apatite closure temperature ($\sim 100°C$) around 10 Ma. Turbidites in the southern part of the Chichibu Belt may have experienced a similar thermal history, as seen in the totally reset apatite age of ~ 10 Ma and zircon age spectrum of pattern I, which together suggest the heating up to the ZPAZ.

Although we cannot determine the precise age at which the peak paleotemperature was attained based on these data, we can constrain it by the youngest peak ages of zircon age spectra and apatite cooling ages. Since zircon grain ages were reduced by partial annealing of tracks, the youngest peak age in the zircon grain age spectrum provides the oldest limit for thermal event, and the apatite age, the youngest limit. Accordingly, the timing of maximum temperature of the Hiwasa unit should have been between ~ 50 Ma and ~ 10 Ma; and that of the southern Chichibu Belt ~ 107 Ma and ~ 10 Ma.

Agar and others (1989) reported K-Ar ages of around 50 Ma on microcrystals obtained from Cretaceous turbidites. They interpret the ages as the timing of cleavage generation, which would be related to the temperature climax (DiTullio and others, this volume). The ages are consistent with the constraints from FT

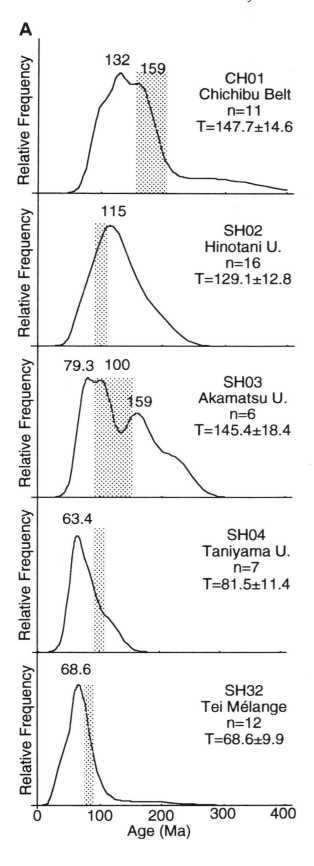

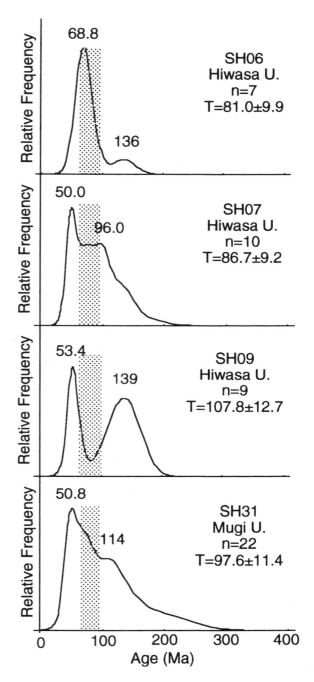

Figure 3 (on this and following two pages). A, FT zircon age spectra of samples from the Chichibu Belt and the Northern Shimanto Belt with crystal number, peak ages (Ma), and pooled age (±2σ, Ma). Dotted region represents the depositional age estimated from fossils (Table 1). B, FT zircon age spectra of samples from the Southern Shimanto Belt with crystal number, peak ages (Ma), and pooled age (±2σ, Ma). Dotted region represents the depositional age estimated from fossils (Table 1). U, unit; A, A member; B, B member; C, C member.

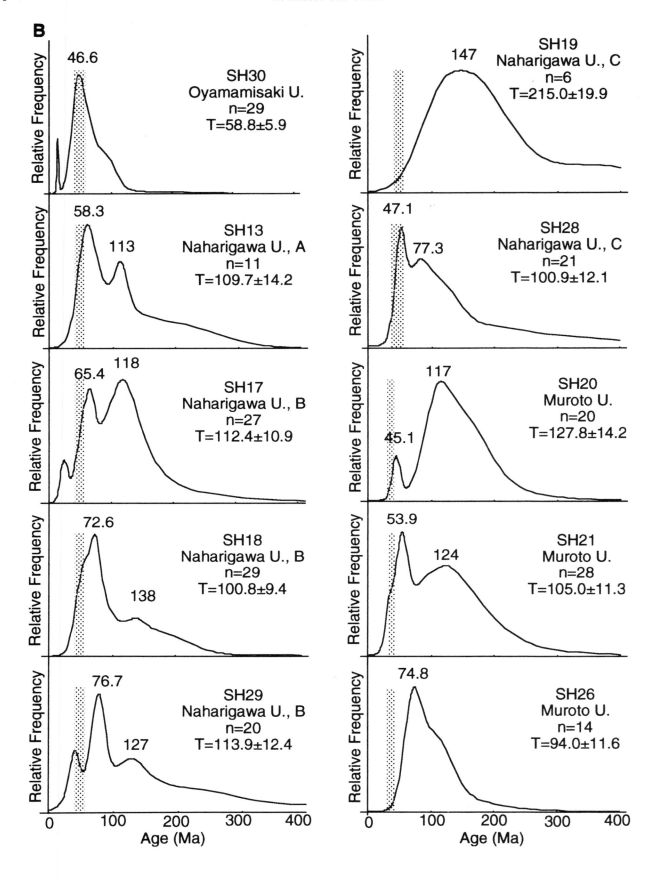

analysis in that they slightly postdate the youngest peak ages in zircon spectra. Hence, we conclude that the turbidites in the Northern Shimanto Belt experienced the maximum temperature of ~200°C, about 15 m.y. after their deposition at trench. The thermal history estimated for the Hiwasa unit is illustrated in Figure 4, as a representative example from the Cretaceous turbidites.

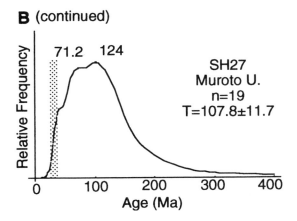

Mélanges. Of the four zircon data from mélange units, two samples from the Tei mélange (SH32) and Taniyama unit (SH04), which are correlated by the geological map published by the Regional Forestry Office of Kochi Prefecture (Katto and others, 1977), indicate higher maximum temperature close to the bottom of ZPAZ (~260°C). This is inferred from relatively greater $P(\chi^2)$ values as well as the narrow and unimodal age spectra, showing pattern I. In particular, SH04 passed the χ^2 test at 5% criterion and, hence, its age variation could be caused only by the radiometric decay processes and may represent when the mélange cooled through the closure temperature of zircon (~225 ± 30°C; Zaun and Wagner, 1985) after the temperature climax at ~260°C. It follows that the timing of the maximum temperature in the Taniyama unit and the Tei mélange would be around their peak ages of 63 and 69 Ma, suggesting a best estimate of ~65 Ma.

Two other samples collected from the Akamatsu and the Mugi units (SH03 and SH31) also show evidence of partial annealing based on the position of youngest peak ages in age spectra, and their $P(\chi^2)$ values of <0.1% in addition to broad age spectra with multiple peaks. This implies the maximum temperature of ~200°C, similar to those of the surrounding turbidites. Furthermore, their times of temperature climax inferred from the youngest peak ages are younger than 79 and 51 Ma, respectively. It appears that the mélanges in the Northern Shimanto Belt sustained variable maximum temperatures at various times. The deduced thermal history of the Tei mélange is illustrated in Figure 4.

The Eocene Shimanto Belt (The Murotohanto subbelt)

In the case of the Murotohanto subbelt, zircon age spectra of three turbidite units provide no indication of heating up into the ZPAZ. The Naharigawa and Muroto units yielded FT apatite cooling ages of ~10 Ma. These facts constrain the maximum temperature of the major part of the Eocene Shimanto Belt above APAZ and probably below ZPAZ, namely ~125 to 190°C, with subsequent cooling through the ~100°C apatite isotherm around 10 Ma. K-Ar microcrystal ages (Agar and others, 1989) of around 30 Ma from Eocene turbidites, correlative with the Oyamamisaki and Naharigawa units, may date the timing of temperature climax. The thermal history of the Murotohanto subbelt is illustrated in Figure 4, as represented by the Naharigawa unit.

The relatively high zircon $P(\chi^2)$ values of SH19 and SH26 could reflect some localized thermal events up to the ZPAZ because other samples from the same unit (the Naharigawa and Muroto units, respectively) show no evidence of heating. The timing of peak heating is less controlled: ~50 to 10 Ma for the Naharigawa unit and ~35 to 10 Ma for the Muroto unit, based on the depositional age and apatite cooling ages. The hypothetical thermal events localized around the two sites appear to be reflected by relatively high vitrinite reflectance values as well (Mori and Taguchi, 1988; see also DiTullio, and others, this volume; Laughland and Underwood, this volume).

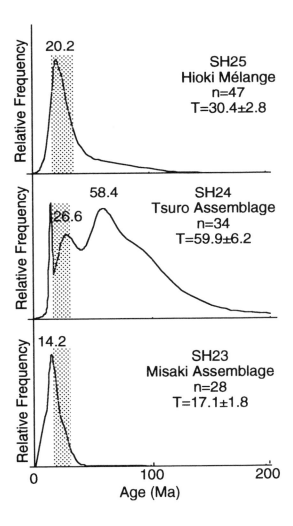

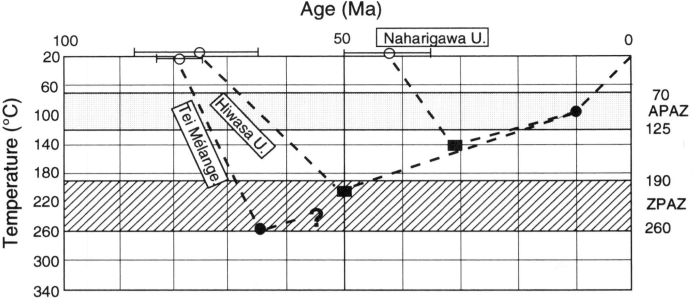

Figure 4. Thermal history of the Tei mélange, Hiwasa unit, and Naharigawa unit, which are representative examples of Cretaceous mélange, Cretaceous turbidite, and Eocene turbidite, respectively. Open circles show the depositional ages with possible depositional duration. Solid circles show time-temperature points estimated from FT ages. Solid squares show the maximum possible temperature derived from FT zircon data, and their timing based on K-Ar cleavage ages (Agar and others, 1989).

The Miocene Shimanto Belt (The Nabae subbelt)

One zircon age spectrum from the Misaki assemblage (SH23) shows pattern I, and two from Hioki mélange (SH25) and Tsuro assemblage (SH24) show pattern II. No apatite data were obtained from the Nabae subbelt.

SH23, ~30 m from the contact with the Muroto-Misaki gabbroic intrusion, has a unimodal age spectrum with a peak age of 14.2 Ma (Fig. 3b) showing the evidence of partial resetting. This result, in conjunction with the geologic evidence of temperature elevation caused by the gabbroic intrusion that produced a hornfels aureole of ~40 m width (Yoshizawa, 1953), would be best explained by the temperature elevation of SH23 up to ZPAZ caused by the gabbroic intrusion. In addition, the peak age of 14.2 Ma is concordant with the Rb-Sr whole-rock biotite isochron age (14.4 ± 0.4 Ma; Hamamoto and Sakai, 1987). The low $P(\chi^2)$ of SH23 and the multimodal zircon age spectrum of SH24 suggest that the thermal effect of the gabbroic intrusion was localized to the tip of the Muroto Peninsula. SH25 shows no evidence of heating events, which also supports this interpretation.

The vitrinite reflectance data of the Nabae subbelt suggest that the maximum paleotemperature was ~200°C for the whole Hioki mélange and increased progressively up to ~270°C toward Cape Muroto in the Misaki assemblage (Hibbard and others, this volume; Laughland and Underwood, this volume). This trend also supports the localized thermal effect of the gabbroic intrusion.

TECTONIC EVOLUTION OF THE CRETACEOUS TO EOCENE SHIMANTO BELT

In reconstructing the evolution of an accretionary wedge on the basis of thermal history, some constrains need to be known or assumed reasonably, such as (1) shape of the accretionary wedge, (2) thermal structure of the wedge, and (3) fundamental pattern of material migration in the wedge. The shape of an accretionary wedge can vary through time in response to various factors, including subduction rate and angle, rheology of décollement, material input or output, and thermal structure. However, the accretionary wedge tends to be kept more or less constant as long as the subducting oceanic plates are of similar characteristics (e.g., Davis and others, 1983; Platt, 1986). The ancient slabs subducting beneath southwest Japan had similar ages of the order of 10 Ma from Late Cretaceous to the present in addition to rather constant convergence rates of ~5 to 10 cm/year (Engebretson and others, 1985; Maruyama and Seno, 1986), except during the subduction of the newly opened Shikoku Basin from ~26 to ~15 Ma (Chamot-Rooke and others, 1987). The present shape of the Southwest Japan forearc system is thus adopted here as a first-order approximation of the Shimanto accretionary wedge.

With respect to thermal structure, modeling based on reasonably assumed subduction parameters predicts subnormal gradients within a wedge (e.g., Dumitru, 1991). Accordingly, a geothermal gradient of 20°C/km is assumed for the Cretaceous to Eocene Shimanto wedge, based on the pressure-temperature path of the Shimanto metamorphism (Toriumi and Teruya, 1988) as well as that of the Sambagawa (Takasu and Dallmeyer, 1990). Note this value is also consistent with the Dumitru's (1991) modelling, assuming a slab age of ~10 m.y. and convergence rate of ~5 cm/year.

Material migration at shallow levels of accretionary wedges has been better understood in recent years by drilling of prisms and seismic reflection profiles (e.g., Karig, 1986). The trench-fill turbidites accrete as offscraped imbricate packages, which subsequently thicken by thrusting in addition to forearc slope sedimentation and move away from trench due to succeeding frontal accretion (e.g., Moore, 1989). This growth pattern of imbricate thrust slices might be maintained at deeper levels as well. Meanwhile, the pelagic and hemipelagic sediments and oceanic basalts accrete mainly at depth, probably due to a step in the décollement associated with the formation of imbricate duplex structures (Moore, 1989). Those underplated materials may progressively migrate away from the subducting plate and be incorporated in the overlying wedge as a result of successive underplating.

Given these guidelines, a plausible evolution of the Cretaceous to Eocene Shimanto Belt is reconstructed schematically in Figure 5. The Northern Shimanto Belt, represented by the Hiwasa unit, accreted at the trench at ~75 Ma and was dragged

down to ~10 km depth at ~50 Ma. It subsequently rebounded to migrate from the slab with gradual exhumation, followed by rapid unroofing since ~10 Ma to the present. The Murotohanto subbelt, represented by the Naharigawa unit, yields a similar pattern of evolution since the accretion at 43 Ma to the exposure at present, except that the path stays consistently at a shallower depth. The Tei mélange was subducted at ~80 Ma prior to accretion, and underplated below offscraped imbricated packages and dragged down to ~12 km depth at ~65 Ma. Then it rebounded to leave the slab, although the subsequent unroofing history remains ambiguous. The notable features of accretionary processes found in Figure 5 are threefold: (1) continuous migration of imbricated turbidites, (2) depth of underplating, (3) rapid unroofing around 10 Ma, which will be discussed below.

Continuous migration of imbricated turbidites

As illustrated in Figure 5, the evolution of the Cretaceous and Eocene Shimanto Belt (NSB and MHSB, respectively) could have followed similar paths of downward convex shape with approximately the same migration rates. It appears that formation and growth of offscraped imbricate packages could have been rather constant through the Cretaceous to Eocene time on the basis of these paths. In addition, two paths exhibit a systematic difference in depth, and this would be manifestation of a layered

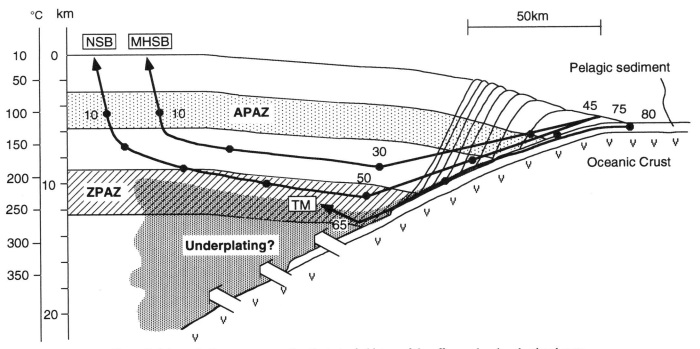

Figure 5. Schematic diagram representing the tectonic history of the offscraped and underplated materials under the frameworks of constant wedge shape and 20°C/km thermal gradient with surface temperature of 10°C. NSB, path for the offscraped units in the Northern Shimanto Belt; MHSB, path for the offscraped units in the Murotohanto subbelt; TM, path for the Tei mélange, an underplated unit in the Northern Shimanto Belt; APAZ, apatite partial annealing zone; ZPAZ, zircon partial annealing zone. Solid circles on trajectories depict positions every 10 Ma. Figures in the diagram indicate timings inferred from thermal histories. Note that the wedge is described here so that the position of the trench has been fixed through time, yielding the evolutionary growth of wedge toward the left.

pattern of material migration in wedge. It is thus likely that the migration of turbidites associated with growing imbrication played a dominant role in the shallow part of an accretionary wedge, implying that other factors, such as out of sequence thrusts, would act in a more localized scale. The tectonic "rebound" of turbidites from a depth of ~10 km could have been caused by increased underplating at deeper levels.

Depth of underplating

Although the evolution of mélanges seems highly complicated due to their various geneses, mélanges comprised of pelagic sediments and oceanic basalts are presumably formed by underplating processes (Moore, 1989). As suggested in the last section, the temperature maximum and its timing should have been different between such underplated mélanges of similar depositional ages, suggesting the variation in underplating process. Those mélanges show, however, slightly higher maximum temperatures than surrounding turbidites, and were underplated probably beneath the bottom of imbricated turbidites. The timing of maximum burial followed by tectonic rebound is well estimated for the Tei mélange, as shown in Figure 5. A notable fact in its underplating process is that it took ~15 m.y. from subduction at the trench to maximum burial. If we assume a plate convergence rate of ~5 cm/year, as well as a migration pattern of the Tei mélange similar to that of slab prior to underplating and to those of turbidites thereafter, then the average calculated migration rate from trench to rebound for the Tei mélange would be ~1 cm/year. This rate is too slow to be explained by the depth of underplating close to rebound. A much shallower depth of underplating would be more probable to allow such a low average migration rate. Hence, we suggest that the tectonic rebound of the Tei mélange at ~12 km depth significantly postdates its underplating. The rebound of mélange and turbidites may be related to accelerated underplating at deeper levels that effectively interrupts the invasion of materials from shallower levels (Fig. 5). This hypothetical zone of accelerated underplating could be the source of high-pressure/low-temperature metamorphic rocks present in other accretionary prisms, which are mainly derived from pelagic and hemipelagic sediments as well as mafic igneous rocks.

Rapid unroofing (or cooling) around 10 Ma

In the context of continuous migration of imbricated turbidites, the rapid unroofing (or cooling) around 10 Ma covering the whole Cretaceous to Eocene Shimanto Belt and the Chichibu Belt could be attributable to some episodic event in the Miocene. Since the subduction parameters reconstructed for the Miocene Southwest Japan forearc are far from the normal state, due to the newly opened Shikoku Basin, here we have to reexamine the frameworks and constraints that control the Miocene accretionary wedge. The Shikoku Basin, located between the Izu-Bonin arc and the Kyushu-Palau Ridge, was formed by back-arc rifting and spreading from ~26 to ~15 Ma (e.g., Chamot-Rooke and others, 1987). It is thus expected that after ~26 Ma, the parameters that control subduction tectonics were subjected to drastic

change, producing a thermal perturbation in wedge, assuming Philippine Sea Plate had been subducted, rather than Pacific Plate (Seno and Maruyama, 1984). Although the rate and angle of subduction from ~26 to ~15 Ma are unknown, the hot Shikoku Basin was involved in rapid subduction beneath southwest Japan when the Southwest Japan block was rotated clockwisely ~50° at ~15 Ma (Otofuji and others, 1985). Hence it is estimated with confidence that the thermal structure of Shimanto accretionary wedge experienced increased thermal input from the descending slab. In such a tectonic framework, the widely concordant apatite age of ~10 Ma can be attributed to (1) increased thermal gradient at ~15 Ma, followed by a recovery to a normal thermal state; (2) tectonic uplift caused by hot, buoyant slab with a shallow dip angle; and/or (3) uplift by accelerated underplating due to an anomalously large convergence rate. Of these three possibilities, the third factor may be the least probable because the hemipelagic and pelagic sediments deposited on the young Shikoku Basin should have been too thin to enable accelerated underplating. With respect to the first factor, the subduction of a young slab can effectively elevate the thermal gradient in an accretionary wedge (Dumitru, 1991). Because heat transfer by thermal diffusion may require a time lag of ~4 m.y. for the geometry shown in Figure 5 (Dumitru, 1990), the studied field would have been cooled uniformly at ~10 Ma. Factor 2 is based on the empirical relationship that a younger slab tends to have a lower angle of subduction (e.g., Jarrard, 1986). A buoyant slab with shallow angle would have uplifted the wedge. As a result, the present surface of the studied region could have been exhumed from ~6 km depth. As the subducting slab cooled down, exhumation would have ceased. It is likely that the combined influence of factors (1) and (2) are responsible for the concordant apatite age profile. We finally stress that the increased thermal gradient should have not overprinted completely the pre-Miocene thermal history, as suggested by our FT zircon data of SH04 and SH32 that preserve cooling at ~65 Ma subsequent to the heating up to the bottom of the ZPAZ, and by the vitrinite reflectance data of Eocene Shimanto Belt (DiTullio and others, this volume; see also Laughland and Underwood, this volume) consistent with deformation structures related to accretionary processes.

CONCLUSIONS

1. The turbidites in the Northern Shimanto Belt and southern Chichibu Belt were heated into the ZPAZ (190 to 260°C) while those in the Southern Shimanto Belt show limited evidence of such heating. This suggests systematic differences in the maximum temperature and thus evolutionary paths during the accretionary process. The age-temperature paths estimated for individual tectonic units imply successive accretion, growth and uplift of offscraped imbricate packages.

2. Mélanges in the Northern Shimanto Belt sustained variable maximum temperatures, presumably reflecting their different evolutionary processes. They were probably underplated at the bottom of offscraped imbricated packages and subsequently dragged down to within or below ZPAZ.

3. The Cretaceous to Eocene Shimanto Belt was cooled through ~100°C isotherm at ~10 Ma, probably due to increased geothermal gradient and/or exhumation related to subduction of the newly formed Shikoku Basin.

ACKNOWLEDGMENTS

We would like to express our gratitude to Masayuki Torii, Rasoul B. Sorkhabi, Gaku Kimura, Simon R. Wallis, and Asahiko Taira for their constructive suggestions on the study. We thankfully acknowledge Mike B. Underwood, Mark T. Brandon, and Trevor A. Dumitru for their critical reviews of the manuscript. We also thank Rieko Arai for improving the manuscript. We thank Takaaki Matsuda, Teruyuki Honda, and Sataro Nishikawa for their help with irradiation processes. This work has been performed by using facilities of the TRIGA II Reactor at Musashi Institute of Technology and the KUR-1 at the Research Reactor Institute, Kyoto University.

REFERENCES CITED

Agar, S. M., Cliff, R. A., Duddy, I. R., and Rex, D. C., 1989, Short paper: Accretion and uplift in the Shimanto Belt, SW Japan: Journal of the Geological Society of London, v. 146, p. 893–896.

Banno, S., and Sakai, C., 1989, Geology and metamorphic evolution of the Sambagawa metamorphic belt, Japan, *in* Daly, J. S., Cliff, R. A., and Yardley, B.W.D., eds., Evolution of metamorphic belts: Geological Society Special Publication, v. 43, p. 519–532.

Barker, C. E., 1988, Geothermics of petroleum systems: Implications of the stabilization of kerogen thermal maturation after a geologically brief heating duration at peak temperature, *in* Magoon, L., ed., Petroleum systems of the United States: U.S. Geological Survey Bulletin 1870, p. 26–29.

Chamot-Rooke, N., Renard, V., and LePichon, X., 1987, Magmatic anomalies in the Shikoku Basin: A new interpretation: Earth and Planetary Science Letters, v. 83, p. 214–228.

Cowan, D. S., 1985, Structural styles in Mesozoic and Cenozoic mélanges in the western Cordillera of North America: Geological Society of America Bulletin, v. 96, p. 451–462.

—— , 1990, Kinematic analysis of shear zones in sandstone and mudstone of the Shimanto Belt, Shikoku, SW Japan: Journal of Structural Geology, v. 12, p. 431–441.

Davis, D., Suppe, J., and Dahlen, F. A., 1983, Mechanics of fold-and-thrust belts and accretionary wedge: Journal of Geophysical Research, v. 88, p. 1153–1172.

DiTullio, L., and Byrne, T., 1990, Deformation paths in the shallow levels of an accretionary prism: The Eocene Shimanto belt of southwest Japan: Geological Society of America Bulletin, v. 102, p. 1420–1438.

Dumitru, T. A., 1990, Subnormal Cenozoic geothermal gradients in the extinct Sierra Nevada magmatic arc: Consequences of Laramide and post-Laramide shallow-angle subduction: Journal of Geophysical Research, v. 95, p. 4925–4941.

—— , 1991, Effects of subduction parameters on geothermal gradient in forearcs, with an application to Franciscan subduction in California: Journal of Geophysical Research, v. 96, p. 621–641.

Engebretson, D. C., Cox, A., and Gordon, R. G., 1985, Relative motions between oceanic and continental plates in the Pacific Basin: Geological Society of America Special Paper, v. 206, 59 p.

Galbraith, R. F., 1981, On statistical models for fission track counts: Mathematical Geology, v. 13, p. 471–488.

Gleadow, A.J.W., 1981, Fission-track dating methods: What are the real alternatives?: Nuclear Tracks, v. 5, p. 3–14.

Gleadow, A.J.W., Duddy, I.R., and Lovering, J. F., 1983, Fission track analysis: A new tool for the evaluation of thermal histories and hydrocarbon potential: Australian Petroleum Exploration Association Journal, v. 23, p. 93–102.

Green, P. F., 1981, A new look at statistics in fission-track dating: Nuclear Tracks, v. 5, p. 77–86.

—— , 1985, Comparison of zeta calibration baselines for fission-track dating of apatite, zircon and sphene: Chemical Geology, Isotope Geoscience Section, v. 58, p. 1–22.

Green, P. F., Duddy, I. R., Laslett, G. M., Hegarty, K. A., Gleadow, A.J.W., and Lovering, J. F., 1989, Thermal annealing of fission tracks in apatite 4; Quantitative modeling techniques and extension to geological timescales: Chemical Geology, Isotope Geoscience Section, v. 79, p. 155–182.

Hamamoto, R., and Sakai, H., 1987, Rb-Sr age of granophyre associated with the Muroto-misaki gabbroic complex: Kyusyu University, Memoirs of the Faculty of Science, v. 15, p. 131–135 (in Japanese).

Harland, W. B., Armstrong, R. L., Cox, A. V., Craig, L. E., Smith, A. G., and Smith, D. G., 1989, A geologic time scale 1989: Cambridge, Cambridge University Press, 263 p.

Harrison, T. M., and McDougall, I., 1980, Investigations of an intrusive contact, northwest Nelson, New Zealand—1, Thermal, chronological and isotopic constraints: Geochimica et Cosmochimica Acta, v. 44, p. 1985–2003.

Hibbard, J. P., and Karig, D. E., 1987, Sheath-like folds and progressive fold deformation in Tertiary sedimentary rocks of the Shimanto accretionary complex, Japan: Journal of Structural Geology, v. 7, p. 845–857.

—— , 1990a, Alternative plate model for the early Miocene evolution of the southwest Japan margin: Geology, v. 18, p. 170–174.

—— , 1990b, Structural and magmatic responses to spreading ridge subduction: An example from southwest Japan: Tectonics, v. 9, p. 207–230.

Hurford, A. J., 1986, Cooling and uplift patterns in the Lapontine Alps, south central Switzerland and an age of vertical movement on the Insubric fault line: Contributions to Mineralogy and Petrology, v. 92, p. 413–427.

—— , 1990, Standardization of fission track dating calibration: Recommendation by the Fission Track Working Group of the I.U.G.S. Subcommission on Geochronology: Chemical Geology, Isotope Geoscience Section, v. 80, p. 171–178.

Hurford, A. J., and Green, P. F., 1983, The zeta age calibration of fission-track dating: Isotope Geoscience, v. 1, p. 285–317.

Hurford, A. J., and Watkins, R. T., 1987, Fission-track age of the tuffs of the Buluk Member, Bakate Formation, northern Kenya: A suitable fission-track age standard: Chemical Geology, Isotope Geoscience Section, v. 66, p. 209–216.

Hurford, A. J., Fitch, F. J., and Clarke, A., 1984, Resolution of the age structure of the detrital zircon populations of two Lower Cretaceous sandstones from the Weald of England by fission track dating: Geological Magazine, v. 121, p. 269–396.

Jarrard, R. D., 1986, Relations among subduction parameters: Reviews of Geophysics, v. 24, p. 217–284.

Kagami, H., Shiono, K., and Taira, A., 1983, Subduction of plate at the Nankai Tough and formation of accretionary prism: Kagaku, v. 51, p. 429–438 (in Japanese).

Kamp, P.J.J., Green, P. F., and White, S. H., 1989, Fission track analysis reveals character of collisional tectonics in New Zealand: Tectonics, v. 8, p. 169–195.

Karig, D. E., 1986, Physical properties and mechanical state of accreted sediments in the Nankai Trough, Southwest Japan Arc: Geological Society of America Memoir, v. 166, p. 117–133.

Kasuya, M., and Naeser, C. W., 1988, The effect of a-damage on fission-track annealing in zircon: Nuclear Tracks and Radiation Measurements, v. 14, p. 477–480.

Katto, J., and others, 1977, Geological map of Shikoku: Regional Forestry Office of Kochi Prefecture, scale 1:200,000.

Katto, J., and Tashiro, M., 1979, A study on the molluscan fauna of the Shimanto Terrain southwest Japan; Part 2, Bivalve fossils from the Murotohanto Group in Kochi Prefecture, Shikoku: Research Report of the Kochi University, v. 28, p. 1–11.

Kimura, G., and Mukai, A., 1991, Underplated units in an accretionary complex: Mélange of the Shimanto Belt of eastern Shikoku, southwest Japan: Tectonics, v. 10, p. 31–50.

Kumon, F., 1983, Coarse clastic rocks of the Shimanto Supergroup in eastern Shikoku and Kii Peninsula, southwest Japan: Kyoto University, Memoirs of the Faculty of Science, Series of Geology and Mineralogy, v. 50, p. 63–109.

Maruyama, S., and Seno, T., 1986, Orogeny and relative plate motions: Example of the Japan Island: Tectonophysics, v. 127, p. 305–329.

Matsugi, H., Yasuda, H., and Okamura, M., 1987, Geologic processes of Lower Cretaceous Shimanto Belt, Shikoku [abs.]: Geological Society of Japan, Annual Meeting, v. 94, p. 235 (in Japanese).

Miller, D. S., Duddy, I. R., Green, P. F., Hurford, A. J. and Naeser, C. W., 1985, Results of interlaboratory comparison of fission track age standards, Fission Track Workshop, 1984: Nuclear Tracks, v. 10, p. 383–391.

Miyake, Y., 1983, Muroto-Misaki gabbroic complex formed in forearc basin: Magma, v. 69, p. 10–14 (in Japanese).

Moore, G. F., Karig, D. E., Shipley, T. H., Taira, A., Stoffa, P. L., and Wood, W. T., 1991, Structural framework of the ODP Leg 131 area, Nankai Trough, *in* Taira, A., Hill, I., Firth, J., and others, eds., Initial Report of the Deep Sea Drilling Project: Washington, D.C., U.S. Government Printing Office, v. 131, p. 15–20.

Moore, J. C., 1989, Tectonics and hydrogeology of accretionary prism: Role of the décollement zone: Journal of Structural Geology, v. 11, p. 95–106.

Moore, J. C., and Watkins, J. S., 1981, Summary of accretionary processes, Deep Sea Drilling Project Leg 66: Offscraping, underplating, and deformation of the slope apron, *in* Watkins, J. S., Moore, J. C., and others, eds., Initial Report of the Deep Sea Drilling Project: Washington, D.C., U.S. Government Printing Office, v. 66, p. 825–836.

Mori, K., and Taguchi, K., 1988, Examination of the low-grade metamorphism in the Shimanto Belt by vitrinite reflectance: Modern Geology, v. 12, p. 325–339.

Naeser, C. W., 1976, Fission track dating: U.S. Geological Survey Open-File Report, 58 p.

—— , 1979, Fission-track dating and geologic annealing of fission tracks, *in* Jager, E., and Hunziker, J. C., eds., Lectures in isotope geology: Berlin, Springer-Verlag, p. 154–169.

—— , 1981, The fading of fission tracks in the geologic environment: Nuclear Tracks, v. 5, p. 248–250.

Okada, M., and Okamura, M., 1980, Calcareous nannofossils obtained from the Shimanto Belt in Kochi Prefecture, *in* Taira, A., and Tashiro, M., eds., Geology and paleontology of the Shimanto Belt: Kochi, Japan, Rinyakosaikai Press, p. 147–152 (in Japanese).

Okamura, M., 1980, Radiolarian fossils from the Northern Shimanto Belt (Cretaceous) in Kochi Prefecture, Shikoku, *in* Taira, A., and Tashiro, M., eds., Geology and paleontology of the Shimanto Belt: Kochi, Japan, Rinyakosaikai Press, p. 153–178 (in Japanese).

Otofuji, Y., Hayashida, A., and Torii, M., 1985, When was the Japan Sea opened? Paleomagnetic evidence from Southwest Japan, *in* Nasu, N., and others, eds., Formation of active ocean margins: Tokyo, Japan, Terra Scientific Publishing Company, p. 551–566.

Ozawa, T., Taira, A., and Kobayashi, F., 1985, How was the geologic structure of SW Japan formed?: Kagaku, v. 55, p. 4–13 (in Japanese).

Platt, J. P., 1986, Dynamics of orogenic wedges and the uplift of high-pressure metamorphic rocks: Geological Society of America Bulletin, v. 97, p. 1037–1053.

Saito, T., 1980, An early Miocene (Aquitanian) planktonic foraminiferal fauna from the Tsuro Formation, the youngest part of the Shimanto Super Group, Shikoku, Japan, *in* Taira, A., and Tashiro, M., eds., Geology and paleontology of the Shimanto Belt: Kochi, Japan, Rinyakosaikai Press, p. 227–234 (in Japanese).

Sakai, H., 1988, Origin of the Misaki Olistostrome Belt and reexamination of the Takachiho orogeny: The Journal of the Geological Society of Japan, v. 94, p. 945–961 (in Japanese).

Sakai, T., and Kanmera, K., 1981, Stratigraphy of the Shimanto Terrain and tectono-stratigraphic setting of greenstones in the northern part of Miyazaki Prefecture, Kyusyu: Kyusyu University, Memoirs of the Faculty of Science, v. 14, p. 31–48.

Seno, T., and Maruyama, S., 1984, Paleogeographic reconstruction and origin of the Philippine Sea: Tectonophysics, v. 102, p. 53–84.

Shibata, K., 1978, Contemporaneity of Tertiary granites in the Outer Zone of Southwest Japan: Bulletin Geological Survey of Japan, v. 29, p. 551–554 (in Japanese).

Sumii, T., Tagami, T., and Nishimura, S., 1987, Anisotropic etching character of spontaneous fission tracks in zircon: Nuclear Tracks and Radiation Measurements, v. 13, p. 275–277.

Suyari, K., 1986, Restudy of the Northern Shimanto Subbelt in eastern Shikoku: University of Tokushima Journal of Science, v. 19, p. 45–54 (in Japanese).

Tagami, T., 1987, Determination of zeta calibration constant for fission track dating: Nuclear Tracks and Radiation Measurements, v. 13, p. 127–130.

Tagami, T., Lal, N., Sorkhabi, R. B., Ito, H., and Nishimura, S., 1988, Fission track dating using external detector method: A laboratory procedure: Kyoto University, Memoirs of the Faculty of Science, Series of Geology and Mineralogy, v. 53, p. 14–30.

Tagami, T., Ito, H., and Nishimura, S., 1990, Thermal annealing characteristics of spontaneous fission tracks in zircon: Chemical Geology, Isotope Geoscience Section, v. 80, p. 159–169.

Taira, A., and Tashiro, M., 1987, Late Paleozoic and Mesozoic accretion tectonics in Japan and eastern Asia, *in* Taira, A., and Tashiro, M., eds., Historical biogeography and plate tectonic evolution of Japan and eastern Asia: Tokyo, Japan, Terra Scientific Publishing Company, p. 1–43.

Taira, A., Katto, J., Tashiro, M., and Okamura, M., 1980a, The geology of the Shimanto Belt in Kochi Prefecture, Shikoku, Japan, *in* Taira, A., and Tashiro, M., eds., Geology and paleontology of the Shimanto Belt: Kochi, Japan, Rinyakosaikai Press, p. 319–389 (in Japanese).

Taira, A., and 8 others, 1980b, Lithofacies and geologic age relationship within mélange zones of Northern Shimanto Belt (Cretaceous), Kochi Prefecture, Japan, *in* Taira, A., and Tashiro, M., eds., Geology and paleontology of the Shimanto Belt: Kochi, Japan, Rinyakosaikai Press, p. 179–214 (in Japanese).

Taira, A., Katto, J., Tashiro, M., Okamura, M., and Kodama, K., 1988, The Shimanto Belt in Shikoku—Evolution of Cretaceous to Miocene accretionary prism: Modern Geology, v. 12, p. 5–46.

Taira, A., Tokuyama, H., and Soh, W., 1989, Accretion tectonics and evolution of Japan, *in* Abraham, Z. B., ed., The evolution of the Pacific Ocean margins: New York, Oxford University Press, p. 100–123.

Takahashi, M., 1986, Magmatic activity of island arcs just before and after the opening of the Japan Sea: Kagaku, v. 56, p. 103–111 (in Japanese).

Takasu, A., and Dallmeyer, R. D., 1990, $^{40}Ar/^{39}Ar$ mineral age constraints for the tectonothermal evolution of the Sambagawa metamorphic belt, central Shikoku, Japan: A Cretaceous accretionary prism: Tectonophysics, v. 185, p. 111–139.

Toriumi, M., and Teruya, J., 1988, Tectono-metamorphism of the Shimanto Belt: Modern Geology, v. 12, p. 303–324.

Wagner, G. A., Reimer, G. M., and Jager, E., 1977, The cooling ages derived by apatite fission track, mica Rb-Sr, and K-Ar dating: The uplift and cooling history of the Central Alps: University of Padova, Memoirs of Institute of Geology and Mineralogy, v. 30, p. 1–27.

Westbrook, G. K., and Smith, M. J., 1983, Long décollements and mud volcanoes: Evidence from the Barbados Ridge Complex for the role of high pore-fluid pressure in the development of an accretionary complex: Geology, v. 11, p. 279–283.

Yoshizawa, H., 1953, On the gabbro of the Cape Muroto, Shikoku Island, Japan; Part 1: University of Kyoto, Memoirs of the College of Science, Series B, v. 20, p. 271–284.

Zaun, P. E., and Wagner, G. A., 1985, Fission-track stability in zircons under geological conditions: Nuclear Tracks, v. 10, p. 303–307.

MANUSCRIPT ACCEPTED BY THE SOCIETY APRIL 24, 1992

Printed in U.S.A.

Geological Society of America
Special Paper 273
1993

Thermal structure of the Nankai accretionary prism as inferred from the distribution of gas hydrate BSRs

Juichiro Ashi and Asahiko Taira
Ocean Research Institute, University of Tokyo, 1-15-1, Minamidai, Nakano-ku, Tokyo 164, Japan

ABSTRACT

Closely spaced seismic reflection profiles obtained from the Nankai accretionary prism provide high resolution images of the prism internal structure. They also provide indirect information about the heat flow estimated from the depths of bottom-simulating reflectors (BSRs) that originate at the gas hydrate phase transition. A careful error evaluation for heat-flow estimation was conducted for this study. More than 40 seismic lines reveal a landward decrease of heat flow closely associated with thickening of the sedimentary section above oceanic basement during prism growth. BSR-estimated heat flow is consistent with heat-flow values obtained by probe as well as expected from the oceanic basement crustal age (15 Ma), except at the prism toe. These data suggest that the thermal structure of the prism is mostly conductive and the effect of fluid flow is relatively small. The oceanic basement surface temperature, extrapolated from thermal gradients and conductivities, is relatively constant (100 to 140°C) for a 30-km-wide zone beneath the seaward part of the prism. Regionally high heat flow occurs at the prism toe, suggesting local advective heat flow transfer. Sidescan sonar images indicate that the advective zone corresponds to regions of rugged topography, where the most active tectonic deformation is presumed to occur by thrust faulting and no slope sediment cover exists. Thus, the locally high heat flow may be caused by localized fluid expulsion related to rapidly increasing tectonic overburden. The existence of mud volcanoes on the trough floor suggests that some pore fluid expelled from the prism migrates seaward through permeable layers.

INTRODUCTION

Formation of an accretionary prism at a subduction zone is an initial step of continental crustal growth and many geologic features observed in orogenic belts are related to this process. Probably the most important factor in determining the evolutionary pattern of an accretionary prism is the thermal structure and the history of heat transfer within the prism. Development and evolution of diagenesis, metamorphism, structural fabric, and physical properties are closely related to the thermal history of the sediment involved (e.g., Langseth and Moore, 1990).

The thermal structure of an accretionary prism in relatively shallow structural levels is apparently controlled by two main factors: age of oceanic lithosphere (magnitude of heat source) and

mode of heat transfer (role of fluid expulsion). Relationships between heat-flow values and oceanic crustal ages, which may control the thermal structure of the prism, have been proposed by several authors (e.g., Lister, 1977; Parsons and Sclater, 1977). Landward decrease of heat flow in an accretionary prism, probably due to prism thickening and the cooling effect of the subducting plate, was reported from the Nankai Trough (Yamano and others, 1982), the northern Cascadia Basin (Davis and others, 1990), and the Barbados Ridge (Langseth and others, 1990). Gas-hydrate BSR (bottom-simulating reflector) studies revealed fluid expulsion resulting in the dewatering of a 10- to 20-km-wide zone landward of the deformation front (Davis and others, 1990) and fluid expulsion through thrust faults near the prism toe (Minshull and White, 1989).

Ashi, J., and Taira, A., 1993, Thermal structure of the Nankai accretionary prism as inferred from the distribution of gas hydrate BSRs, *in* Underwood, M. B., ed., Thermal Evolution of the Tertiary Shimanto Belt, Southwest Japan: An Example of Ridge-Trench Interaction: Boulder, Colorado, Geological Society of America Special Paper 273.

Significant questions—such as, What is the mode of prism heat transfer (conductive versus advective)? How does tectonic thickening affect the prism thermal structure? What is the relationship between the internal variation of thermal structure to prism diagenetic and structural evolution?—need to be addressed with a careful study of modern accretionary prisms.

The modern Nankai forearc slope is one of the best examples of the clastic-dominated accretionary prism. Surface heat flow measurements, drillings from the Deep Sea Drilling Project (DSDP) and Ocean Drilling Program (ODP), and gas-hydrate bottom-simulating reflectors (BSR) have provided preliminary estimate of the heat-transfer process within the prism (Yamano and others, 1982; Kinoshita and Yamano, 1986; Taira and others, 1991). This study is a continuation of these efforts and we present a comprehensive reevaluation of the BSR distribution. Based on this data set, we analyze two problems of the Nankai accretionary prism thermal structure: (1) What is the cross-section variation in prism thermal structure? and (2) What is the major heat transfer mechanism within the prism, conductive or advective?

TECTONIC SETTING AND PREVIOUS WORKS

The Nankai Trough is the convergent plate boundary between the Philippine Sea and Eurasian Plates. Estimates of the convergence rate vary from 1- to 2 cm/yr (Ranken and others, 1984) through 3- to 4 cm/yr (Seno, 1977). The crustal age of the subducting Shikoku Basin is between 14 to 25 Ma (Kobayashi and Nakada, 1978; Shih, 1980; Chamot-Rooke and others, 1987). The Kinan Seamount Chain, situated in the center of the Shikoku Basin, is a series of seamounts trending north-northwest to south-southeast interpreted to be the product of an aborted stage of sea-floor spreading (Chamot-Rooke and others, 1987) later modified by widely distributed off-ridge volcanism (Klein and Kobayashi, 1980). An accretionary prism is well developed landward of the Nankai Trough, especially off Shikoku, in southwest Japan (Fig. 1). The internal structure of the prism, which is mainly characterized by landward-dipping imbricate thrusts, has been revealed by seismic reflection studies (Nasu and others, 1982; Aoki and others, 1983; Moore and others, 1990).

The IAZNAGI sidescan sonar includes a coregistered 11- to 12-kHz acoustic backscattering measurement and a bathymetric mapping system in a single towed body. The IZANAGI backscattering image shows continuous rugged topographic features along the frontal part of the prism (Fig. 2). Based on seismic reflection profiles, these topographic features correspond to anticlinal ridges caused by thrust faults with wavelengths of a few kilometers (Fig. 3). To the landward of the prism, fine-grained slope sediments with low reflectivity overlie and partly mask the fault-bend fold undulations. A small trench fan, supplied by slope sediments, occurs across the deformation front and is involved in deformation front thrusting. These observations suggest that deformation is more active in the frontal part of the prism than in the landward part.

Heat flow in the Nankai Trough is conspicuously high compared to that at other subduction zones (Uyeda, 1972; Watanabe and others, 1977; Yoshii, 1979). Independent from the conventional method of heat-flow measurement, Yamano and others (1982) estimated heat flow from the depth distribution of gas hydrate as detected by BSRs in seismic reflection surveys from the Nankai accretionary prism. The heat-flow contour map compiled by Yamano and others (1984) shows that the highest heat flow occurs on the trough floor. Outside of the trough, heat flow gradually decreases toward the Japanese Island, and Shikoku Basin values are highly scattered. Sediment temperatures were also obtained at three drill sites near the prism toe. Estimated heat flow values, 63 to 67 mW/m^2 in the western area (Kinoshita and Yamano, 1986) and 126 to 129 mW/m^2 in the eastern area (Taira and others, 1991), are in good agreement with the observed values by thermal probe.

Some specific causes of high heat flow on the trough have been proposed by several authors. First, high heat flow on the trough floor is attributed to subduction of a young hot plate (Yamano and others, 1984; Kinoshita and Yamano, 1986). Second, heat flow may be raised by the migration of pore fluid extruded from the sediments during underthrusting (Yamano and others, 1984; Kinoshita and Yamano, 1986). Third, heat rebound accompanied by buried hydrothermal circulation may provide the recovery of high heat flow in the trough (Nagihara and others, 1989). In contrast, highly scattered heat flow values in the Shikoku Basin are interpreted to be due to fluid circulation within the sediment (Nagihara and others, 1989). Recent ODP Leg 131 drilling result indicates that the thermal regime of the prism toe is dominated by conductive heat transfer. These studies reveal that the relative importance of advective versus conductive heat transfer process within the Nankai accretionary prism is quite debatable.

ESTIMATION OF THERMAL GRADIENT AND HEAT FLOW BASED ON BSRs

Gas hydrate BSR

Anomalous acoustic reflectors are observed on most seismic reflection profiles obtained from the Nankai accretionary prism (Fig. 3). These reflectors are generally subparallel to the sea floor and cut across the reflectors of sedimentary layers. These bottom simulating reflectors (BSRs) are considered to represent the phase boundary between solid gas hydrates and gas-soluble fluid (Yamano and others, 1982). Gas hydrate compounds have an icelike crystalline structure with large amounts of gas molecules trapped within the water molecule lattice. Free gas can be trapped beneath the hydrate layer, which forms a barrier to upward migrating fluids and gases. Experimental studies suggest that artificially formed gas hydrate in sediments causes a marked increase in acoustic wave velocity (Stoll and others, 1971). In addition, a concentration of gas below the hydrate, if present, causes a considerable decrease in velocity and density at the boundary (Bryan, 1974). Consequently, a BSR is often associated with reversed

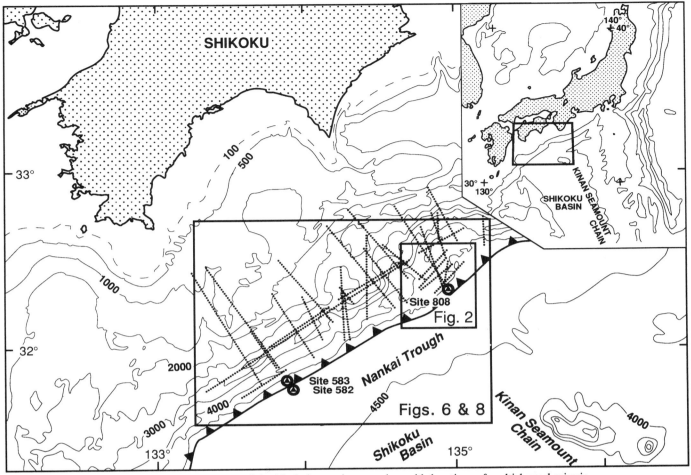

Figure 1. Bathymetric map of the Nankai accretionary prism with locations of multichannel seismic lines. Only part of the seismic lines used in this study are shown because the profile density would result in an illegible map. Bathymetric contour interval is 500 m. Rectangles show locations of Figures 2, 6, and 8.

polarity seismic data, indicating an acoustic impedance decrease from gas hydrate to fluid phase in the sedimentary sequence.

The stability of hydrate with various gas compositions depends more strongly on temperature than on pressure (e.g., Tucholke and others, 1977). The depth distribution of the gas hydrate layer as inferred from the presence of a BSR, therefore, provides an estimate of the in-situ temperature at the BSR. Thermal gradients have been estimated on this basis (Shipley and others, 1979; Yamano and others, 1982). In this study, we used 47 seismic profiles obtained by JAPEX, ORI, and R/V Fred Moore multichannel profiles and Kaiko project single-channel profiles for mapping sea-floor structure and depth to BSRs (Fig. 1). The heat-flow contour map was constructed using more than 4,000 data points on these seismic reflection profiles.

Method for heat-flow estimation

The method of BSR heat-flow estimation is discussed in detail by Yamano and others (1982). Seismic reflection profiles exhibiting the depth to a BSR in two-way travel time must first be depth converted. Travel time to depth conversion is based on an average velocity law (Fig. 4) constructed using published interval velocity data (Nasu and others, 1982; Stoffa and others, 1992) and the correlation between seismic reflection profiles and ODP Leg 131 drilling results (e.g., Moore and others, 1990; Taira and others, 1991). The pressure at the BSR is calculated by the combining sediment and seawater columns overburdens. The average sediment density obtained from drilling results is about 1.9 g/cm^3 (Bray and Karig, 1986; Taira and others, 1991). Seawater density is about 1.025 g/cm^3. We used a density of 1.5 g/cm^3 for calculating the overburden between sea floor and BSR, because the pressure at the BSR ranges from hydrostatic to lithostatic pressures. The temperature at the BSR can then be found using the phase diagram (Fig. 5) and the pressure estimate, assuming a fixed gas composition. Gas extracted from continental margin sediments is often composed of methane and small amounts of gas mixtures such as ethane and CO_2. The ODP Leg 131 drilling

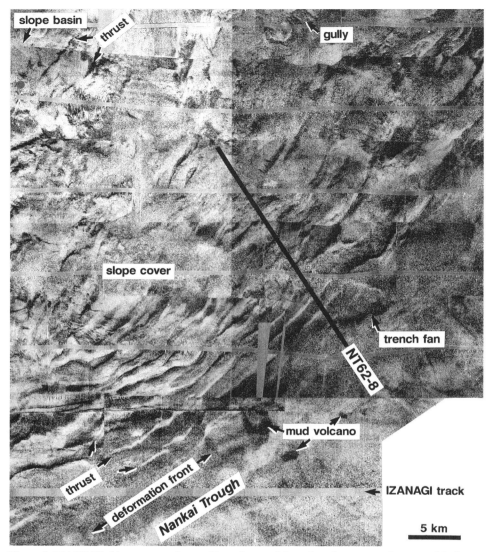

Figure 2. IZANAGI sidescan sonar image. Frontal part of the prism shows rugged topographic feature caused by fault-bend folds. Dashed line is a seismic reflection line shown in Figure 3.

data shows an absence of ethane gas in the sediment above 700 m (Taira and others, 1991). In this study, we used the boundary condition for pure water plus pure methane system, because a small addition of NaCl reduces the influence of CO_2 gas mixtures (Kvenvolden and Barnard, 1983). The mean geothermal gradient is then estimated from the difference between temperatures at the BSR and the sea floor. The sea-floor temperature is assumed to be the bottom-water temperature obtained by CTD (Ocean Research Institute, University of Tokyo, unpublished data).

The determination of heat flow requires the knowledge of the thermal conductivity as well as the geothermal gradient. According to Horai (1982), there are good correlations between thermal conductivity and other physical properties such as compressional wave velocity (V_p). Relations between conductivity,

porosity, and velocity, strongly dependent on lithology, are also based on ODP Leg 131 results (Taira and others, 1991). Even when the average velocity between the sea floor and BSRs is constant, differences in the thicknesses in each velocity layer change the average conductivity value. Average conductivity, therefore, is calculated from the various interval velocities and layer thicknesses obtained by the two-ship seismic reflection experiment (Stoffa and others, 1992). Locations of split-spread profiles (SSPs) and expanding-spread profiles (ESPs) are shown in Figure 6. Figure 7 indicates the average conductivity law constructed by curve fitting. We excluded the data from the Shikoku Basin and the Nankai Trough (SSP1, SSP2, ESP7), prism toe (SSP3) and landward slope of the prism (SSP23) for the reasons mentioned below.

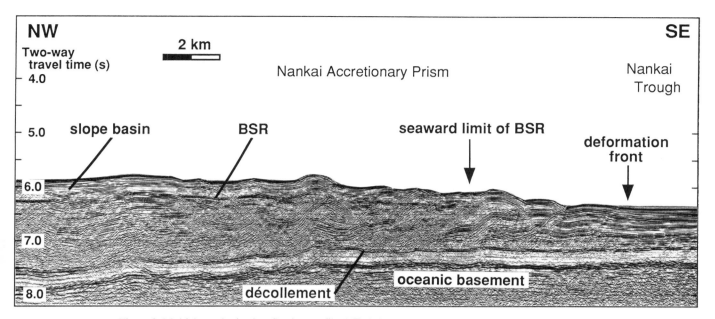

Figure 3. Multichannel seismic reflection profile (NT62-8) showing bottom-simulating reflectors (BSRs) in the Nankai accretionary prism. Reflector indicates bottom of gas-hydrate layer.

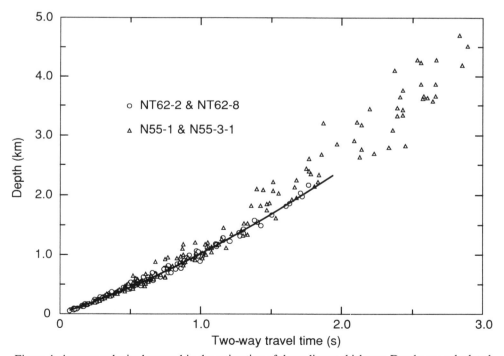

Figure 4. Average velocity law used in the estimation of the sediment thickness. Depths are calculated from two-way travel times using interval velocities from the multichannel seismic reflection studies (Nasu and others, 1982; Stoffa and others, 1992) and drilling results compared with a seismic reflection line (Taira and others, 1991).

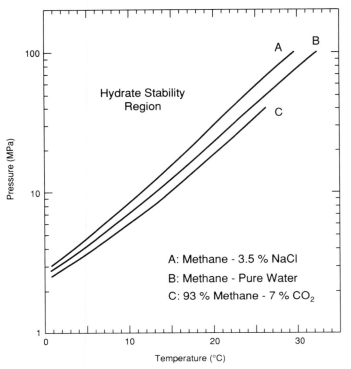

Figure 5. The stability field of methane hydrate showing phase boundary changes due to variation in the composition of gas-bearing fluid (from Tucholke and others, 1977).

Seaward of the deformation front, conductivity is inferred to be relatively low due to slowly increasing velocity with depth suggesting less consolidation (Fig. 7). These values may be suitable for areas with thick slope sediments. But we exclude them since they lack BSRs. At the site of SSP3 where no BSRs are observed, high conductivity inferred from the high velocity seems to indicate rapid dewatering within the protothrust zone. High conductivity observed at the shallow part of SSP23 may result from the presence of highly consolidated offscraped sediments near the sea floor. Such a high conductivity is an inadequate value to use for shallow depths within the prism because the BSR depth increases to more than 500 m below sea floor as heat flow decreases toward the upper slope of the prism.

Error estimation

Yamano and others (1982) argued that errors in estimated heat flow could be up to 25%, and suggested that the main cause of error is poor constraints on thermal conductivity values and velocity structure. Our estimation of maximum cumulative error reaches up to 30% for extreme conditions such as regional high velocity or high thermal conductivity. We believed that this estimation of error is unrealistic and thus, we discuss below about errors derived from indiscriminate adoption of theoretical relations.

Water-bottom temperatures vary only between 1.5 and 2°C since most depths in the survey area are deeper than 2,000 m.

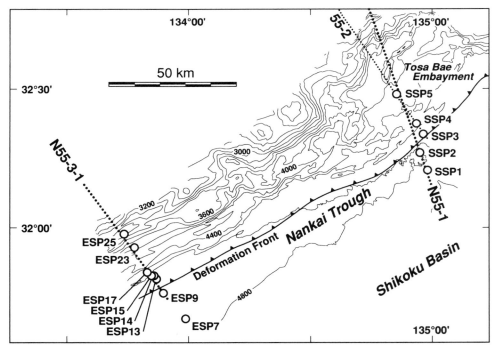

Figure 6. Circles show source-receiver midpoints of expanding-spread profiles (ESPs) and source points of split-spread profiles (SSPs) from the two-ship seismic-reflection experiment (Stoffa and others, 1992). The accretionary prism thermal structures for multichannel seismic reflection lines N55-1 and N55-3-1 (dotted lines) are shown on Figure 11. Bathymetric contour interval is 200 m.

Error caused by water-bottom temperature is inferred to be insignificant.

Phase diagrams used for temperature estimation from BSR depth is controlled by pore fluid chemistry. We adopted a phase diagram of pure water–pure methane for calculating heat-flow value. Pure methane in 3.5% salt water leads to 10% overestimate, and pore fluid with methane and 7% CO_2 creates a 10% underestimate in comparison to our heat-flow values. ODP Leg 131 drilling at the toe of the Nankai accretionary prism shows that there are no major pore fluid chemical anomalies (e.g., existence of pure methane and CO_2, although precise amounts of these are not known), which supports our assumption (Taira and others, 1991).

Seismic reflection profiles provide two-way travel times of both the sea floor and the BSRs. Two-way travel time picking errors are ±0.02%, respectively. Error of sea-floor depth picking gives less than ±1% heat-flow error. In the case of BSR picking, ±3% error occurs, although it reaches ±5% in only 10% of the survey area where BSRs are less than 300 m below the sea floor.

Conversion from two-way travel time to depth is the most common cause of heat flow estimation errors. If the interval velocity between the sea floor and the BSR is larger than that of an average velocity law adopted in our calculation, temperature at the BSR will increase with rising pressure, and geothermal gradient will decrease with increasing BSR depth. Furthermore, heat-flow values increase due to the increasing thermal conductivity associated with the porosity-velocity relation. We evaluated the error in heat-flow values using high and low limits of the average velocity law derived from SSP and ESP data (Stoffa and others, 1992). This suggests errors mostly within ±8%.

Thermal conductivity is related to sediment composition and porosity. The Nankai accretionary prism is mainly composed of turbidite and hemipelagic sediments and compositional variations cause a maximum of ±10% error.

Pressure at a BSR should be between hydrostatic and lithostatic pressures, and abnormally high pore pressures are unlikely to occur at the shallow burial depth where gas hydrates form. Two extreme pressure values show that errors gradually increase from ±1% at 4,500 m water depth to ±5% at 1,000 m water depth. In areas shallower than 1,000 m, this error rapidly increases to ±10%.

Careful examination of the individual data set suggests that the estimated maximum error reaches up to 30% in the extreme. These extreme values, however, are quite limited and we used only the data with less than 15% total error.

THERMAL STRUCTURE OF THE PRISM

Heat-flow distributions

Heat-flow estimated obtained in this study show almost the same general distribution as in previous heat-flow studies (Yamano and others, 1984; Kinoshita and Kasumi, 1988). Heat-flow

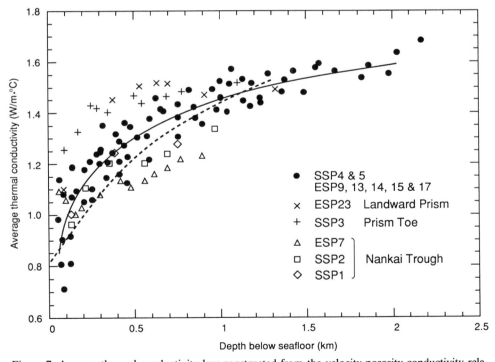

Figure 7. Average thermal conductivity law constructed from the velocity-porosity-conductivity relation. Acoustic velocity data is from a two-ship multichannel reflection study (Stoffa and others, 1992). Symbols represent average conductivities over various depths. Solid line is a curve fitting the solid circles. Dashed line is an average conductivity estimated from the Hole 808 data (Taira and others, 1991).

estimates increase toward the trough floor as described by Yamano and others (1982). However, conspicuously high heat flow
occurs in the topographically low region named "Tosa Bae Embayment" (see Fig. 6) by Leggett and others (1985).

The cross-sectional heat-flow patterns have been explained
using two-dimensional models (Yamano and others, 1984; Kinoshita and Kasumi, 1988; Nagihara and others, 1989). Kinoshita
and Kasumi (1988) showed that the landward heat flow decrease
resulted from the cooling effect of the subducting plate and the
landward thickening of the accretionary prism. Similar results
were obtained by heat-flow observations in other accretionary
prisms (Langseth and others, 1990; Davis and others, 1990).
Based on these results, we expect heat flow to change with prism
thickness. Seismic reflection profiles show that the prism in the
Tosa Bae embayment tapers at a lower angle than in surrounding
areas. This may be the cause of the observed high heat flow.

The sediment thickness (prism thickness) above oceanic
basement is estimated from seismic reflection profiles and the
average velocity relationship. Sixty-seven seismic profiles were
used in mapping the sediment thickness. The sediment thickness
contour map is characterized by northwest-southeast–trending
topographic highs, which follow the same orientation as presumed subducted fossil ridges formed during spreading in the
Shikoku Basin (Chamot-Rooke and others, 1987). Le Pichon and
others (1987) attributed the distribution of trench-fill thickness to
the relief of the oceanic basement.

A combined contour map of heat flow and sediment thickness is shown in Figure 8. A good correlation is observed between
heat flow and sediment thickness, especially between the 2,500-m
contour in thickness and 70-mW/m² contour in heat flow. Figure 9 illustrates the relationship between heat flow and sediment
thickness in this study area. The heat flow increase in proportion
to decrease of sediment thickness appears to extend from the toe
to the area where the sediment thickness exceeds 5,000 m. Lower
heat flow values in slope basins (solid triangles) than in the offscraped sediments (open circles) are interpreted to result from the
cooling effect of sedimentation. If a lower conductivity is used for
heat-flow estimation (because of less consolidated basin sediment), heat-flow values might be even more reduced. Three solid
curves represent theoretical conductive heat flow using the
temperature at the top of oceanic basement as a parameter and
using the average thermal conductivity. Temperatures between
100 and 140°C at the top of the oceanic basement may account
for heat-flow variations in the relatively shallow areas where
BSRs occur.

Heat transfer process

A comparison between observed heat flow and a model of
conductive thermal structure may be used to determine whether
or not heat transfer through the prism is conductive. Kinoshita
and Kasumi (1988) proposed a two-dimensional, steady-state
model based only on conductive heat flow. They calculated the
heat-flow variation along Line 55-2 (see Fig. 6) by treating the

basal heat flux as a parameter. Figure 10 shows the comparison
between the theoretical surface heat flow and the observed heat
flow including our new data. The 110-mW/m² and 150-
mW/m² curves represent the surface heat flow calculated from
the basal heat fluxes derived from the estimated age (20 Ma) of
the Shikoku Basin and the mean heat flow observed on the
trough floor, respectively. Kinoshita and Kasumi (1988) suggested that the observed values for the prism landward slope are
roughly consistent with the 110-mW/m² curve: toward the
trough floor the values gradually approach the 150-mW/m²
curve. Our new data (Fig. 10) generally confirm Kinoshita and
Kasumi's (1988) observations and supports their inference of
more advective heat transfer in the prism toe. ODP Site 808
results document an oceanic basement age of about 15 Ma for
this site (Taira and others, 1991). Heat-flow values of 120 to 130
mW/m², therefore, are a reasonable estimation based on heat
flow–crustal age relations (Lister, 1977; Parsons and Sclater,
1977). The possibility of fluid expulsion at the prism toe is indirectly supported by the presence of a single shell of *Calyptogena*
observed by a deep-towed camera during the KH86-5 cruise
(Taira and others, 1988) and the presence of mud volcanoes near
the deformation front (Fig. 2).

In Figure 11, we show thermal structural cross sections
along seismic reflection lines N55-1 and N55-3-1. These profiles
were made by extrapolating temperatures determined from estimated thermal gradients calculated from BSR depths based only
on conduction. Conductivities were estimated using interval velocities (Nasu and others, 1982) and the velocity–thermal conductivity relation. These estimates suggest that the surface
temperature of oceanic basement is relatively constant, between
100 and 140°C, along both sections. In the eastern area (Line
N55-1), however, the thermal gradient increases at the frontal
part of the prism. The large thermal gradient at the toe may be
caused by up-dip pore fluid migration and this interpretation is
further discussed in the next section.

DISCUSSION

Figures 9 and 11 illustrate that the overall thermal structure
of the Nankai accretionary prism is basically conductive and
mainly controlled by the temperature of the subducting oceanic
basement. Fluid flow along thrust faults, even if present, has little
effect on the thermal structure of the prism as a whole, because
the observed BSRs do not bend up at the fault planes. In contrast,
Minshull and White (1989) reported shallowing BSRs near thrust
faults suggesting warm fluid up-dip migration. The possibility of
heat flow higher than that estimated from the conductive model
must be considered for the eastern area of the Nankai accretionary prism, particularly within 15 km of the deformation front. In
such regions, the heat transfer may be advective due to active
fluid migration.

The boundary between conductive and advective zones appears to coincide with the transition zone on the IZANAGI
image from well-defined lineated regions to low reflectivity re-

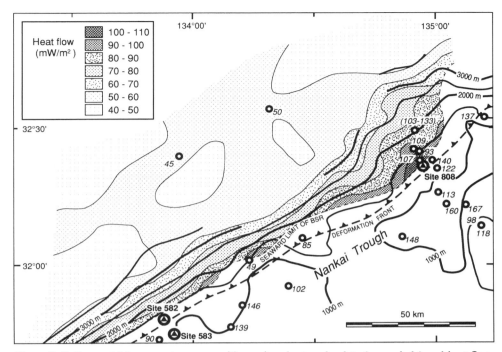

Figure 8. Sediment thickness contour map with previous heat probe data (open circle) and heat-flow distribution estimated by BSR depth. Heat-flow values decrease with the increasing sediment thickness.

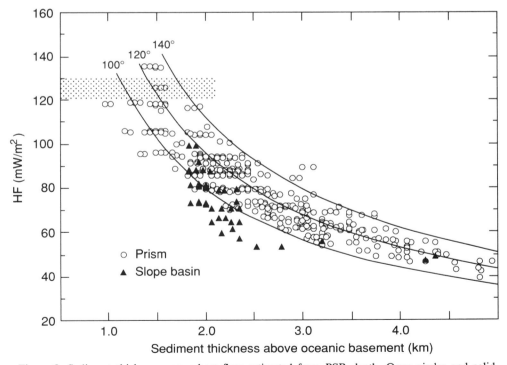

Figure 9. Sediment thickness versus heat flow estimated from BSR depth. Open circles and solid triangles represent heat-flow values in the offscraped sediments (prism) and the slope-basin sediments, respectively. The average probe heat-flow value observed in the trough is indicated by the hatched area. Solid curves are calculated heat-flow variations for various temperatures (100, 120, and 140°C) at the top of oceanic basement and using the estimated thermal conductivities.

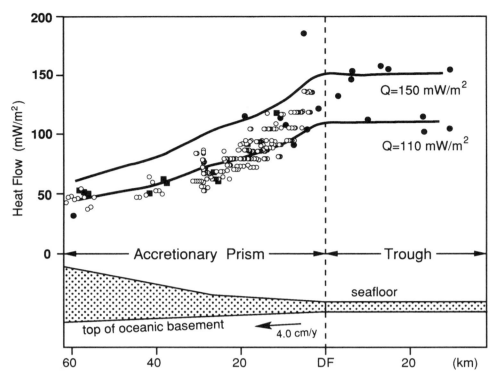

Figure 10. Calculated surface heat flow (solid lines) according to the steady-state model along Line 55-2 (Kinoshita and Kasumi, 1986) with observed heat-flow values in the vicinity of the line (see Fig. 6 for location). Heat-flow values are obtained by heat probe (solid circles) and BSR (solid squares, Yamano and others, 1982; open circles, this study).

gions covered by slope sediments (Fig. 2). At the toe, pore fluids may be easily expelled from the prism because there is no sedimentary cap on the slope. In addition, the toe region is regarded as the most active dewatering zone due to the rapid increase in tectonic overburden during frontal accretion (Bray and Karig, 1986). Such tectonic thickening induces decreased heat flow due to repetition of the same sedimentary sequence by thrusting or folding (e.g., Shi and others, 1988). Therefore, influence of fluid expulsion on heat flow is a more convincing explanation. Furthermore, the accreted terrigenous sedimentary layers are tilted, resulting in steeper-dipping strata favorable for preferred fluid migration. Advective heat flow may result from the rapid increase in tectonic overburden, the presence of tilted permeable layers and an absence of a sedimentary cap at the toe region. The apparent lack of an advection zone in the western area (Fig. 11) may be attributed to the absence of data: no heat-flow estimates are made due to a lack of BSRs.

We infer that the Nankai accretionary prism heat transfer process is dominated by conductive heat transfer. This, however, does not exclude large-scale fluid migration out of the prism. The fluid must be expelled from the prism to accommodate the porosity reduction within the prism sediment as deduced from landward increase of the seismic velocity (Stoffa and others, 1992). We, thus, believe that the main mode of fluid migration within

the prism is diffusive, and does not affect the conductive thermal structure. Transient rapid fluid migration, however, may occur within limited fluid conduits.

In the northern Cascadia accretionary prism, differences of heat-flow values between probe measurements and estimations from BSRs were reported by Davis and others (1990). The probe rsults always exceeded the BSR results within a 10- to 20-km-wide zone landward of the deformation front. This phenomenon was interpreted to be caused by regional fluid expulsion due to compaction of sediments at the prism toe. In the Nankai Trough, there are few available heat probe data directly above BSRs to estimate upward fluid flow. The BSR data and theoretical calculation shown above, however, suggest that the over-all heat transfer within the prism is mostly conductive, except in the toe region.

BSR distribution around the Nankai Trough is restricted to the accretionary prism and no BSRs are observed at the prism toe. In addition, sometimes no BSRs are observed within the slope basins (Fig. 3). Gas hydrates are not always accompanied by the occurrence of BSRs. In the Middle American trench off Guatemala, drilling results provided abundant evidence for the presence of gas hydrate in the sediments, although BSRs are not readily discernible on the seismic reflection profiles (Harrison and Curiale, 1982). The variation in BSR distribution among various

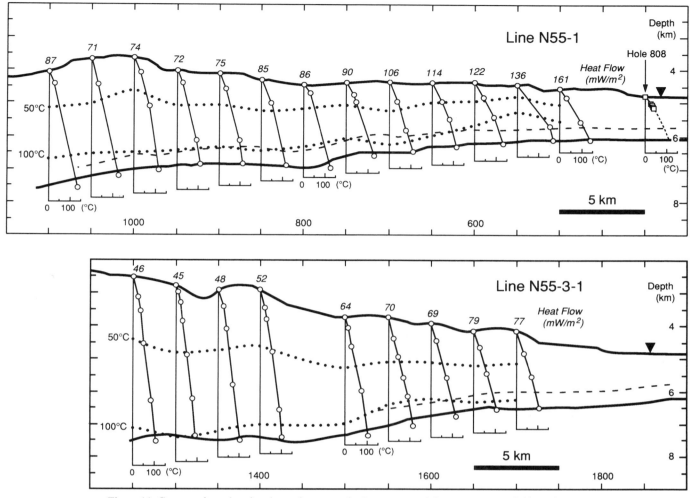

Figure 11. Cross sections showing thermal structure in the eastern and the western part of this study area (see Fig. 6 for location). Upper and lower solid lines represent the sea floor and top of oceanic basement, respectively. Dashed line is the décollement. Geotherm is extrapolated from heat-flow value (number above sea floor) estimated from the BSR depth using an assumed average conductivity only considering conduction. The geotherm contours (50°C and 100°C) are indicated as dotted lines. The rightmost graph in line N55-1 (top) shows a geotherm obtained from ODP Leg 131 Hole 808 for reference (Taira and others, 1991). Triangles indicate deformation front.

sedimentary and tectonic settings in the Nankai accretionary prism is a further topic that we continue to investigate and will present in a separate paper.

SUMMARY AND CONCLUSIONS

Estimation of heat-flow value using BSRs provides wide and continuous thermal structural data of active margins. Detailed heat-flow mapping using the BSR method in the Nankai accretionary prism reveals that the prism heat flow correlates with the thickness of sediments above oceanic basement and heat flow decreases with the increasing sediment thickness. This relationship extends from the toe of the prism to where sediment thickness exceeds 5,000 m.

Based on the observed heat flow values and a conductive steady state model (Kinoshita and Kasumi, 1988), surface heat flow is dominated by conductive heat transfer, except at the toe of the prism. Isotherms based on estimated conductivities and thermal gradients indicate that the oceanic basement surface temperature is between 100 and 140°C beneath the lower slope of the prism. The toe of the eastern area is characterized by higher than predicted heat flow; probably this anomaly is due to pore-fluid expulsion.

The prism's high heat flow region is associated with the well-developed fault-bend fold zone revealed in IZANAGI side-scan sonar images. We propose a model that explains this correlation: First, tectonic thickening by frontal accretion causes a rapid dewatering at the prism toe. Expelled pore fluids may flow upward through tilted permeable layers and/or migrate seaward of the deformation front, as suggested by the presence of mud volcanoes on the trough floor (Fig. 2). In addition, slope sediments that can prevent fluid expulsion are very thin at the prism toe.

This study shows that the thermal structure of the Nankai accretionary prism is primarily dependent on the oceanic basement temperature and the heat transfer within the prism is largely conductive, except for a part of the prism toe. Overall fluid migration within the prism, thus, must be diffusive, except for localized and transient fluid expulsion.

ACKNOWLEDGMENTS

We wish to thank Hidekazu Tokuyama and Adam Klaus for valuable advice and constructive reviews. We also thank Michael Underwood and an anonymous GSA reviewer for their criticism. Discussions with Wonn Soh, Masataka Kinoshita, and Timothy Byrne helped to clarify various aspect of an earlier version of this manuscript. Ashi was funded through the Japan Society for the Promotion of Science (JSPS) postdoctoral fellowship from the Ministry of Education, Science, and Culture.

REFERENCES CITED

Aoki, Y., Tamano, T., and Kato, S., 1983, Detailed structure of the Nankai Trough from migrated seismic sections: American Association of Petroleum Geology Memoir, v. 34, p. 309–322.

Bray, C. J., and Karig, D. E., 1986, Physical properties of sediments from the Nankai Trough, Deep Sea Drilling Project Leg 87A, sites 582 and 583, *in* Kagami, H., Karig, D. E., and others, eds., Initial Reports of the Deep Sea Drilling Project: Washington, D.C., U.S. Government Printing Office, v. 87, p. 827–842.

Bryan, G. M., 1974, In situ indications of gas hydrates, *in* Kaplan, I. R., ed., Natural gases in marine sediments: New York, Plenum, p. 298–308.

Chamot-Rooke, N., Renard, V., and Le Pichon, X., 1987, Magnetic anomalies in the Shikoku Basin: A new interpretation: Earth and Planetary Science Letters, v. 83, p. 214–228.

Davis, E. E., Hyndman, R. D., and Villinger, H., 1990, Rates of fluid expulsion across the northern Cascadia accretionary prism: Constraints from new heat flow and multichannel seismic reflection data: Journal of Geophysical Research, v. 95, p. 8869–8890.

Harrison, W. E., and Curiale, J. A., 1982, Gas hydrates in sediments of holes 497 and 498A, Deep Sea Drilling Project Leg 67, *in* Aubouin, J., von Huene, R., and others, eds., Initial Reports of the Deep Sea Drilling Project: Washington, D.C., U.S. Government Printing Office, v. 67, p. 591–594.

Horai, K., 1982, Thermal conductivity of sediments and igneous rocks recovered during Deep Sea Drilling Project Leg 60, *in* Hussong, D., Uyeda, S., and others, eds., Initial Reports of the Deep Sea Drilling Project: Washington, D.C., U.S. Government Printing Office, v. 60, p. 807–834.

Kinoshita, H., and Yamano, M., 1986, The heat flow anomaly in the Nankai Trough area, *in* Kagami, H., Karig, D. E., and others, eds., Initial Reports of the Deep Sea Drilling Project: Washington, D.C., U.S. Government Printing Office, v. 87, p. 737–743.

Kinoshita, M., and Kasumi, Y., 1988, Heat flow measurement in the Nankai Trough area: Preliminary report of the Hakuho Maru cruise KH86-5: University of Tokyo Ocean Research Institute, p. 190–206.

Klein, G deV., and Kobayashi, K., 1980, Geological summary of the North Philippine Sea based on Deep Sea Drilling Project Leg 58 results, *in* Klein,

G. deV., and Kobayashi, K., and others, eds., Initial Reports of the Deep Sea Drilling Project: Washington, D.C., U.S. Government Printing Office, v. 58, p. 951–961.

Kobayashi, K., and Nakada, M., 1978, Magnetic anomalies and tectonic evolution of the Shikoku interarc basin: Journal of Physics of the Earth, v. 26 (Supplement), p. 391–402.

Kvenvolden, K. A., and Barnard, L. A., 1983, Gas hydrates of the Blake Outer Ridge, site 553, Deep Sea Drilling Project Leg 76, *in* Sheridan, R. E., and Gradstein, F. M., eds., Initial Reports of the Deep Sea Drilling Project: Washington, D.C., U.S. Government Printing Office, v. 76, p. 353–365.

Langseth, M. G., and Moore, J. C., 1990, Introduction to special section on the role of fluids in sediment accretion, deformation, diagenesis, and metamorphism in subduction zones: Journal of Geophysical Research, v. 95, p. 8737–8742.

Langseth, M. G., Westbrook, G. K., and Hobart, M., 1990, Contrasting geothermal regimes of the Barbados ridge accretionary complex: Journal of Geophysical Research, v. 95, p. 8829–8844.

Leggett, J. K., Aoki, Y., and Toba, T., 1985, Transition from frontal accretion to underplating in a part of the Nankai Trough accretionary complex off Shikoku (SW Japan) and extensional features on the lower trench slope: Marine Petroleum Geology, v. 2, p. 131–141.

Le Pichon, X., and 16 others, 1987, Nankai Trough and the fossil Snikoku Ridge, results of Box 6 Kaiko survey: Earth and Planetary Science Letters, v. 83, p. 186–198.

Lister, C.R.B., 1977, Estimates for heat flow and deep rock properties based on boundary layer model: Tectonophysics, v. 41, p. 157–171.

Minshull, T., and White, R., 1989, Sediment compaction and fluid migration in Makran accretionary prism: Journal of Geophysical Research, v. 94, p. 7387–7402.

Moore, G. F., and 7 others, 1990, Structural of the Nankai Trough accretionary zone from multichannel seismic reflection data: Journal of Geophysical Research, v. 95, p. 8753–8765.

Nagihara, S., Kinoshita, H., and Yamano, M., 1989, On the high heat flow in the Nankai Trough area—A simulation study on a heat rebound process: Tectonophysics, v. 161, p. 33–41.

Nasu, N., and 17 others, 1982, Multichannel seismic reflection data across Nankai Trough: University of Tokyo, Ocean Research Institute, IPOD-Japan Basic Series 4, 34 p.

Parsons, B., and Sclater, J. G., 1977, An analysis of the variation of ocean floor bathymetry and heat flow with age: Journal of Geophysical Research, v. 82, p. 803–827.

Ranken, B., Cardwell, R. K., and Karig, D. E., 1984, Kinematics of the Philippine Sea Plate: Tectonics, v. 3, p. 555–575.

Seno, T., 1977, The instantaneous rotation vector of the Philippine Sea Plate relative to the Eurasian Plate: Tectonophysics, v. 42, p. 209–226.

Shi, Y., Wang, C., Langseth, M. G., Hobart, M., and von Huene, R., 1988, Heat flow and thermal structure of the Washington-Oregon accretionary prism: Geophysical Research Letter, 15, p. 1113–1119.

Shih, T. C., 1980, Magnetic lineations in the Shikoku Basin, *in* Klein, G. deV., Kobayashi, K., and others, eds., Initial Reports of the Deep Sea Drilling Project: Washington, D.C., U.S. Government Printing Office, v. 58, p. 783–788.

Shipley, T. H., and 6 others, 1979, Seismic evidence for widespread possible gas hydrate horizons on continental slopes and rises: American Association of Petroleum Geologists Bulletin, v. 63, p. 2204–2213.

Stoffa, P. L., Wood, and 8 others, 1992, Deepwater high-resolution expanding spread and split spread seismic profiles in the Nankai Trough: Journal of Geophysical Research, v. 97, p. 1687–1713.

Stoll, R. D., Ewing, J., and Bryan, G. M., 1971, Anomalous wave velocities in sediments containing gas hydrate: Journal of Geophysical Research, v. 76, p. 2090–2095.

Taira, A., Nishiyama, E., Tokuyama, H., Kinoshita, M., Soh, W., and Ashi, J., 1988, Hydrogeologic framework of the Nankai Trough accretionary prism: Preliminary Report of the Hakuho Maru cruise KH86-5: Ocean Research

Institute ODP Site Survey, University of Tokyo, p. 207–221.

Taira, A., Hill, I., Firth, J., and the Leg 131 Shipboard Scientific Party, 1991, Proceedings of the Ocean Drilling Program, Initial Report, v. 131: College Station, Texas (Ocean Drilling Program), p. 71–269.

Tucholke, B. E., Bryan, G. M., and Ewing, J. I., 1977, Gas-hydrate horizons detected in seismic-profiler data from the western North Atlantic: American Association of Petroleum Geologists Bulletin, v. 61, p. 698–707.

Uyeda, S., 1972, Heat flow, *in* Miyamura, S., and Uyeda, S., eds., The crust and upper mantle of the Japanese area; Part 1, Geophysics: Tokyo, University of Tokyo, Earthquake Research Institute, p. 97–105.

Watanabe, T., Langseth, M., and Anderson, R. N., 1977, Heat flow in back-arc basins of the western Pacific, *in* Talwani, M., and Pitman, W. C., III, eds.,

Island arcs, deep sea trenches, and back arc basins: Washington, D.C., Transactions of the American Geophysical Union, Maurice Ewing Series, v. 1, p. 137–167.

Yamano, M., Uyeda, S., Aoki, Y., and Shipley, T. H., 1982, Estimates of heat flow derived from gas hydrates: Geology, v. 10, p. 339–343.

Yamano, M., Honda, S., and Uyeda, S., 1984, Nankai Trough: A hot trench?: Marine Geophysical Researches, v. 6, p. 187–203.

Yoshii, T., 1979, Compilation of geophysical data around the Japanese Islands (1): Bulletin of Earthquake Research Institute, University of Tokyo, v. 54, p. 75–117.

MANUSCRIPT ACCEPTED BY THE SOCIETY APRIL 24, 1992

Geological Society of America
Special Paper 273
1993

The effects of ridge subduction on the thermal structure of accretionary prisms: A Tertiary example from the Shimanto Belt of Japan

Michael B. Underwood
Department of Geological Sciences, University of Missouri, Columbia, Missouri 65211
Tim Byrne
Department of Geology and Geophysics, University of Connecticut, Storrs, Connecticut 06268
J. P. Hibbard
Department of Marine, Earth, and Atmospheric Sciences, Box 8208, North Carolina State University, Raleigh, North Carolina 27695
Lee DiTullio*
Department of Geology, Kochi University, Kochi 780, Japan
Matthew M. Laughland*
Department of Geological Sciences, University of Missouri, Columbia, Missouri 65211

ABSTRACT

The Shimanto accretionary complex on the Muroto Peninsula of Shikoku comprises two major units of Tertiary strata: the Murotohanto subbelt (Eocene-Oligocene) and the Nabae subbelt (Oligocene-Miocene). Field-based structural analyses and laboratory measurements of thermal maturity show that both subbelts were affected by thermal overprints long after the initial stages of accretion-related deformation. Paleotemperatures for the entire Tertiary section range from about 140 to 315°C, based upon mean vitrinite reflectance values of 0.9 to 5.0%R_m. In general, older rocks of the Murotohanto subbelt display higher levels of thermal maturity and better cleavage than rocks of the Nabae subbelt. The Murotohanto subbelt also displays a spatial decrease in thermal maturity from south to north, and this pattern probably was caused by regional-scale differential uplift following peak heating. Conversely, the paleothermal structure exposed within the Nabae subbelt is fairly uniform, except for the local effects of mafic intrusions at the tip of Cape Muroto. The most important structural discontinuity to display a large component of post-metamorphic vertical offset is the Shiina-Narashi fault; this out-of-sequence thrust forms the boundary between hanging-wall rocks of the Murotohanto subbelt and the younger Nabae subbelt.

Rapid heating of the Shimanto Belt evidently occurred immediately after a middle Miocene reorganization of the subduction boundary of southwest Japan. Whether or

---

*Present addresses: DiTullio, 326 12th St., Apt. 2L, Brooklyn, New York 11215; Laughland, Mobil Research and Development, Exploration, and Producing Technical Center, 3000 Pegasus Park Dr., Dallas, Texas 75265-0232.

Underwood, M. B., Byrne, T., Hibbard, J. P., DiTullio, L., and Laughland, M. M., 1993, The effects of ridge subduction on the thermal structure of accretionary prisms: A Tertiary example from the Shimanto Belt of Japan, *in* Underwood, M. B., ed., Thermal Evolution of the Tertiary Shimanto Belt, Southwest Japan: An Example of Ridge-Trench Interaction: Boulder, Colorado, Geological Society of America Special Paper 273.

not all portions of the inherited (Eocene-Oligocene) paleothermal structure were over-printed during the middle Miocene remains controversial. Subduction prior to middle Miocene time probably involved the Kula, or fused Kula/Pacific, Plate. Hot oceanic lithosphere from the Shikoku Basin back-arc spreading ridge first entered the subduction zone at approximately 15 Ma; this event also coincided with the opening of the Sea of Japan and the rapid clockwise rotation of southwest Japan. Middle Miocene thermal overprints elsewhere in the Shimanto Belt are spatially related to and, therefore, probably synchronous with widespread acidic magmatism, local mafic intrusions, and formation of anthracite coals in the forearc. The Kii Peninsula of Honshu appears to be the hottest spot along the strike of the subduction zone; this locality may well correspond to the collision point between the Shimanto accretionary complex and the central ridge of the Shikoku Basin. However, the spreading ridge probably was unstable and disorganized at the time of the collision, which allowed widespread, strike-parallel flux of mantle heat and magma into the accretionary prism. Apatite fission-track data show that most of the Eocene-Oligocene Shimanto strata experienced differential uplift and cooling through the 100°C isotherm shortly after the ridge-trench collision (11 to 8 Ma).

The Shimanto Belt provides an excellent example of an ancient accretionary complex that does not follow the paradigm of high-pressure, low-temperature metamorphism. Although the thermal history of southwest Japan was punctuated by an unusual burst of middle Miocene thermal and tectonic activity, the residual effects of interaction between the trench and a young subducting slab persist today, as evidenced by high heat flow within submerged portions of the Nankai accretionary prism.

INTRODUCTION

The Shimanto Belt of southwest Japan represents the youngest subaerial portion of a long-lived accretionary margin (Taira and others, 1982, 1988, 1989). These strata make up a large part of the Outer Zone of Japan and range in age from Cretaceous to Miocene (Hada and Suzuki, 1983). In general, metamorphic conditions within the Outer Zone range from zeolite facies to prehnite-pumpellyite facies (Toriumi and Teruya, 1988). Conceptual and stratigraphic correlations have been made between the offshore accretionary domain (modern Nankai Trough) and the Shimanto Belt (Taira, 1985; Taira and others, 1988). The history of interaction between the Japanese accretionary margin and subducting plates of the Pacific Basin has been complicated (Uyeda and Miyashiro, 1974; Maruyama and others, 1989; Taira and others, 1989). For example, the Miocene epoch, alone, was punctuated by the cessation of sea-floor spreading in the Shikoku Basin, the opening of the Sea of Japan, clockwise rotation of southwest Japan, widespread anomalous near-trench magmatism, and incipient collision between the Izu-Bonin Arc and the Honshu Arc (Seno and Maruyama, 1984; Niitsuma and Akiba, 1985; Otofuji and Matsuda, 1987; Niitsuma, 1988).

The two principal objectives that motivated our collective study of the Tertiary Shimanto Belt were: (1) to document the thermal history of a well-studied and well-exposed accretionary complex; and (2) to evaluate the thermal consequences of subducting young oceanic crust. We initially considered these objectives to be fairly straightforward because previous reconnaissance-level investigations had already documented anomalously high paleotemperatures within the Tertiary strata (Mori and Ta-

guchi, 1988); in addition, plate-tectonic reconstructions suggested subduction of an active spreading center during middle Tertiary time (Seno and Maruyama, 1984). Numerical simulations of accretionary prisms indicate that the contrast between "normal" subduction of a cold slab and subduction of young lithosphere should be substantial (DeLong and others, 1979; James and others, 1989), so we expected the paleothermal anomalies to be easy to locate and to evaluate. As our work progressed, however, we realized that the plate-tectonic history of southwest Japan is not that well constrained in either time or space, and many ambiguities exist in the absolute timing of local geologic events. Consequently, regional-scale syntheses of thermal maturity and geochronology, both along and across the structural grain of the Shimanto Belt, are required before all of the important questions can be answered more effectively.

Thermal models

As reviewed by Underwood, Hibbard, and DiTullio (this volume), several workers have demonstrated through numerical models that the thermal structure of an accretionary prism is affected profoundly by the amount of heat conducted across the base of the wedge (Wang and Shi, 1984; James and others, 1989; Dumitru, 1991). These theoretical results are heavily dependent on the selected values of several boundary conditions, but the well-established decrease in heat flow as a function of the increasing age of oceanic crust (Lister, 1977; Parsons and Sclater, 1977) means that the thermal regimes of convergent plate boundaries will vary in response to the age of the subducting slab. The most extreme perturbation of the normal thermal condition develops when an active spreading ridge approaches or collides with a subduction front.

The geologic and bathymetric manifestations of ridge subduction have been documented and discussed by several workers (DeLong and Fox, 1977; DeLong and others, 1978; Herron and others, 1981; Weissel and others, 1982; Barker and others, 1984; Forsythe and Nelson, 1985; Cande and Leslie, 1986; Cande and others, 1987; Nelson and Forsythe, 1989; Hibbard and Karig, 1990a). The subduction of young oceanic crust, for example, may trigger a flip in the vergence direction of thrust faults, folds, and cleavage (Byrne and Hibbard, 1987). Igneous activity, including the intrusion of S-type granites, may take place seaward of the normal volcanic arc, that is, within the forearc basin and/or accretionary prism (Marshak and Karig, 1977; Forsythe and others, 1986). Magmatism, in turn, generally leads to superposition of local contact-metamorphic aureoles upon a regional environment of relatively high-temperature, low-pressure metamorphism.

Many variables exert control over the magnitude, spatial extent, and temporal decay of thermal perturbations within accretionary prisms (van den Beukel and Wortel, 1988; Dumitru, 1991). The orientation of the spreading ridge with respect to the strike of the accretionary prism obviously is important. The width of the thermal overprint should decrease as the incidence angle between the ridge and the trench increases. Another consideration is whether or not ridge segments are offset along intervening fracture zones in a direction toward or away from the trench (Cande and others, 1987). Most heat-flow transects and theoretical models are oriented perpendicular to strike, and in both cases, thermal gradients increase substantially with proximity to the subduction front (DeLong and others, 1979; Cande and others, 1987; James and others, 1989; Davis and others, 1990; Dumitru, 1991). Shallow-level temperatures can be increased further by the effects of frictional heat flux, plus the vertical and lateral transfer of heat by fluids (van den Beukel and Wortel, 1988; Reck, 1987; Moore, 1989; Fisher and Hounslow, 1990). Most of the heat-conduction models lead to the conclusion that a subducting slab must be very young to achieve a thermal gradient conducive to the development of greenschist-facies metamorphism. For example, using what they considered to be realistic boundary conditions, James and others (1989) argued that the subducting oceanic lithosphere must be 1 m.y. or less in age to produce a geothermal gradient greater than 30°C/km at a distance of approximately 100 km landward of the trench. The validity of these numerical simulations must be tested with empirical data sets.

As documented in preceding contributions within this volume, the level of thermal maturity displayed by Tertiary strata of the Shimanto Belt is anomalously high when compared to most coeval accretionary terranes of the Pacific Rim that have been exposed to similar depths of burial. Values of mean vitrinite reflectance ($\%R_m$) for the Shimanto strata of the Muroto Peninsula range from 0.9 to 5.0%R_m, and this rank of organic metamorphism corresponds to a paleotemperature range of approximately 140 to 315°C (Laughland and Underwood, this volume). In comparison, average %R_m values for Paleogene pelites within the Franciscan Complex of northern California (Yager and Coast-

al terranes) are only 0.7%R_m, which equates to an approximate paleotemperature of only 110°C (Underwood and others, 1988). Similarly, most of the Tertiary rocks exposed on the Kodiak Islands, in the western Gulf of Alaska, fall below 1.0%R_m (Moore and Allwardt, 1980). The Tertiary Shimanto, Franciscan, and Kodiak sequences all lack high-pressure mineral assemblages (such as lawsonite), so burial pressures were evidently less than about 3 kbar. Therefore, geothermal gradients appear to have been higher in southeast Japan during the phase of peak heating, as compared to California and Alaska. The cause of this abnormal heating, in a general sense, must be related to the subduction of young oceanic lithosphere. When examined in detail, however, several unknowns remain in our reconstructions of thermal history, including the absolute timing of peak heating event(s) and the identity of the lithospheric plate(s) responsible for the thermal overprint.

Objectives of this study

The purposes of this paper are to reiterate the critical observations presented in the previous chapters, and to integrate these data into a regionally comprehensive and consistent interpretation. Readers should refer to the other papers in this volume for in-depth discussions and citations regarding the regional geology and analytical techniques. Collectively, we have presented a great deal of field-based and laboratory data in this publication, but unfortunately, much of the data allow for nonunique interpretations. This has led to healthy differences of opinion among virtually all of the co-authors. In recognition of this uncertainty and absence of uniformity, we present and critique several scenarios to help explain the occurrence of off-axis magmatism, the generation of high geothermal gradients in southwest Japan, and the relationships between peak paleotemperature and specific structural fabrics and faults. This paper should not be viewed as a "grand finale" to the study of thermal history within the Tertiary Shimanto Belt; instead, we hope to stimulate additional debate and establish focal points for future research.

SHIMANTO METAMORPHISM

Methods

Analyses of vitrinite reflectance on the Muroto Peninsula (Laughland and Underwood, this volume) amplify and confirm the results of previous research (Mori and Taguchi, 1988), which covered a much larger portion of the Shimanto Belt at a reconnaissance scale and included Cretaceous sections located to the north of our study area (Fig. 1). Paleotemperature estimates (140 to 315°C), derived from vitrinite reflectance data, fall within the regional constraints imposed by inorganic mineral phases assigned to the zeolite and prehnite-pumpellyite facies (Toriumi and Teruya, 1988). In addition, statistical correlations between mean vitrinite reflectance and illite crystallinity (Underwood, Laughland, and Kang, this volume) match the limits established

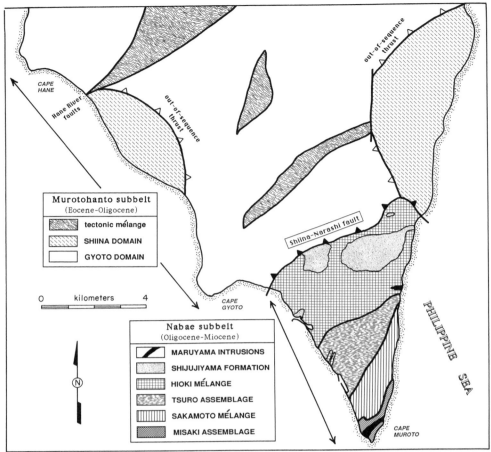

Figure 1. Geologic framework of the Tertiary Shimanto Belt, Muroto Peninsula, Shikoku. Stratigraphic nomenclature and structural boundaries are based upon field mapping by Hibbard (1988), DiTullio (1989), DiTullio and Byrne (1990), and Hibbard and others (1992). See Underwood, Hibbard, and DiTullio (this volume) for summary descriptions of all rock units.

for advanced diagenesis and anchimetamorphism (that is, the transition into lowermost greenschist facies).

Most of our paleotemperature estimates agree favorably with constraints imposed by the annealing temperatures of apatite and zircon fission tracks; the bulk of the Tertiary Shimanto Belt appears to have been heated below the upper limit of the zircon partial annealing zone. This boundary corresponds to a temperature maximum of approximately 260°C, assuming a rate of heating of 10°C/m.y. (Hasebe and others, this volume). With faster rates of heating and/or smaller durations of effective heating time, this limit to the annealing zone would be higher (Zaun and Wagner, 1985; Naeser and others, 1989). We emphasize here that the apparent temperature constraint of 260°C is equivalent to a value of $\%R_m$ of about 3.0%, using the model of organic metamorphism proposed by Barker (1988); values as high as 3.7%R_m, moreover, still fall within an error bar of ±20°C with respect to the limit of the zircon partial annealing zone adopted by Hasebe and others (this volume). Additional discussion of errors inherent

in the paleotemperature estimates appears in Laughland and Underwood (this volume).

Murotohanto subbelt

The Murotohanto subbelt (Eocene-Oligocene Shimanto Belt) has been subdivided into two structural domains (Shiina and Gyoto) and several fault-bounded zones of tectonic mélange (Fig. 1). DiTullio and Byrne (1990) showed that early stage-1 accretion-related deformation features within these rocks were overprinted by younger structures. Intermediate stage-2 deformation, which was caused by intraprism shortening, involved the successive development of (1) seaward-vergent folding and cleavage development within the Shiina domain, (2) out-of-sequence thrusting of the Gyoto domain over the Shiina domain, and (3) folding and cleavage development in the Gyoto domain. The similarity of thermal-maturity values in adjacent domains and across map-scale folds leads us to conclude that the paleothermal structure preserved within the Murotohanto subbelt is not a function of the earliest structural architecture of the accretionary prism (DiTullio and others, this volume).

Paleotemperature estimates for the Eocene-Oligocene strata range from 180 to 315°C (Fig. 2). Using mica b_0 lattice dimensions as an indirect measure of maximum burial pressure, Underwood, Laughland, and Kang (this volume) concluded that the maximum burial depth for the Murotohanto subbelt was about 9 km, and the geothermal gradient at the time of peak heating was at least 33°C/km. In terms of regional-scale spatial changes, a limited number of analyses from fault-bounded units in the rugged interior of the peninsular suggest that the interior has been heated slightly more than the coastal sections (Fig. 2). Thermal maturity on the east coast decreases steadily toward the north. If comparisons are restricted to localities near the Shiina-Narashi fault, there are no significant differences in thermal maturity from the west coast to the east (Fig. 2). On the other hand, farther north of the Shiina-Narashi fault, the west-coast exposures of the Shiina domain are significantly higher in organic rank than comparable rocks of the Shiina domain on the east coast (Fig. 2). Overall, the average rank of organic metamorphism is higher in the Gyoto domain than the Shiina domain (Fig. 3), but some of this statistical difference is due to sampling bias (more high-rank Gyoto specimens were collected).

The intensity of cleavages in the Shiina and Gyoto domains clearly correlates in a spatial sense with the zone of highest-rank rocks located just north of the Shiina-Narashi fault, so it seems logical to conclude that peak heating of the Murotohanta subbelt was contemporaneous with the formation of the D_1 and D_2 cleavages (DiTullio and others, this volume). The isoreflectance surfaces are not folded with the limbs of map-scale F_2 folds (DiTullio and others, this volume), which means that most or all of the organic metamorphism postdated the culmination of stage-2 deformation. In addition, neither the Shiina domain nor the Gyoto domain south of Cape Hane (west coast) displays consistent tilting or offset of isoreflectance surfaces (Fig. 2). This uni-

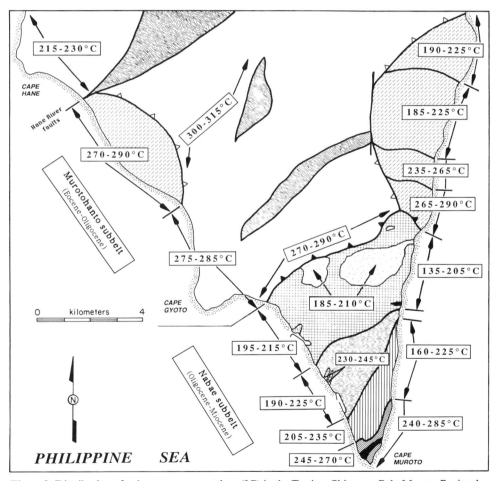

Figure 2. Distribution of paleotemperature values (°C) in the Tertiary Shimanto Belt, Muroto Peninsula, Shikoku. See Figure 1 for identification of tectonostratigraphic units. The equation [T (°C) = 104 (ln %R_m) + 148] was used to calculate paleotemperature based on values of mean random vitrinite reflectance (from Barker, 1988).

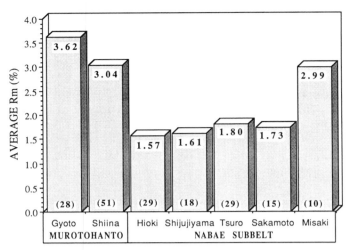

Figure 3. Bar diagram showing average values of mean random vitrinite reflectance for each of the tectonostratigraphic units of the Tertiary Shimanto Belt, Muroto Peninsula, Shikoku. Small numbers in parentheses indicate the number of analyses for each unit. See Laughland and Underwood (this volume) for a complete listing of data.

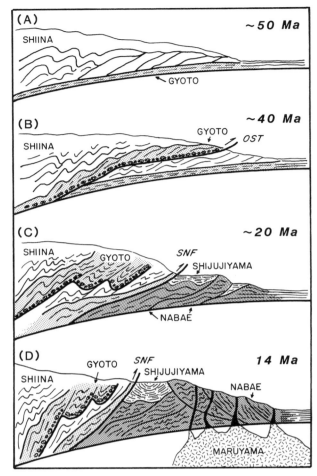

Figure 4. Schematic illustration of the tectonostratigraphic evolution of the Tertiary Shimanto Belt on the Muroto Peninsula of Muroto. OST, out-of-sequence thrust (Shiina-Gyoto domain boundary); SNF, Shiina-Narashi fault (Murotohanto-Nabae subbelt boundary). Based largely upon the structural interpretations of DiTullio and Byrne (1990) and Hibbard and others (1992).

formity of $\%R_m$ values is likewise consistent with a late-stage thermal overprint.

The lowest-rank rocks of the Murotohanto subbelt coincide with a sandstone-rich sedimentary facies on both the east and west coasts of the Muroto Peninsula (Figs. 1 and 2). Although one might argue that the lowest-rank successions represent part of a slope-apron or slope-basin depositional facies (which would explain the shallow burial depths and lower temperatures), structural analyses of one such section tend to refute that interpretation (DiTullio and Byrne, 1990). Regardless of the primary depositional setting, the paleothermal data certainly support the contention that these sandstone-rich sections remained at the shallowest levels of the accretionary prism during the peak heating event. Hasebe and others (this volume) used apatite fission-track data to show that the northern part of the Eocene-Oligocene section passed through the 100°C isotherm at about 14.6 Ma, which was earlier than the coeval higher-rank rocks farther to the south (8 to 11 Ma). Presumably, this difference in cooling history is a combined response to contrasts in burial position at the time of peak heating (that is, deeper to the south) and to differential uplift following organic metamorphism.

Out-of-sequence thrust. Perhaps the most important structural feature within the Murotohanto subbelt is an out-of-sequence thrust that places the Gyoto domain over the Shiina domain (Fig. 4). The eastern half of the Muroto Peninsula displays the same high ranks of organic metamorphism in both the hanging wall and the southern part of the footwall (Fig. 2). The same relationship exists along most of the west coast (Fig. 2). Only at Cape Hane, on the west coast, did we document a clear contrast between the two domains, and the Gyoto hanging wall there is actually lower in rank than the Shiina footwall (DiTullio and others, this volume). Several scenarios involving post-

metamorphic out-of-sequence thrusting could explain this relationship (e.g., two phases of post-peak thermal faulting, or folding of isoreflectance surfaces followed by thrusting). However, the simplest interpretation calls for a north-side-down sense of fault offset where the Shiina-Gyoto contact intersects the coastline. A late-stage, northeast-striking, high-angle fault (Hane River fault zone) has been mapped through this same locality (Taira and others, 1980; Sakai, 1987; see Underwood, Hibbard, and DiTullio, this volume, their Fig. 5), and this fault may be responsible for a late-stage offset of the inherited paleothermal structure.

The apatite fission-track data of Hasebe and others (this volume) show that both of the structural domains were uplifted and cooled through the 100°C isotherm at roughly the same time. Some additional deformation and/or differential uplift could have affected the domains after they passed above the 100°C isotherm, but this would be difficult to detect using fission-track techniques. Based upon the available data, we therefore conclude that organic metamorphism in the Murotohanto subbelt occurred after the culmination of Gyoto-Shiina out-of-sequence thrusting; furthermore, thrust displacement was not responsible for any detectable offsets of the paleothermal structure, as defined by isoreflectance contours.

Nabae subbelt

The Oligocene-Miocene Nabae subbelt of the Shimanto Belt contains two units of mélange (Sakamoto and Hioki), two assemblages of complexly deformed sedimentary rocks (Misaki and Tsuro), inferred slope-basin deposits of the Shijujiyama Formation, and mafic intrusions of the Maruyama igneous suite (Fig. 1). Just as in the Murotohanto subbelt to the north, deformed rocks within the Nabae complex show clear evidence that the early stage structural fabrics (due to sediment offscraping) were overprinted by intermediate-stage intraprism deformation (Hibbard and others, 1992). It is important to note, however, that the first two stages of deformation in the Nabae subbelt occurred after stage-1 and stage-2 deformation in the Murotohanto subbelt.

Overall, the Nabae subbelt is lower in thermal maturity than the older Murotohanto rocks (Fig. 3). Moreover, there is no obvious relationship between the degree or style of stage-1 stratal disruption and the magnitude of thermal maturity (Hibbard and others, this volume). Peak burial temperatures within most of the Nabae complex were approximately 140 to 230°C (Fig. 2), and most of the structural and stratigraphic units display similar levels of thermal maturity (Fig. 3). Higher-rank windows of mélange (230 to 245°C) are exposed beneath the Tsuro assemblage (190 to 225°C) on the west coast of Cape Muroto (Fig. 2), and these anomalies may have been retained from an early phase of hydrothermal activity within the mélange (Hibbard and others, this volume).

Thermal overprints during late-stage deformation of the Nabae subbelt are obvious, particularly near the southern tip of Cape Muroto (Fig. 2), where abnormally high %R_m values for the Misaki assemblage (Fig. 3) are genetically related to the emplacement of the Maruyama intrusive suite at approximately 14 Ma (Hamamoto and Sakai, 1987). The maximum rock temperatures adjacent to the intrusions were at least 285°C, based upon %R_m values of 3.7%. Hasebe and others (this volume) showed that detrital zircons in the Misaki sandstone beds were completely reset and then passed through the annealing window immediately after the intrusion (14.1-Ma cooling age). We believe that peak heating and organic metamorphism through most of the Nabae subbelt was probably synchronous with this magmatic activity (Fig. 4). However, the spatial extent of the contact aureole is restricted to roughly twice the outcrop width of the intrusion (Laughland and Underwood, this volume).

Based upon several arguments and empirical comparisons, Hibbard and others (this volume) concluded that the paleothermal gradient within inferred slope-basin deposits of the Shijujiyama Formation was probably around 70°C/km at the time of peak heating. This estimate is qualitatively consistent with the existence of interfingered mafic volcanic rocks and siliciclastic mudstones near the base of the section. Paleotemperature gradients calculated for other rock units within the Nabae complex are conservatively estimated to range from 60 to 90°C/km (Hibbard and others, this volume).

At first glance, one might expect the Shijujiyama basin-fill to be lower in thermal maturity than the underlying accreted basement (Fig. 4). However, this anticipated paleothermal contrast with respect to the adjacent Hioki mélange (inferred accretionary basement) cannot be substantiated. In fact, %R_m values for several of the mélange sites, located within only a few 100 m of the contact, are significantly lower (down to 0.9%R_m) than those of the Shijujiyama Formation, which are consistently around 1.6%R_m (Laughland and Underwood, this volume). We believe this is because the primary tectonostratigraphic architecture (slope basin over mélange basement) has been structurally overprinted. The slope-basin hypothesis, therefore, must also include some element of fault dislocation along the Shijujiyama-Hioki contact. One possibility is that mélange-cored ridges bordering the slope basin were uplifted sufficiently along thrust faults prior to peak heating; this syndepositional uplift could have prevented resetting of %R_m values even if the regional geothermal gradient subsequently increased and isotherms rose higher in the structural section. Alternatively, vertical offset along the slope-basin margin might have occurred after peak heating, but then the Shijujiyama strata must have been uplifted with respect to the Hioki mélange.

Shiina-Narashi fault

The most conspicuous discontinuity in thermal maturity on the Muroto Peninsula is associated with the Shiina-Narashi fault (Figs. 1 and 4). Eocene-Oligocene (Murotohanto) rocks immediately to the north of the fault display a well-developed cleavage and yield %R_m values that average about 3.7% (285°C), whereas younger and weakly cleaved Nabae rocks to the south average only 1.6%R_m (195°C). This difference in organic metamorphism corresponds to an average contrast in paleotemperature across the thrust of about 90°C. The apatite cooling ages (8 to 11 Ma) documented for the Eocene-Oligocene rocks by Hasebe and others (this volume) almost certainly are a response to uplift from late Miocene through Quaternary time. These data are extremely important, because they confirm the occurrence of substantial differential offset between the two subbelts after the culmination of organic metamorphism within both belts.

We acknowledge that vertical displacement along the Shiina-Narashi fault may have started prior to peak heating in the footwall. The truncation and offset of isoreflectance surfaces across the fault, however, requires substantial post-metamorphic displacement. Furthermore, there is no evidence of anomalously high %R_m values immediately to the south of the fault, which might be expected if there had been conductive transfer of heat from the hanging wall into the footwall aureole (Underwood, Hibbard, and DiTullio, this volume). Consequently, either the hanging wall was relatively cool at the time of dislocation, or fault-zone temperatures were moderated by high rates of hanging-wall erosion. Given the evidence for neotectonic activity on the Muroto Peninsula, including the exposure and deformation of upper Pliocene marine sediments and the formation of prominent marine terraces (Sakai, 1987), there is a strong possibility of protracted offset through much of late Cenozoic time. This contention of late-stage offset within the Shimanto Belt is also supported by the obvious post-metamorphic slip along the Hane River fault zone, which is superimposed on the Gyoto-Shiina out-of-sequence thrust south of Cape Hane (Fig. 2). For both faults (Hane River and Shiina-Narashi), the structural levels observed at the surface today (on opposite sides of the fault) were juxtaposed after both fault blocks had reached their maximum burial temperatures.

If one accepts the idea that the regional geothermal gradient was as high as 70°C/km during the middle Miocene (Hibbard and others, this volume), and that both Murotohanto and Nabae strata attained their maximum burial temperatures at the same time, then the minimum amount of post-metamorphic vertical offset on the Shiina-Narashi fault equates to approximately 1,200 m. In other words, this amount of slip is required to account for the contrast in vitrinite reflectance. As discussed subsequently, peak heating of the Murotohanto section could have occurred during Oligocene time under the influence of a lower geothermal gradient. If so, then the cumulative vertical offset along the Shiina-Narashi fault may be significantly greater than our minimum estimate of 1,200 m.

Muroto flexure

The Muroto flexure is a major north-trending cross fold that affects all of the strata analyzed on the Muroto Peninsula (Hibbard and Karig, 1990a). Within the Eocene-Oligocene part of the section, the structure appears to maintain a subhorizontal axis (DiTullio and Byrne, 1990). Based on mapping within the Nabae subbelt, Hibbard and Karig (1990a) concluded that the fold axis is steeper toward the south and the limbs tighten. The flexure may be responsible for the pronounced bent that appears in the surface trace of the Shiina-Narashi fault (Fig. 1), assuming that the fault trace was originally straight. In addition, there is widespread evidence of quartz-vein precipitation and displacement on a system of high-angle, mesoscale faults and tension fractures during the same stage of regional deformation (Hibbard and Karig, 1990a).

Thermal-maturity data fail to support the contention of monotonic post-metamorphic warping of paleotemperature gradients across the core of the fold or its limbs. For example, %R_m values within the Nabae subbelt are actually somewhat higher on the west coast (on the limb of the flexure) than on the east coast near the core (Hibbard and others, this volume); this spatial variation in thermal maturity is opposite to that expected for a post-metamorphic anticline, particularly if the plunge of the axial trace is truly subhorizontal. In addition, the uniformity of

paleotemperature values along an east-west transect just north of the Shiina-Narashi fault suggest little or no modification by the flexure (Fig. 2). On the other hand, farther to the north, especially on the east coast within the Shiina domain, the gradual northward reduction of thermal-maturity values can be broken up into subtle steplike shifts; this pattern has been explained through small offsets along high-angle tear faults on the limbs of the Muroto flexure (DiTullio and others, this volume). Some of the fault-bounded blocks display fairly uniform levels of thermal maturity, whereas others display an obvious tilting of isoreflectance surfaces. In most cases, the offsets of organic rank across the faults are subtle, at best. Finally, if one compares the Shiina rocks south of Cape Hane (on the west coast and in the interior) with rocks on the east coast at equivalent distances from the Shiina-Narashi fault, there is an obvious contrast in thermal maturity (Fig. 2). DiTullio and others (this volume) have attributed this spatial change to an important north-south high-angle fault (east-side down); this fault is located near the axis of, and presumably is related to, the Muroto flexure.

In conclusion, the Muroto flexure does not exert consistent control on the orientation of isoreflectance surfaces within the Tertiary Shimanto Belt, even though the fold appears to be post-peak thermal. One explanation is that the magnitude of vertical offset of isoreflectance surfaces, from the core of the flexure to either limb, was modest, particularly close to and south of the Shiina-Narashi fault. With a subvertical fold axis, we would expect the east-west warping of paleotemperature horizons to be negligible. Another possibility, at least within the Nabae subbelt, is that the flexure formed during peak heating (Hibbard and others, this volume).

Coeval rocks in southwest Japan

Several critical observations and data sets must be reiterated to summarize the effects of Oligocene and Miocene thermal events elsewhere in the Japanese Islands (see also Underwood, Hibbard, and DiTullio, this volume). In particular, we consider radiometric dates on cleavage in other parts of the Tertiary Shimanto Belt, evidence for igneous activity throughout the Outer Zone of southwest Japan during the middle Miocene, and chronologic constraints on local uplift and cooling histories from fission-track analyses.

Hata Peninsula. Beginning the discussion with the Hata Peninsula of Shikoku (Fig. 5), illite crystallinity data define a dramatic gradient toward Miocene granitic bodies at Cape Ashizuri, where values are consistent with greenschist-facies metamorphism (DiTullio and Hada, this volume). Agar and others (1989) dated early (accretion-related) cleavages within some of these lower-rank strata using K-Ar radiometric methods; their dates range from approximately 43 to 18 Ma within the Eocene-Oligocene part of the Shimanto section (Fig. 5). Most of the dates cluster between 34 and 26 Ma. In addition, for strata that were heated above the apatite partial annealing zone (~125°C), the resultant fission-track cooling ages for both Late Cretaceous and

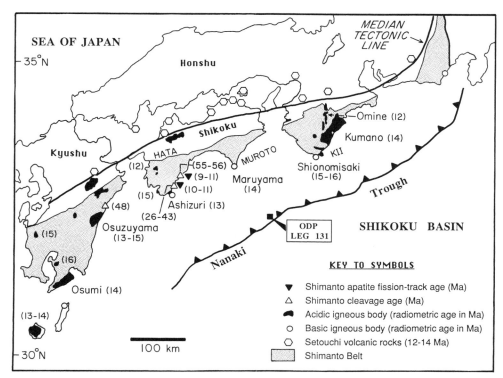

Figure 5. Map showing the distributions of anomalous basic and acidic igneous bodies of middle Miocene age within the Outer Zone of Japan, together with complementary radiometric age dates (Ma). Occurrences of the somewhat younger Setouchi volcanic series are shown north of the Median Tectonic Line. Also depicted are control points and corresponding absolute K-Ar ages (Ma) of cleavage formation within the Shimanto Belt, plus uplift/cooling ages (Ma) based on apatite fission tracks. Data for this figure were obtained from the following sources: Shibata and Nozawa (1967), Oba (1977), Miyake (1985), Hamamoto and Sakai (1987), Terakado and others (1988), Agar and others (1989), Hibbard and Karig (1990b), and MacKenzie and others (1990).

Paleogene strata range from approximately 11 to 9 Ma. These dates are virtually identical to the apatite cooling ages documented by Hasebe and others (this volume) on the Muroto Peninsula, and they collectively point to an uplift event of regional extent that involved all stratigraphic components of the accretionary prism.

Igneous activity on the Hata Peninsula (Ashizuri granites) has been dated at approximately 13 Ma (Shibata and Nozawa, 1967; Oba, 1977), and the emplacement of these granitic intrusions must have coincided, at least locally, with organic metamorphism and recrystallization of illite/mica (DiTullio and Hada, this volume). The spatial limit of intrusion-related overprints of the inherited accretion-related thermal structure remains unresolved. It is possible that some or most of the dated Oligocene cleavages on the Hata Peninsula formed 5 to 20 m.y. before the attainment of maximum paleotemperatures at 13 Ma. Conversely, the cycle of uplift and/or cooling of strata through the apatite 100°C isotherm definitely occurred immediately after the 13-Ma magmatic overprint.

The effects of significant uplift of the Hata Peninsula prior to the early Miocene epoch needs to be considered when evaluating the possibility of a subsequent thermal overprint. The best evidence for an early phase of uplift is the unconformity that separates the basal Misaki Group (~18 Ma) from the underlying Shimanto Belt (Kimura, 1985; Sakai, 1988). This contact probably is a submarine unconformity, however, separating the accretionary prism from overlying slope-basin deposits (similar to the basal Shijujiyama deposits of Muroto). Consequently, subaerial uplift and erosion of Shimanto overburden prior to the early Miocene are not required. Furthermore, many of the coeval (early Miocene) slope-basin deposits along the strike of the Outer Zone grade up stratigraphic section from middle bathyal deposits into shallow-water microfossils and depositional facies (Chijiwa, 1988; Hisatomi, 1988; Osozawa, 1988). These data strongly support the idea of regional uplift of a submerged accretionary prism and progradation of the Miocene shoreline (Hibbard and Karig, 1990b), but not subaerial erosion.

Eastern Kyushu. K-Ar age dates for cleavage in the Shimanto Belt of eastern Kyushu (Fig. 5) are approximately 48 Ma, which is less than 10 m.y. after than the fossil-controlled age of deposition (Mackenzie and others, 1990). Although the uplift and cooling history of the Shimanto Belt on Kyushu has not yet been quantified using fission-track analysis, this area (just as within southern Shikoku) likewise was affected by intrusive activity during a time interval of 13 to 15 Ma (Shibata and Nozawa, 1967; Oba, 1977). The Paleogene strata there yield $\%R_m$ values as high as 4.0%; in addition, vertical maturation gradients are on the order of $0.1\%R_m/100$ m, and spatial patterns of organic metamorphism clearly overprint the structural architecture of the accreted section (Aihara, 1989). Values of illite crystallinity for these rocks have been reported by DiTullio and Hada (this volume), and they are generally consistent with conditions of advanced diagenesis and incipient greenschist-facies metamorphism.

Stratigraphically above the Shimanto strata of eastern Kyushu, there is a pronounced decrease in $\%R_m$ values (to <0.5%), as well as a sharp deflection in the vertical $\%R_m$ gradient at the unconformable contact with upper Miocene sequences of the Miyazaki Group (Aihara, 1989). The gap in $\%R_m$ values corresponds to a paleotemperature contrast across the unconformity of approximately 75°C. Deposition of the Miyazaki Group began at approximately 11 Ma (Sakai, 1988). Thus, we conclude that peak heating, uplift, and erosion of at least 1 km of the underlying Shimanto stratigraphy must have occurred before 11 Ma. Presumably, at least some of the thermal overprinting of the Shimanto rocks was contemporaneous with the middle Miocene igneous activity and organic metamorphism on Shikoku, and the uplift event on Kyushu (Miyazaki unconformity) probably was synchronous with the late Miocene uplift of Shikoku.

Kii Peninsula. On the Kii Peninsula of Honshu (Fig. 5), there are pronounced angular unconformities between the intensely deformed Shimanto strata and overlying slope-basin and forearc-basin deposits (Chijiwa, 1988; Hisatomi, 1988; Kumon and others, 1988). Cleavage within the accreted strata evidently formed prior to the development of the unconformities, because overlying strata (of middle Miocene or late Miocene age) are not cleaved. We have no data, however, to establish the absolute age of cleavage in this region. On the other hand, maximum burial temperatures over much of the Kii Peninsula must have coincided with the emplacement of plutonic and volcanic rocks at 14 to 15 Ma (Shibata and Nozawa, 1967; Oba, 1977; Miyake, 1988; Terakado and others, 1988), rather than with the formation of cleavage. This contention is supported by $\%R_m$ values documented by Chijiwa (1988) and Aihara (1989) in the Miocene forearc-basin succession (up to $6.0\%R_{max}$); the forearc coals overlap ranks of dispersed organic matter within the underlying Shimanto rocks (up to nearly $7.0\%R_{max}$ in the subsurface). Thus, unlike the example from Kyushu, the Miocene unconformity on the Kii Peninsula predated peak heating.

Nankai Trough analog

Reconstructions of the thermal history of southwest Japan should take into account the modern analog represented by the Nankai accretionary prism (Fig. 5). As summarized by Yamano and others (1984) and Kinoshita and Yamano (1986), geother-

mal gradients within the upper kilometer of the structural pile range from 41 to 66°C/km, and near-surface heat flow ranges from 70 to 130 mW/m². Horizontal gradients in heat flow define a regional-scale decrease toward the shoreline. Temperatures recently recorded in the borehole at ODP Site 808 are consistent with a local geothermal gradient of 110°C/km (Shipboard Scientific Party, 1991), but this site is located at the toe of the prism where heat flow is at a maximum (Fig. 5). Detailed documentation and interpretation of bottom-simulating reflectors by Ashi and Taira (this volume) suggest that the thermal structure may be affected by fluid flux; in addition, along-strike variations within the prism are a response to the age and structure of the subducting slab, with some modification by indentations (seamount collision, etc.).

The data cited above point to a persistence of high heat flow in the Nankai accretionary prism, even within midslope bathymetric domains. Interpretations of sea-floor magnetic anomalies imply that volcanism in the subducting Shikoku Basin ended at approximately 12 Ma (Chamot-Rooke and others, 1987). Heat flux from the subducting slab should reach maximum values of 136 to 145 mW/m², given an age of 12 Ma (Lister, 1977; Parsons and Sclater, 1977). Heat-flow data agree, in part, with the framework established through numerical models of subduction (e.g., James and others, 1989; Dumitru, 1991), in that the flux of heat through the overriding accretionary prism does decrease toward the rear of the wedge. On the other hand, based upon the empirical data base from Nankai Trough, we stress that abnormally high geothermal gradients (>40°C/km) are maintained within the outermost 50 km of the accretionary prism over a range of slab-age conditions that is much lengthier than those indicated in the numerical simulations. In other words, the subducting slab does not have to be exceptionally young to produce a significant regional-scale thermal overprint of the accretionary wedge.

INTERPRETATION AND SPECULATION

The following discussion stresses what we think we know about the regional geothermal history of southwest Japan, together with the many uncertainties and ambiguities. The discussion begins with an evaluation of the Nabae subbelt, because that part of the tectonostratigraphy is linked with more confidence to the present-day plate-tectonic regime and the offshore record of sea-floor magnetic anomalies.

Oligocene-Miocene accretionary prism

Plate models. At the present time, plate convergence along the Nankai Trough involves the Eurasian and Philippine Sea Plates (Ranken and others, 1984; Seno and Maruyama, 1984). The initiation of rifting in the Shikoku Basin cannot be dated directly through analysis of sea-floor magnetic anomalies, because the older intervals of oceanic crust have been consumed. However, if one assumes that the openings of the Shikoku and

Parece Vela Basins were approximately synchronous, then rifting started at 30 Ma, based on the identification of anomaly 10 in the central Parece Vela Basin (Shih, 1980). The oldest observable anomaly in Shikoku Basin is anomaly 7, which has been dated at 26 Ma (Watts and Weissel, 1975). Several phases of spreading and realignment of the ridge can be recognized, and the last volcanic activity evidently occurred between 14 Ma and 12 Ma (Chamot-Rooke and others, 1987).

According to the retreating trench model of Seno and Maruyama (1984), subduction beneath Nankai Trough during late Oligocene time (30 Ma) involved older lithosphere of the Pacific Plate (Fig. 6). Westward migration of the Izu-Bonin arc-trench system eventually brought the central ridge of the Shikoku Basin into a position almost directly offshore of the Muroto Peninsula by approximately 17 to 15 Ma. A recent model by Otsuki (1990) is similar except that the central ridge is located farther to the east in the middle Miocene and then migrated to the west (see Underwood, Hibbard, and DiTullio, this volume). Abnormally high geothermal gradients quite clearly affected the Shimanto forearc region across its entire length at about 15 Ma, as shown by the widespread and simultaneous occurrence of anthracite-rank coals, acidic volcanic rocks, and granitic intrusions (Fig. 5). The formation of a back-arc spreading center in the Shikoku Basin, together with subsequent subduction of this newly created oceanic crust, must have contributed to the thermal alteration of the Nabae subbelt, but the timing of its arrival remains a topic of debate.

An important modification to the Seno and Maruyama (1984) plate reconstruction was proposed recently by Hibbard and Karig (1990b). The critical difference is the addition of a transform boundary between the Pacific and Philippine Sea Plates with a protrusion of older Pacific crust separating the Shimanto subduction front from the active Shikoku Basin spreading ridge (Fig. 6). This configuration is very appealing because it allows for a lower geothermal gradient through Oligocene and early Miocene time, with a rapid transition to higher geothermal gradients when the Shikoku Basin lithosphere entered the subduction zone. The radiometric dates summarized on Figure 5 prove that Miocene magmatic activity did not sweep gradually across the Shimanto forearc in the wake of a migrating Shikoku Basin spreading center, as would be expected with the Seno and Maruyama (1984) reconstruction. Instead, high regional heat flow and widespread near-trench magmatism began suddenly across the entire forearc, within a time window of about 15 to 13 Ma.

Paleomagnetic evidence shows that the Outer Zone magmatic activity occurred immediately after the culmination of 42 to 56° of rapid clockwise rotation of southwest Japan, which itself was caused by the opening of the Sea of Japan (Otofuji and

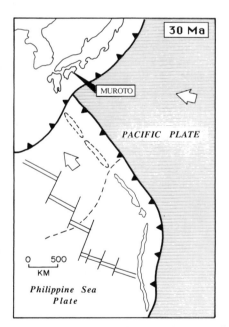
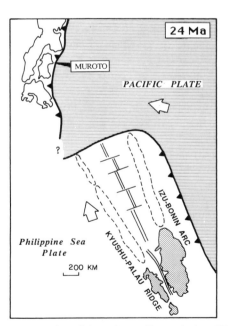
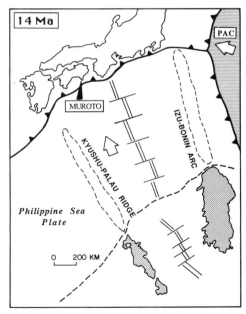

Figure 6. Plate-tectonic reconstructions for southwest Japan during Oligocene, earliest Miocene, and middle Miocene time. PAC, Pacific Plate. Dashed line in the Shikoku Basin represents the approximate limit of oceanic crust that has been subducted since 14 Ma. Present-day Isu-Bonin and Kyushu-Palau Ridges are highlighted by stippled pattern. Arrows indicate directions of motion for the Pacific and Philippine Sea Plates relative to Eurasia. For the 24-Ma reconstruction, an allowance has been made for opening of the Sea of Japan and 50° of clockwise block rotation. At 14 Ma, the locus of collision between the central ridge of Shikoku Basin and the subduction front is placed near the Kii Peninsula of Honshu. The 24-Ma and 14-Ma reconstructions are modified from Hibbard and Karig (1990b). The 30-Ma reconstruction is modified from Seno and Maruyama (1984) and does not take into account the block rotations associated with opening of the Sea of Japan.

Matsuda, 1984, 1987; Otofuji and others, 1985; Nakajima and Hirooka, 1986). We believe that the thermal regime throughout the Shimanto Belt changed abruptly at this same time because of a shift from subduction of Pacific Plate to subduction of very young and unusually hot Philippine Sea Plate (Fig. 6). As discussed subsequently, the Pacific Plate just prior to 15 Ma must have been at least 28 m.y. old, because Kula-Pacific spreading ended at 43 Ma (Lonsdale, 1988). At the time of the plate reorganization, early Miocene strata within the accretionary wedge would have been in very close proximity to the subduction front, so background geothermal gradients of 70°C/km, or more, certainly would be expected (see, for a modern analog, Cande and others, 1987). Even higher gradients must have occurred within the contact aureoles produced by granitic and mafic intrusions.

Location of ridge collision. According to Hibbard and Karig (1990a), the Muroto flexure formed soon after reorganization of the subduction boundary as a mechanical response to collision between the Shimanto accretionary prism and a rigid indenter of significant size. Although the indenter may have been the central ridge of Shikoku Basin (which would also explain the occurrence of cogenetic mafic magmatism), active seamounts located off the ridge axis could have accomplished the same thing in terms of both deformation style and magmatic activity. Descriptions of the mechanical effects of seamount subduction on deformation of the present-day Nankai trench slope have been presented by Yamazaki and Okamura (1989). In addition, the Muroto flexure is only one of several so-called "megakinks" mapped within the Outer Zone of southwest Japan; superficially similar structures have been documented in Kyushu, the Hata Peninsula, and the Kii Peninsula (Murata, 1987; Kosaka and others, 1988; Yanai, 1989; Kano and others, 1990), where they have been attributed to horizontal compression at shallow crustal levels during opening of the Sea of Japan. Future investigations might be able to test whether or not the other cross folds and flexures display unusual gradients in thermal maturity, but at the present time, the combination of structural geometry and coeval mafic magmatism does make the Muroto flexure unique with respect to the other megafolds.

Based solely on regional trends in thermal maturity, one might argue that the locus of collision between the Shikoku Basin ridge crest and the Shimanto subduction front was located to the east of Cape Muroto at about 15 Ma, in the general vicinity of the Kii Peninsula (Fig. 6). Our provisional identification of this "hot spot" is based on three lines of evidence: (1) values of organic metamorphism within the Tertiary Shimanto section (Aihara, 1989), which are significantly higher than values for coeval rocks at Cape Muroto; (2) coal rank within the Miocene Kumano Group (the forearc basin succession on the Kii Peninsula), which is likewise extremely high (Chijiwa, 1988); and (3) the widespread distribution of both extrusive and intrusive magma bodies. The remarkably high levels of coal rank on the Kii Peninsula cannot be attributed to comparatively larger amounts of uplift and erosion into deeper stratigraphic levels of the accretionary forearc, because the coal-bearing successions of the Kumano

Group are associated with shallow-marine and nonmarine facies, which should be the first deposits lost to subaerial erosion. According to Chijiwa (1988), the middle Miocene paleogeothermal gradient was 80 to 110°C/km, but we view these numbers as conservative estimates because they are based on an outdated time-dependent model of thermal maturation. If the Barker (1988) model of vitrinite maturation is applied to the data from the Kumano coal fields, then an estimate for a typical geothermal gradient increases to 140°C/km, which is substantially higher than comparable estimates made by Hibbard and others (this volume) for the Muroto Peninsula.

Compared to the Maruyama intrusive suite at Cape Muroto, the Kii Peninsula igneous rocks are much more voluminous and chemically more diverse. The igneous rocks range from MORB-type ophiolites to high-magnesian andesite, rhyolitic tuff, and S-type granite (Murata, 1984; Miyake, 1985, 1988; Torii and Ishikawa, 1986; Terakado and others, 1988). As with other igneous bodies of the Outer Zone, chemical data, occurrences of xenoliths, and zones of migmatite suggest that the granitoid bodies probably originated deep in the crust as upwelling mantle convection cells passed beneath the Nankai accretionary prism (Oba, 1977; Nakada and Takahashi, 1979; Shibata and Ishihara, 1979; Nakada, 1983; Terakado and others, 1988). Mixing between igneous and sedimentary components evidently occurred deep in the source region, prior to chemical differentiation of the magma (Terakado and others, 1988). In comparison, the mantle heat flux beneath the Muroto Peninsula evidently was not sufficient to melt the lower crust. Instead, the local MORB-type bodies of the Maruyama intrusive suite probably were injected directly from the upper mantle into the accreted sedimentary rocks (Hibbard and Karig, 1990a).

To link all of the Miocene igneous bodies of the Outer Zone to a single point of ridge-trench intersection leads to an enigma because magmatism was nearly simultaneous along a strike-length of 700 km (Fig. 5). One way to widen the spatial extent of magmatism within a narrow time window (15 to 12 Ma) is to allow for several segments of the spreading axis in the Shikoku Basin, together with east-west offsets along intervening transform faults. For example, one ridge segment may have collided with the accretionary prism at the Kii Peninsula at 15 Ma, with a second segment providing an intersection point at Cape Muroto at 14 Ma. A better explanation, however, involves widespread plate instability and disorganization of the magmatic locus as the spreading ridge approached the trench at about 15 Ma. The unusual magnetic anomaly pattern for the central Shikoku Basin, which has been described and interpreted by Chamot-Rooke and others (1987), probably was caused by a phase of north-south spreading between 15 Ma and 12 Ma. This deviation in the direction of sea-floor spreading may have been a response to either simple rotation of the ridge axis or a ridge propagation/jumping process. Among the many possible effects of ridge reorganization is the high near-surface heat flow recorded within the present-day central trench segment (Chamot-Rooke and others, 1987). Because all of the direct evidence has been subducted, we have no means of producing an accurate reconstruction of the ridge-transform geometry during the inception of ridge-trench collision. Nevertheless, we suggest that protracted adjustment of an unstable spreading ridge, coupled with extensive off-axis magmatism, provides the most logical explanation for what must have been an unusually large zone of mantle upwelling and heat flux into the Shimanto accretionary prism.

Eocene-Oligocene accretionary prism

Plate models. Interpretations of the thermal history of Eocene-Oligocene rocks are even more tenuous than those for the Miocene, in part because the remaining sea-floor record of earlier plate motions is much less complete. Among the recognized complications, events in the late Eocene and the Oligocene took place before middle Miocene clockwise rotation of southwest Japan (Fig. 6). In addition, the position of the Kula-Pacific spreading ridge prior to the cessation of sea-floor spreading remains unknown. Most plate reconstructions place the Kula-Pacific boundary well to the north of the Japanese Islands at about 43 Ma (e.g., Engebretson and others, 1985; Lonsdale, 1988; Jolivet and others, 1989), but DiTullio (1989) and Byrne and DiTullio (1992) speculated that the triple junction may have been close to the Shimanto subduction zone. We simply do not know which of several possible plates was subducted beneath Shikoku during latest Eocene and early Oligocene time. Possibilities include the Pacific Plate, the Kula Plate, the Western Philippine Sea Plate, and (after the 43-Ma plate reorganization) the fused Pacific-Kula Plate.

A major issue within the context of our study is whether or not the early Oligocene epoch was a time of unusually high heat flow, and if so, what was the source of the heat transfer through the Murotohanto accretionary prism? One line of evidence favoring the subduction of relatively young lithosphere towards the end of Eocene time is the small time gap between turbidite fossil ages and the ages of basalts within the Murotohanto subbelt (Taira, 1985). These temporal gaps by themselves, however, do not help us determine which plate was subducted. According to Sakai (1988), the rocks at Cape Gyoto contain microfossils whose range extends into the early Oligocene (perhaps as young as 34 Ma); fossils as young as 33 Ma also have been identified within the Oligocene Shimanto section of the Hata Peninsula (Agar and others, 1989). The minimum depositional age for this portion of the Shimanto Belt, therefore, seems to be significantly younger than the plate reorganization described by Lonsdale (1988). Thus, even if the Kula-Pacific ridge had been in close proximity to southwest Japan when spreading ceased at 43 Ma, some of the subsequent Murotohanto subduction history must have involved the fused Kula-Pacific crust. In addition, at least within the Oligocene section, the slab age must have been older than 10 m.y. at the time of sediment accretion. By analogy with the present-day Nankai accretionary prism, a slab age of 10 to 15 m.y. should have been young enough to produce a background geothermal gradient in excess of 30°C/km, particularly near the toe of the prism.

To establish the absolute timing of peak heating within the Murotohanto subbelt remains a critical objective for future research. DiTullio and others (this volume) have argued that the formation of cleavages during stage 2 of their deformation history coincided with the attainment of maximum burial temperatures. Furthermore, DiTullio and Byrne (1990) contended that the divergence in cleavage orientations within the Gyoto and Shiina domains is related to a change in shortening direction (from north-south to northwest-southeast, relative to present-day geographic coordinates). The inferred rotation of convergence vectors also caused the intermediate stage of out-of-sequence thrusting. DiTullio and Byrne (1990) assigned the realignment of shortening directions to the reorganization in plate motions at about 43 Ma (Kula subduction followed by fused Kula-Pacific subduction). In other words, according to this model, stage-2 deformation began during late middle Eocene time.

One weakness in the above argument involves the magnitude of changes in plate trajectories at 43 Ma. According to Lonsdale (1988), the angular change in convergence vectors only amounted to about 35° of counterclockwise rotation, which is less than the ~50° of rotation in shortening direction recognized by DiTullio and Byrne (1990). Another discrepancy is the occurrence of early Oligocene fossils in this part of the section which, again, are up to 10 m.y. younger than the Eocene plate reorganization. In addition, most of the cleavage dates for correlative Shimanto rocks of Eocene-Oligocene age on the Hata Peninsula are between 34 Ma and 26 Ma (Agar and others, 1989). Consequently, even if one accepts the hypothesis that cleavage formation and organic metamorphism were synchronous, the timing of the thermal overprint appears to postdate Kula-Pacific Plate reorganization by a considerable amount.

Thermal overprint of cleavage. In order to retain the Eocene and/or early Oligocene thermal structure of the Murotohanto subbelt (imparted, e.g., by a geothermal gradient of roughly 30°C/km), the rocks must have been uplifted considerably prior to middle Miocene time. Otherwise, paleotemperature values at a given structural position within the prism should have been reset during the Miocene as the regional geothermal gradient increased (Fig. 7). The key to this relationship, for any particular unit-volume of rock, is the absolute magnitude of uplift with respect to the sea floor (or ground surface) balanced against the absolute vertical shift in isotherms. We point out here that the southern limit of the Murotohanto subbelt was within at least 50 km of the subduction front during the Miocene reorganization at 15 Ma (Taira, 1985). By analogy with the Chile Rise collision zone, conductive geothermal gradients at such a position would likely be in excess of 70°C/km (Cande and others, 1987), even in the absence of near-trench magmatic activity. This estimate for the geothermal gradient is also consistent with the calculations of Hibbard and others (this volume). Therefore, a parcel of Eocene-Oligocene rock at depths of 6 to 7 km would have to be uplifted at least 3 km with respect to the sea floor and/or subaerial ground surface to prevent an overprint of $\%R_m$ during the middle Miocene (Fig. 7). As outlined previously, regional submarine

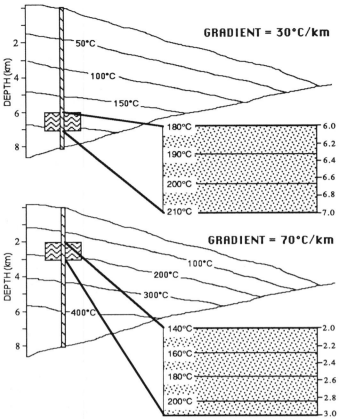

Figure 7. Schematic cross section showing the thermal structure of an accretionary prism and the balance between uplift/erosion and temporal increases in geothermal gradient. Note that the isotherms wedge out slightly with distance from the subduction front, but they are not based on rigorous application of any particular thermal model. The gradients of 30°C/km and 70°C/km are linear averages and apply only to the section encountered by a hypothetical borehole (diagonal pattern). A parcel of accreted sediment residing at prism depths of 6 to 7 km (geothermal gradient of 30°C/km) would have to be uplifted at least 3 km with respect to the sea floor to avoid overprints of the thermal structure, assuming a subsequent geothermal gradient of 70°C/km.

Japan (~50° at 15 Ma) are perfectly consistent with the measured deflection in shortening direction inferred from fabric analysis (DiTullio and Byrne, 1990). Even if the change in the intraprism shortening direction really did occur during the Paleogene, the angle of Miocene block rotation still must be added in. In other words, the D_1 shortening direction (inferred Kula subduction) would rotate back to northwest-southeast, and D_2 shortening (inferred Pacific subduction) would be restored to roughly east-west. In both cases, these restored directions deviate from the Pacific-Kula Plate motions described by Lonsdale (1988), although the differences may not be enough to discount the hypothesis.

One of the most important pieces of evidence to refute the idea of Miocene cleavage formation within the Murotohanto subbelt comes from the K-Ar ages of illites on the Hata Peninsula, which are 28 to 3 m.y. older than the onset of Outer Zone magmatic activity (Agar and others, 1989). One pitfall associated with radiometric dating of illite/mica in low-grade pelitic rocks, however, is the possibility of mixing among detrital populations and one or more populations of authigenic illite and/or K-mica. The critical temperature required to reset argon completely in these minerals is approximately $260 \pm 30°C$ (Hunziker and others, 1986). Several variables control the resulting K-Ar dates, including the particle size, detrital provenance, the relative proportions of $2M_1$ and $1Md$ mica polymorphs, and burial temperature (Hunziker and others, 1986). Because of these factors, inverse correlations often exist between illite K-Ar ages and metamorphic grade (that is, higher-grade rocks yield younger cleavage ages). Composite-particle K-Ar ages from the Shimanto Belt do not necessarily represent the peak heating event unless all of the crystals were heated beyond the illite/muscovite blocking temperature. This important boundary condition did not exist within the study area of Agar and others (1989), as they intentionally avoided the highest-grade zones closer to the Ashizuri granitic intrusions; in fact, many of their specimens did not even pass through the temperature window for total annealing of apatite fission tracks. In conclusion, the illite K-Ar dates only show that the cleavage is no older than 43 to 18 Ma.

As a final alternative, one might hypothesize that the respective D_1 and D_2 cleavages in the Shiina and Gyoto domains were progressively intensified at progressively deeper levels of the prism during stage-2 deformation, but the organic constituents were still thermally overprinted during the middle Miocene. Without a major phase of uplift and erosion of overburden, the more intensely cleaved strata should have remained at deeper structural levels of the prism as the regional geothermal gradient suddenly increased at about 15 Ma (Fig. 7). Additional growth of illite along the existing cleavage surfaces might have been enhanced with a rise of isotherms, or a new generation of illite might have crystallized. At the same time, values of vitrinite reflectance would be reset in response to the upward migration of isotherms. Deeper zones would not only retain the most intense cleavage, they would also attain the highest rank of organic metamorphism, even though the fabric developed before peak heating.

uplift during the Miocene did occur, but removal of 3 km or more of overburden appears excessive. Consequently, the likelihood of a regional thermal overprint of the Murotohanto strata seems strong.

Perhaps organic metamorphism and cleavage formation within the Murotohanto subbelt both occurred during the middle Miocene thermal event. If so, then the change in shortening direction documented by DiTullio and Byrne (1990) may be a response to prerotation versus postrotation convergence, and a shift from Pacific-Eurasia subduction to Eurasia–Philippine Sea subduction at 15 Ma, instead of a change in the Kula-Pacific convergence vectors. Among the many complications to consider is the possibility of oblique alignment of intraprism shortening fabrics with respect to the subduction vector. We point out here that the clockwise sense and magnitude of block rotation in southwest

This hypothesis is difficult to test, unfortunately, because the earlier generations of authigenic illite/mica would not experience total outgassing of argon at 15 Ma unless the temperatures rose above 260 ± 30°C (Hunziker and others, 1986). Therefore, unless the youngest population of authigenic illite can be isolated from the rest of the sample prior to age dating, we would expect the radiometric data to shift toward younger dates, but still not match the true age of peak heating.

Uplift/cooling history. Unfortunately, unambiguous age dates for all generations of cleavage and illite crystallization are not available, so the critical evidence required to test the completing hypotheses comes from apatite fission-track analyses. Although regional uplift may have started earlier, rocks of the Murotohanto subbelt were cooled below the 100°C isotherm at roughly the same time as coeval rocks on the Hata Peninsula, which was shortly after magmatic activity at Cape Muroto (Hasebe and others, this volume). Moreover, one specimen from the Nabae subbelt (near the mafic dikes) yielded a reset detrital-zircon fission-track peak age of 14 Ma; this datum is identical to the age of the intrusion and consistent with wall-rock temperatures above 260°C. An earlier uplift/cooling history for the Murotohanto strata, at temperatures greater than the apatite partial annealing zone, remains a possibility. However, except for the basal unconformity of the Shijujiyama Formation (Sakai, 1988), which appears to be entirely submarine (Hibbard and others, 1992), no evidence exists to support the idea of a widespread, pre-Miocene phase of subaerial uplift and erosion.

Additional apatite fission-track data from the Tertiary Shimanto Belt certainly would be helpful to constrain the regional cooling history. In the absence of such data, the provisional interpretation favored by most of the co-authors is that large portions of the Murotohanto and Nabae subbelts experienced peak heating, uplift, and cooling together. At the same time, we stress that the competing hypotheses are not mutually exclusive, especially if applied over the entire spatial regime of the study area. Some spatial overlap is likely, for example, between the temperature windows associated wtih the Oligocene and the middle Miocene temperature regimes, respectively. When viewed in three dimensions, the geothermal gradient at 14 Ma probably decreased with distance from the subduction front. In other words, it is possible that some portions of the Murotohanto subbelt were reset during the Miocene, while other parts of the section (farther from the trench, perhaps) retained their inherited (Eocene-Oligocene) values of thermal maturity. At the present time, we are unable to differentiate between the overlapping paleothermal regimes with complete confidence.

CONCLUSIONS

Comparisons among regional data sets from the Outer Zone of Japan show that paleotemperature patterns are fairly consistent for hundreds of kilometers along strike. This compilation of data demonstrates that our results from the Muroto Peninsula of Shikoku are representative of the Tertiary accretionary prism in general. We have not documented an along-strike anomaly.

Values of mean vitrinite reflectance for strata within the Murotohanto subbelt (Eocene to early Oligocene) range between 1.4 and 5.0%R_m; corresponding estimates of peak paleotemperature are 180 to 315°C. Rocks with higher ranks of organic metamorphism generally display better pressure-solution cleavage, and thermal maturity decreases from south to north. The geothermal gradient at the time of peak heating within this subbelt was at least 33°C/km.

Vitrinite reflectance values for the Nabae subbelt (late Oligocene to early Miocene) range from 0.9 to 3.7%R_m; paleotemperature estimates for these rocks are 140 to 285°C. The paleothermal structure within the Nabae section overprints all but the final stage of regional deformation. The highest-rank rocks occur in close proximity to gabbroic bodies of the Maruyama intrusive suite, which yield a radiometric age of 14 Ma. Local rock temperatures (near intrusions or related hydrothermal discharge) were substantially higher than the regional Miocene gradient, which itself was probably 70°C/km or higher.

The thermal peak in the Murotohanto subbelt postdated the main phase of deformation in the Eocene-Oligocene accretionary prism, including the out-of-sequence thrust offset between the Shiina and Gyoto structural domains. This conclusion is consistent with some models of evolving mountain belts that predict a lag in peak thermal metamorphism relative to peak burial and deformation (e.g., England and Thompson, 1984; Barr and Dahlen, 1989).

The qualitative correlation between intensities of pressure-solution cleavage and ranks of organic metamorphism indicates that the two are temporally related. The illite K-Ar ages from coeval units on the Hata Peninsula demonstrate that the cleavages are no older than 43 to 18 Ma. However, the absolute timing of organic metamorphism and cleavage formation within the Murotohanto subbelt remains unknown. Because of this, we offer two alternative scenarios: (1) a progressive heating event, ranging from early Eocene to late Oligocene, followed by uplift and erosion of overburden prior to 15 Ma; (2) a single, relatively short-lived thermal overprint at 15 Ma that coincided with anomalous near-trench igneous activity throughout the Outer Zone of southwest Japan. With option 1, the documented rotation in shortening direction within the Murotohanto subbelt would be linked to a change in the convergence vector of the Pacific-Kula lithosphere; with scenario 2, the shortening direction shifted because of the opening of the Sea of Japan and clockwise block rotation of southwest Japan.

The heat source responsible for the Eocene-Oligocene thermal structure (whether subsequently overprinted, or not) likewise remains unconstrained. Subduction of relatively young crust from the Kula Plate or fused Kula-Pacific Plate may have generated moderately high geothermal gradients within the Eocene-Oligocene section. However, with respect to the Muroto subduction front, the exact position of the Kula-Pacific spreading ridge at 43 Ma remains unknown.

On a regional scale, the dramatic middle Miocene perturbation of geothermal gradients was punctuated by widespread and voluminous near-trench acidic volcanism, emplacement of S-type granites, local mafic intrusions, and the formation of high-rank coals within forearc-basin sequences such as the Kumano Group of the Kii Peninsula. At Cape Muroto, this anomalous event involved collision between the accretionary prism and a rigid indentor of significant size. In addition to mafic magmatism, the mechanical responses to the collision included high-angle faulting, differential uplift of the Murotohanto subbelt, late-stage mesoscopic faulting and veining within the Nabae complex, and formation of a major cross fold (the Muroto flexure).

A final controversy involves the heat source responsible for the middle Miocene thermal overprint. We offer four possible explanations: (1) subduction of an active spreading segment directly beneath Cape Muroto; (2) subduction of an active spreading ridge beneath the Kii Peninsula, with off-axis magmatism (and seamount collision) at Cape Muroto; (3) development of an unstable and disorganized Shikoku Basin spreading system, with several off-axis mantle convection cells passing beneath the subduction front; and (4) a sudden change from Pacific Plate convergence to subduction of very young oceanic crust of the Philippine Sea Plate; this change occurred as the crustal block of southwest Japan rotated and the margin of Shikoku Basin changed from strike-slip to consuming. These scenarios are not mutually exclusive.

The only faults to show unequivocal evidence of postmetamorphic displacement are the Shiina-Narashi fault (Nabae-Murotohanto boundary), with more than 1 km of post-peak-thermal offset, and the Hane River fault zone. A late Miocene phase of regional uplift is consistent with apatite fission-track cooling ages of 11 to 8 Ma. Cooling occurred soon after the cessation of middle Miocene magmatic activity. Neotectonic features, including high-angle faults and uplifted marine terranes, show that vertical offset of the Muroto Peninsula has continued into Quaternary time.

Low-temperature, high-pressure metamorphic conditions, which typically are associated with blueschist-facies subduction complexes, obviously did not develop in this case. In addition, the existing theoretical models of accretionary-prism thermal structure are inadequate to explain the thermal history of southwest Japan, including the present-day offshore thermal structure. Even though many uncertainties and ambiguities remain, the combination of structural analyses and laboratory measurements of thermal maturity proves that the Tertiary Shimanto Belt of southwest Japan deviated significantly from the paradigm of a cold subduction zone.

ACKNOWLEDGMENTS

Funding for field mapping and the lab-based elements of this research was provided through a wide variety of sources listed in the individual papers in this volume. We direct special attention to the Donors of the Petroleum Research Fund, administered by the American Chemical Society, for partial support of this research (Grant No. 19187-AC2 to Underwood). Reviews by Dan Karig and Dan Orange helped clarify both the form and substance of the manuscript. Finally, Underwood acknowledges the patience of his co-authors and the difficulty of presenting all of their diverse opinions in a coherent and even-handed way.

REFERENCES CITED

Agar, S. M., Cliff, R. A., Duddy, I. R., and Rex, D. C., 1989, Accretion and uplift in the Shimanto Belt, SW Japan: London, Journal of the Geological Society, v. 146, p. 893–896.

Aihara, A., 1989, Paleogeothermal influence on organic metamorphism in the neotectonics of the Japanese Islands: Tectonophysics, v. 159, p. 291–305.

Barker, C. E., 1988, Geothermics of petroleum systems: Implications of the stabilization of kerogen thermal maturation after a geologically brief heating duration at peak temperature, *in* Magoon, L. B., ed., Petroleum systems of the United States: U.S. Geological Survey Bulletin 1870, p. 26–29.

Barker, P. F., Barber, P. L., and King, E. C., 1984, An early Miocene ridge crest-trench collision on the South Scotia Ridge: Tectonophysics, v. 102, p. 315–332.

Barr, T. D., and Dahlen, F. A., 1989, Brittle frictional mountain building; 2, Thermal structure and heat budget: Journal of Geophysical Research, v. 94, p. 3923–3947.

Byrne, T., and DiTullio, L., 1992, Evidence for changing plate motions in southwest Japan and reconstructions of the Philippine Sea Plate: The Island Arc, v. 1, p. 148–165.

Byrne, T., and Hibbard, J., 1987, Landward vergence in accretionary prisms: The role of the backstop and thermal history: Geology, v. 15, p. 1163–1167.

Cande, S., and Leslie, R., 1986, Late Cenozoic tectonics of the southern Chile Trench: Journal of Geophysical Research, v. 91, p. 471–496.

Cande, S. C., Leslie, R. B., Parra, J. C., and Hobart, M., 1987, Interaction between the Chile Ridge and Chile Trench: Geophysical and geothermal evidence: Journal of Geophysical Research, v. 92, p. 495–520.

Chamot-Rooke, N., Renard, V., and LePichon, X., 1987, Magnetic anomalies in the Shikoku Basin: A new interpretation: Earth and Planetary Science Letters, v. 83, p. 214–228.

Chijiwa, K., 1988, Post-Shimanto sedimentation and organic metamorphism: An example of the Miocene Kumano Group, Kii Peninsula: Modern Geology, v. 12, p. 363–387.

Davis, E. E., Hyndman, R. D., and Villinger, H., 1990, Rates of fluid expulsion across the northern Cascadia accretionary prism: Constraints from new heat flow and multichannel seismic reflection data: Journal of Geophysical Research, v. 95, p. 8869–8889.

DeLong, S., and Fox, P., 1977, Geological consequences of ridge subduction, *in* Pitman, W., and Talwani, M., eds., Island arcs, deep sea trenches, and back arc basins: American Geophysical Union, Maurice Ewing Series, v. 1, p. 221–228.

DeLong, S. E., Fox, P. J., and McDowell, F. W., 1978, Subduction of the Kula Ridge at the Aleutian Trench: Geological Society of America Bulletin, v. 89, p. 83–95.

DeLong, S. E., Schwarz, W. M., and Anderson, R. N., 1979, Thermal effects of ridge subduction: Earth and Planetary Science Letters, v. 44, p. 239–246.

DiTullio, L. D., 1989, Evolution of the Eocene accretionary prism in SW Japan: Evidence from structural geology, thermal alteration, and plate reconstructions [Ph.D. thesis]: Providence, Rhode Island, Brown University, 161 p.

DiTullio, L. D., and Byrne, T., 1990, Deformation paths in the shallow levels of an accretionary prism: The Eocene Shimanto belt of southwest Japan: Geological Society of America Bulletin, v. 102, p. 1420–1438.

Dumitru, T., 1991, Effects of subduction parameters on geothermal gradients in forearcs, with an application to Franciscan subduction in California: Journal of Geophysical Research, v. 96, p. 621–641.

Engebretson, D. C., Cox, A., and Gordon, R. G., 1985, Relative motions between oceanic and continental plates in the Pacific Basin: Geological Society of America Special Paper 206, 59 p.

England, P. C., and Thompson, A. B., 1984, Pressure-temperature-time paths of regional metamorphism; I, Heat transfer during the evolution of regions of thickened continental crust: Journal of Petrology, v. 25, p. 894–928.

Fisher, A. T., and Hounslow, M. W., 1990, Transient fluid flow through the toe of the Barbados accretionary complex: Constraints from Ocean Drilling Program Leg 110 heat flow studies and simple models: Journal of Geophysical Research, v. 95, p. 8845–8858.

Forsythe, R., and Nelson, E., 1985, Geological manifestations of ridge collision: Evidence from the Gulfo de Penas–Taitao Basin, southern Chile: Tectonics, v. 4, p. 477–495.

Forsythe, R., and 7 others, 1986, Pliocene near-trench magmatism in southern Chile: A possible manifestation of ridge collision: Geology, v. 14, p. 23–27.

Hada, S., and Suzuki, T., 1983, Tectonic environments and crustal section of the Outer Zone of southwest Japan, in Hashimoto, M., and Uyeda, S., eds., Accretion tectonics in the circum-Pacific regions: Tokyo, Terra Scientific Publishing Company, p. 207–218.

Hamamoto, R., and Sakai, H., 1987, Rb-Sr age of granophyre associated with the Cape Muroto gabbroic complex: Kyushu University, Department of Geology Science Reports, v. 15, p. 1–5.

Herron, E. M., Cande, S. C., and Hall, B. R., 1981, An active spreading center collides with a subduction zone: A geophysical survey of the Chile margin triple junction: Geological Society of America Memoir 154, p. 683–701.

Hibbard, J. P., 1988, Evolution of anomalous structural fabrics in an accretionary prism; The Oligocene-Miocene portion of the Shimanto Belt at Cape Muroto, southwest Japan [Ph.D. thesis]: Ithaca, New York, Cornell University, 227 p.

Hibbard, J. P., and Karig, D. E., 1990a, Structural and magmatic responses to spreading ridge subduction; An example from southwest Japan: Tectonics, v. 9, p. 207–230.

—— , 1990b, An alternative plate model for the early Miocene evolution of the SW Japan margin: Geology, v. 18, p. 170–174.

Hibbard, J. P., Karig, D. E., and Taira, A., 1992, Anomalous structural evolution of the Shimanto accretionary prism at Murotomisaki, Shikoku Island, Japan: The Island Arc, v. 1, p. 133–147.

Hisatomi, K., 1988, The Miocene forearc basin of southwest Japan and the Kumano Group of the Kii Peninsula: Modern Geology, v. 12, p. 389–408.

Hunziker, J. C., and 8 others, 1986, The evolution of illite to muscovite: Mineralogical and isotopic data from the Glarus Alps, Switzerland: Contributions to Mineralogy and Petrology, v. 92, p. 157–180.

James, T. S., Hollister, L. S., and Morgan, W. J., 1989, Thermal modeling of the Chugach Metamorphic Complex: Journal of Geophysical Research, v. 94, p. 4411–4423.

Jolivet, L., Huchon, P., and Rangin, C., 1989, Tectonic setting of Western Pacific marginal basins: Tectonophysics, v. 160, p. 23–47.

Kano, K., Kosaka, K., Murata, A., and Yanai, S., 1990, Intra-arc deformations with vertical rotation axes: The case of the pre-middle Miocene terranes of southwest Japan: Tectonophysics, v. 176, p. 333–354.

Kimura, K., 1985, Sedimentation and sedimentary facies of the Tertiary Shimizu and Misaki Formations in the southwestern part of Shikoku: Geological Society of Japan Journal, v. 91, p. 815–831.

Kinoshita, H., and Yamano, M., 1986, The heat flow anomaly in the Nankai Trough area, in Kagami, H., Karig, D. E., and Coulbourn, W. T., eds., Initial Reports of the Deep Sea Drilling Project, v. 87: Washington, D.C., U.S. Government Printing Office, p. 737–743.

Kosaka, K., Itoga, H., and Yanai, S., 1988, Macroscopic and mesoscopic kink folds of the Kobotoke Group in the southern Kanto Mountains, central Japan: Journal of the Geological Society of Japan:, v. 94, p. 221–224.

Kumon, F., and 8 others, 1988, Shimanto Belt in the Kii Peninsula, southwest Japan: Modern Geology, v. 12, p. 71–96.

Lister, C.R.B., 1977, Estimates for heat flow and deep rock properties based on boundary layer theory: Tectonophysics, v. 41, p. 157–171.

Lonsdale, P., 1988, Paleogene history of the Kula Plate: Offshore evidence and onshore implications: Geological Society of America Bulletin, v. 100, p. 733–754.

Mackenzie, J. S., Taguchi, S., and Itaya, T., 1990, Cleavage dating by K-Ar isotopic analysis in the Paleogene Shimanto Belt of eastern Kyushu, S.W. Japan: Journal of Mineralogy, Petrology, and Economic Geology, v. 85, p. 161–167.

Marshak, S., and Karig, D. E., 1977, Triple junctions as a cause for anomalously near-trench igneous activity between the trench and volcanic arc: Geology, v. 5, p. 233–236.

Maruyama, S., Liou, J. G., and Seno, T., 1989, Mesozoic and Cenozoic evolution of Asia, in Ben-Avraham, Z., ed., The evolution of the Pacific Ocean margin: New York, Oxford University Press, p. 75–99.

Miyake, T., 1985, MORB-like tholeiites formed within the Miocene forearc basin, southwest Japan: Lithos, v. 18, p. 23–24.

—— , 1988, Petrology of the Shionomisaki igneous complex, southwest Japan and its implication for the ophiolite generation: Modern Geology, v. 12, p. 283–302.

Moore, J. C., 1989, Tectonics and hydrogeology of accretionary prisms: Role of the décollement zone: Journal of Structural Geology, v. 11, p. 95–106.

Moore, J. C., and Allwardt, A., 1980, Progressive deformation of a Tertiary trench slope, Kodiak Islands, Alaska: Journal of Geophysical Research, v. 85, p. 4741–4756.

Mori, K., and Taguchi, K., 1988, Examination of the low-grade metamorphism in the Shimanto Belt by vitrinite reflectance: Modern Geology, v. 12, p. 325–339.

Murata, A., 1987, Conical folds in the Hitoyoshi Bending, south Kyushu, formed by the clockwise rotation of the southwest Japan Arc: Journal of the Geological Society of Japan, v. 93, p. 91–105.

Murata, M., 1984, Petrology of Miocene I-type and S-type granitic rocks in the Ohmine district, central Kii Peninsula: Journal of Japan Association of Mineralogy, Petrology, and Economic Geology, v. 79, p. 351–369.

Naeser, N. D., Naeser, C. W., and McCulloh, T. H., 1989, The application of fission-track dating to the depositional and thermal history of rocks in sedimentary basins, in Naeser, N. D., and McCulloh, T. H., eds., Thermal history of sedimentary basins, methods and case histories: Springer-Verlag, New York, p. 157–180.

Nakada, S., 1983, Zoned magma chamber of the Osuzuyama acid rocks, southwest Japan: Journal of Petrology, v. 24, p. 471–494.

Nakada, S., and Takahashi, M., 1979, Regional variations in chemistry of the Miocene intermediate to felsic magmas in the Outer Zone and the Setouchi Province of southwest Japan: Journal of the Geological Society of Japan, v. 85, p. 571–582.

Nakajima, T., and Hirooka, K., 1986, Clockwise rotation of southwest Japan inferred from paleomagnetism of Miocene rocks in Fukui Prefecture: Journal of Geomagnetism and Geoelectricity, v. 38, p. 513–522.

Nelson, E. P., and Forsythe, R. D., 1989, Ridge collision at convergent margins: Implications for Archean and post-Archean crustal growth: Tectonophysics, v. 161, p. 307–315.

Niitsuma, N., 1988, Neogene tectonic evolution of southwest Japan: Modern Geology, v. 12, p. 497–532.

Niitsuma, N., and Akiba, F., 1985, Neogene tectonic evolution and plate subduction in the Japanese island arc, in Nasu, N., Yueda, S., Kushiro, I., Kobayashi, K., and Kagmi, H., eds., Formation of active ocean margins: Tokyo, Terra Publishing Company, p. 75–108.

Oba, N., 1977, Emplacement of granitic rocks in the Outer Zone of southwest Japan and geologic significance: Journal of Geology, v. 85, p. 383–393.

Osozawa, S., 1988, Accretionary process of the Tertiary Setogawa and Mikasa Groups, southwest Japan: Journal of Geology, v. 96, p. 199–208.

Otofuji, Y., and Matsuda, T., 1984, Timing of rotational motion of southwest Japan inferred from paleomagnetism: Earth and Planetary Science Letters, v. 70, p. 373–382.

—— , 1987, Amount of rotation of southwest Japan—Fan shaped opening of the southwestern part of the Japan Sea: Earth and Planetary Science Letters, v. 85, p. 289–301.

Otofuji, Y., Hayashida, A., and Torii, M., 1985, When was the Japan Sea opened?: Paleomagnetic evidence from southwest Japan, in Nasu, N.,

Uyeda, S., Kushiro, I., Kobayashi, K., and Kagami, H., eds., Formation of active ocean margins: Tokyo, Terra Scientific Publishing Company, p. 551–566.

Otsuki, K., 1990, Westward migration of the Izu-Bonin Trench, northward motion of the Philippine Sea Plate, and their relationships to the Cenozoic tectonics of Japanese island arcs: Tectonophysics, v. 180, p. 351–367.

Parsons, B., and Sclater, J. G., 1977, An analysis of the variation of ocean floor bathymetry and heat flow with age: Journal of Geophysical Research, v. 82, p. 803–827.

Ranken, B., Cardwell, R. K., and Karig, D. E., 1984, Kinematics of the Philippine Sea Plate: Tectonics, v. 3, p. 555–576.

Reck, B. H., 1987, Implications of measured thermal gradients for water movement through the northeast Japan accretionary prism: Journal of Geophysical Research, v. 92, p. 3683–3690.

Sakai, H., 1987, Active faults in the Muroto Peninsula of "non-active fault province": Journal of the Geological Society of Japan, v. 93, p. 513–516.

——, 1988, Origin of the Misaki Olistostrome Belt and re-examination of the Takachiho orogeny: Journal of the Geological Society of Japan, v. 94, p. 945–961.

Seno, T., and Maruyama, S., 1984, Paleogeographic reconstruction and origin of the Philippine Sea: Tectonophysics, v. 102, p. 53–84.

Shibata, K., and Ishihara, S., 1979, Initial $^{87}Sr/^{86}$ ratios of plutonic rocks from Japan: Contributions to Mineralogy and Petrology, v. 70, p. 381–390.

Shibata, K., and Nozawa, T., 1967, K-Ar ages of granitic rocks from the Outer Zone of southwest Japan: Geochemical Journal, v. 1, p. 131–137.

Shih, T. C., 1980, Magnetic lineations in the Shikoku Basin, *in* Klein, G. deV., and Kobayashi, K., eds., Initial reports of the Deep Sea Drilling Project, v. 58: Washington, D.C., U.S. Government Printing Office, p. 783–788.

Shipboard Scientific Party, 1991, Site 808, *in* Taira, A., Hill, I., Firth, J. V., and others, eds., Proceedings of the Ocean Drilling Program, initial reports, v. 131: College Station, TX, Ocean Drilling Program, p. 71–272.

Taira, A., 1985, Sedimentary evolution of Shikoku subduction zone: The Shimanto Belt and Nankai Trough, *in* Nasu, N., Uyeda, S., Kushiro, I., Kobayashi, K., and Kagami, H., eds., Formation of active ocean margins: Tokyo, Terra Scientific Publishing Company, p. 835–851.

Taira, A., Tashiro, M., Okmura, M., and Katto, J., 1980, The geology of the Shimanto Belt in Kochi Prefecture, Shikoku, Japan, *in* Taira, A., and Tashiro, M., eds., Geology and paleontology of the Shimanto Belt, Kochi: Rinya-Kosakai Press, p. 319–389.

Taira, A., Okada, H., Witaker, J.H.McD., and Smith, A. J., 1982, The Shimanto Belt of Japan: Cretaceous to lower Miocene active margin sedimentation, *in* Leggett, J. K., ed., Trench-forearc geology: Geological Society of London, Special Publication 10, p. 5–26.

Taira, A., Katto, J., Tashiro, M., Okamura, M., and Kodama, K., 1988, The Shimanto Belt in Shikoku, Japan—Evolution of Cretaceous to Miocene accretionary prism: Modern Geology, v. 12, p. 5–46.

Taira, A., Tokuyama, H., and Soh, W., 1989, Accretion tectonics and evolution of Japan, *in* Ben-Avraham, Z., ed., The evolution of the Pacific Ocean margins: New York, Oxford University Press, p. 100–123.

Terakado, Y., Shimizu, H., and Masuda, A., 1988, Nd and Sr isotopic variations in acidic rocks formed under a peculiar tectonic environment in Miocene southwest Japan: Contributions to Mineralogy and Petrology, v. 99, p. 1–10.

Torii, M., and Ishikawa, N., 1986, Age estimation of a high-magnesian andesite from the Kii Peninsula, southwest Japan: Possible interpretation of anomalous magnetic direction: Journal of Geomagnetism and Geoelectricity, v. 38, p. 523–528.

Toriumi, M., and Teruya, J., 1988, Tectono-metamorphism of the Shimanto Belt: Modern Geology, v. 12, p. 303–324.

Underwood, M. B., O'Leary, J. D., and Strong, R. H., 1988, Contrasts in thermal maturity within terranes and across terrane boundaries of the Franciscan Complex, northern California: Journal of Geology, v. 96, p. 399–416.

Uyeda, S., and Miyashiro, A., 1974, Plate tectonics and the Japanese islands: A synthesis: Geological Society of America Bulletin, v. 85, p. 1159–1170.

van den Beukel, J., and Wortel, R., 1988, Thermo-mechanical modelling of arc-trench regions: Tectonophysics, v. 154, p. 177–193.

Wang, C. Y., and Shi, Y., 1984, On the thermal structure of subduction complexes: A preliminary study: Journal of Geophysical Research, v. 89, p. 7709–7718.

Watts, A. B., and Weissel, J. K., 1975, Tectonic history of the Shikoku marginal basin: Earth and Planeteary Science Letters, v. 25, p. 239–250.

Weissel, J. K., Taylor, B., and Karner, G. D., 1982, The opening of the Woodlark Basin, subduction of the Woodlark spreading system, and the evolution of northern Melanesia since mid-Pliocene time: Tectonophysics, v. 87, p. 253–277.

Yamano, M., Honda, S., and Uyeda, S., 1984, Nankai Trough: A hot trench?: Marine Geophysical Researches, v. 6, p. 187–203.

Yamazaki, T., and Okamura, Y., 1989, Subducting seamounts and deformation of overriding forearc wedges around Japan: Tectonophysics, v. 160, p. 207–229.

Yanai, S., 1989, A horizontal buckle model as a dynamic mechanism for back arc spreading of the Japan Sea: Journal of Geology, v. 97, p. 569–583.

Zaun, P. E., and Wagner, G. A., 1985, Fission-track stability in zircons under geologic conditions: Nuclear Tracks, v. 10, p. 303–307.

MANUSCRIPT ACCEPTED BY THE SOCIETY APRIL 24, 1992

Index

[Italic page numbers indicate major references]

Typeset by WESType Publishing Services, Inc., Boulder, Colorado
Printed in U.S.A. by Malloy Lithographing, Inc., Ann Arbor, Michigan